RESEARCH COMPANION TO INNOVATION IN CONSTRUCTION

ELGAR COMPANIONS TO THE BUILT ENVIRONMENT

This important and timely series brings together critical and thought-provoking contributions on the most pressing topics and issues in built environment research. Comprising specially-commissioned chapters from leading academics these comprehensive *Companions* feature cutting-edge research, help to define the field and are written with a global readership in mind. Equally useful as reference tools or high-level introductions to specific topics, issues, methods, innovations and debates, these *Companions* will be an essential resource for academic researchers and postgraduate students in building, construction management and related disciplines.

For a full list of Edward Elgar published titles, including the titles in this series, visit our website at www.e-elgar.com.

Research Companion to Innovation in Construction

Edited by

Mohammed Fadhil Dulaimi

Professor of Construction Project Management and Innovation, Leeds Beckett University, UK

ELGAR COMPANIONS TO THE BUILT ENVIRONMENT

Edward Elgar
PUBLISHING

Cheltenham, UK • Northampton, MA, USA

Published by

Edward Elgar Publishing Limited
The Lypiatts
15 Lansdown Road
Cheltenham
Glos GL50 2JA
UK

Edward Elgar Publishing, Inc.
William Pratt House
9 Dewey Court
Northampton
Massachusetts 01060
USA

Authorised representative in the EU for GPSR queries only: Easy Access System Europe – Mustamäe tee 50, 10621 Tallinn, Estonia, gpsr.requests@easproject.com

A catalogue record for this book is available from the British Library

Library of Congress Control Number: 2025935774

This book is available electronically in the **Elgar**online
Geography, Planning and Tourism subject collection
https://doi.org/10.4337/9781035326594

MIX
Paper | Supporting responsible forestry
FSC® C013604
FSC www.fsc.org

ISBN 978 1 0353 2658 7 (cased)
ISBN 978 1 0353 2659 4 (eBook)
ISBN 978 1 0353 7031 3 (ePub)

Printed and bound by CPI Group (UK) Ltd, Croydon, CR0 4YY

Contents

Contributors

Mohammed Abdelmegid is a Lecturer in Engineering Management, School of Civil Engineering, University of Leeds, UK.

Shaima AlHarmoodi is an Associate Professor of Innovation and Project Management, School of Business and Quality Management, Hamdan Bin Mohamed Smart University, UAE.

M.K.S. Al-Mhdawi is a Lecturer (Assistant Professor) in Engineering Project Management at the School of Computing, Engineering, and Digital Technologies, Teesside University, UK.

Suhaib Arogundade has a PhD in Civil Engineering and is founder of Carbonetrix, an AI-powered digital construction carbon management platform based in the UK.

Arun Bajracharya is an Associate Professor of Operations and Project Management at the School of Management in New Uzbekistan University, Tashkent, Uzbekistan.

T.E. Butt is a Senior Lecturer in Environment, Engineering and Management at Northumbria University, UK.

George Denny-Smith is a Project Manager at RPS Australia.

Gerard de Valence retired from the School of the Built Environment, University of Technology Sydney, and the Bartlett School of Sustainable Construction, University College London, in 2020.

Hemanta Doloi is an Associate Professor of Construction and Project Management and Director of the Smart Villages Lab (SVL), Faculty of Architecture, Building and Planning, University of Melbourne, Australia.

Mohammed Fadhil Dulaimi is a Professor of Construction Project Management and Innovation, School of Built Environment, Engineering and Computing, Leeds Beckett University, UK.

Bismark Duodu is a Lecturer at the Department of Construction Technology and Management, Faculty of Built Environment, Kwame Nkrumah University of Science and Technology, Kumasi, Ghana.

Per Erik Eriksson is a Professor of Construction Management at the Division of Industrialized and Sustainable Construction, Luleå University of Technology, Sweden.

Hossein Sadeghzadeh Fasaghandis is a PhD student in Construction Management at the School of Built Environment, Massey University, New Zealand.

Ximena Ferrada is the Director of the Ingeniería Civil en Obras Civiles Program, Faculty of Engineering, Universidad del Desarrollo, Chile.

Vicente Gonzalez is a Professor and Tier 1 Canada Research Chair in Digital Lean Construction, Department of Civil and Environmental Engineering, Faculty of Engineering, University of Alberta, Canada.

Chris Gorse is a Professor of Construction Engineering and Management at the School of Architecture, Building and Civil Engineering, Loughborough University, UK.

Malena Havenvid is an Associate Professor of Construction Project Management, Head of the Division of Construction and Facilities Management, KTH Royal Institute of Technology, Sweden.

Mostafa Babaeian Jelodar is Academic Lead and Senior Lecturer in Construction Engineering and Management at the School of Built Environment, Massey University, New Zealand.

Mohan Kumaraswamy is an Honorary Professor at the University of Hong Kong and the University of Moratuwa; Editor-in-Chief, *Journal of Built Environment Project and Asset Management* (BEPAM); and Editor-in-Chief, World Scientific Book Series on Domain-Specific Bodies of Knowledge in Project Management.

Johan Larsson is an Associate Professor of Construction Management, Head of the Division of Industrialized and Sustainable Construction, Luleå University of Technology, Sweden.

Martin Loosemore is a Distinguished Professor of Project Management at the Faculty of Design, Architecture and Building, University of Technology, Sydney, Australia.

Peter E.D. Love is a John Curtin Distinguished Professor of Infrastructure and Engineering Informatics at Curtin University, Perth, Australia.

C. Mahammedi is a Lecturer in Engineering (Civil & Construction) at the School of Computing, Engineering, and Digital Technologies, Teesside University, UK.

Shoeb Ahmed Memon is an Assistant Professor of Construction Management and Quantity Surveying, Faculty of Society and Design, Bond University, Gold Coast, Queensland, Australia.

George Ofori is a Professor Emeritus in the School of the Built Environment and Architecture at London South Bank University, UK.

Michael O'Sullivan is an Associate Professor at the Department of Engineering Science and Biomedical Engineering, Faculty of Engineering, University of Auckland, New Zealand.

Mani Poshdar is a Senior Lecturer in Built Environment Engineering, School of Future Environments, Faculty of Design and Creative Technologies, Auckland University of Technology, New Zealand.

Steve Rowlinson is Emeritus Professor of Construction Project Management at Hong Kong University and Professor of Construction Project Management (part-time), Faculty of Society and Design, Bond University, Gold Coast, Queensland, Australia.

Ali M. Saad is a Lecturer in Construction Engineering and Project Management at the School of Architecture, Building and Civil Engineering, Loughborough University, UK.

Alfredo Serpell is a Professor of Construction Management, Faculty of Engineering, Ingeniería Civil en Obras Civiles, Universidad del Desarrollo, Chile.

Nipuni Sumanarathna is a Lecturer in the School of the Built Environment, London Metropolitan University, London, UK.

Riza Yosia Sunindijo is an Associate Professor of Construction Management at the School of Built Environment, University of New South Wales, Australia.

Claudia Valderrama-Ulloa is a Research Associate at the Faculty of Engineering, Research Centre on Technologies for Society, Universidad del Desarrollo, Chile.

Cameron Walker is a Professor at the Department of Engineering Science and Biomedical Engineering, Faculty of Engineering, University of Auckland, New Zealand.

Derek H.T. Walker is Emeritus Professor of Project Management at RMIT University, Melbourne, Australia.

Suzanne Wilkinson is Deputy Dean and Professor in the Faculty of Design and Creative Technologies, Auckland University of Technology, Auckland, New Zealand.

Megan Williams is a CEO and Research Lead, Yulang Indigenous Evaluation and University of Sydney, Australia.

Foreword

The construction industry is not often regarded as highly innovative and is considered a slow adopter of new technologies. However, there are numerous innovations that take place on a daily basis and that have helped to transform the construction project delivery process over the years. With these two perspectives in mind, it is imperative to have an in-depth understanding of the state of innovation in construction to ensure that the industry is able to seize emerging opportunities.

Professor Mohammed Dulaimi and his colleagues have sought to explore the dynamics of innovation in construction and have written a comprehensive book that addresses the critical aspects of this subject. It is situated within the context of previous studies on innovation in construction and discusses innovation at different scales – project, industry, national, and global. A major strength of the book is its organisation into four logical parts:

- *Innovation Policy and Industry Strategies:* The chapters in this part discuss innovation at the industry level and cover the role that government policies can play in engendering an innovative climate in the construction industry. The situation in developing countries is also highlighted and contrasted with that in developed countries.
- *Implementing Innovations in Organisations and Projects:* In this part, the chapters focus on the specific approaches to implementing and fostering innovation in the construction industry. These include the need for a 'systems' approach, the importance of innovation diffusion models, and how corporate relationships within the construction industry influence innovation in construction projects.
- *Innovations in Construction Project Procurement:* There have been numerous innovations in the procurement methods adopted in the construction industry, and the chapters in this part discuss these. This is very important as the procurement method adopted has a major impact on the construction supply chain, the project team relationships and dynamics, the level and extent of knowledge-sharing, project culture, and the capacity for innovations at project level.
- *Innovative Technological Systems and Organisations:* Specific innovative technologies, tools, and approaches that have been developed for various aspects of the construction industry are presented in the chapters in this part. These include systems for improved health and safety, measurement of 'corporate progress', and stakeholder integration for sustainability delivery.

Given the continued rapid pace of technological developments, the increasing urgency to ensure sustainability of both the construction process and constructed facilities, and changing demographic and workforce trends, this is a timely book that should be read by construction researchers, industry practitioners, and policymakers. It offers useful insights that will enable the construction industry to leverage emerging technologies and make innovation endemic to its business processes.

Professor Chimay J. Anumba
Dean, College of Design, Construction and Planning
University of Florida, USA

PART I

INNOVATION POLICY AND INDUSTRY STRATEGIES

1. Introduction to *Research Companion to Innovation in Construction*

Mohammed Fadhil Dulaimi

THE CHALLENGE OF INNOVATION IN CONSTRUCTION

The review of the literature on innovation in construction shows that researchers in the construction management area started to develop a specific interest in this topic towards the end of the 1980s. This interest picked up speed in the 1990s, arguably triggered by the publications of several reports into the state of the construction industry, such as the Latham report (1994) and Egan report (1998). Such reports have examined the state of the construction industry, and the challenges and shortcomings of the industry's performance. The construction industry was criticised for failing to be more innovative in addressing the needs, concerns and aspirations of its clients and users, as well the wider network of its stakeholders.

The pressure on the construction industry to develop and adopt new technologies, tools and practices is significant and global (Jelodar et al., 2018). Many researchers have argued that one of the main reasons for the construction industry lagging behind other sectors is that the industry is different, and the nature of its one-off and short-term projects puts it at a disadvantage compared to other sectors (Lloyd-Walker et al., 2014; Walker, 2016; Jacquetta et al., 2016). The delivery of successful innovation may be argued to require clients to accept greater and long-term uncertainty regarding the long-term performance of the acquired facility. Such arguments may have even discouraged, in the past, the construction industry from crossing the boundaries to sectors other than construction to adopt and/or adapt models, theories and innovative tools and practices developed in a completely different environment and market.

First, we need to be clear about what we mean by innovation. Drucker (1985) defined innovation as the effort to create purposeful, focused change in the economic and social potential of an enterprise. His definition linked the importance, in a business environment, of both concepts of invention and entrepreneurship. Innovation requires creative inventions to be implemented and successfully commercialised (Dulaimi, 1995). We can argue that invention can be an individual activity and that, in order to realise innovation, a team needs to be involved in a social and collective activity (Guan and Liu, 2016) that takes place in organisations. The organisation, in this case, will/should provide the necessary knowledge, skills and resources needed to deliver a successful innovation.

So, is innovation important to businesses? Research has linked the success of organisations to their ability to continuously develop processes and products that perform better than their competitors on attributes that their customers consider to be of value (Wittmann et al., 2009). Researchers and practitioners recognise that without investment in innovation, the construction industry cannot meet the challenges it faces in meeting the needs and aspirations of its clients, users and general society.

It was argued that the problem was not the lack of new ideas but the poor understanding of the environment that is conducive to innovation at corporate and, even more challenging, at project levels. Dulaimi (1995) challenged the construction industry to learn from other sectors on how they managed to create the culture that is conducive to innovation. The interest in innovation in the field of business management was evident much earlier, with extensive research and publications examining the dynamics of innovation at corporate, team and individual levels. It is not surprising that the early research into innovation in construction shows that researchers in construction management were influenced by theories, models and tools used in business management literature.

UNDERSTANDING THE DYNAMICS OF INNOVATION

Tatum (1987, 1989) provided pioneering research into construction innovation, investigating the process of innovation in construction projects and analysing the relationships between the key stakeholders. Such research has paved the way and encouraged more researchers to explore this important issue. Research by Gann and Salter (2000) differentiated construction innovation at the project and business levels and emphasised the need for integrating project and business processes within the firm to effectively manage technological capabilities. The linking of project-level issues and corporate-level priorities presents opportunities for researchers to better understand the relationship between short-term, project-level goals and long-term strategic, corporate and industry-level goals.

With a few exceptions, the focus of research into innovation tends to be more concerned with the dynamics of innovation within the boundaries of one organisation, known as closed innovation. However, the nature of innovation in construction projects requires input and effort from several organisations involved in the procurement of that project, and hence the closed innovation model is not applicable. Innovation is a goal-oriented process which, in an open innovation setting such as construction, would create a challenge of aligning the motivation as well as the business processes of the different organisations involved and contributing to innovation. It is suggested, as will be argued later in this book, that the alignment of motivation of different entities within the construction project organisation is one of the necessary conditions for the effective implementation of innovation. Alignment of motivation may be made possible through alliances between the contributing organisations to bridge their differences and motivate greater collaboration. Therefore, the concepts related to alliances are reviewed later in this book to understand the inter-organisational interactions.

Researchers have argued that we can draw parallels between how construction projects are organised and structured and what can be observed in research and development (R&D) settings in other sectors of industry. In construction, Dulaimi et al. (1996) argued, we can find that the project development and execution processes, from concept to detailed design and construction, are comparable with the process of progressing from R&D to successful innovations in manufacturing scenarios. In addition, their research has argued that the structure and organisation of activities in an R&D environment are comparable with existing project organisational models and procurement frameworks deployed in construction projects. In both settings, essentially, the aim and aspiration is to find a solution to a problem or a need.

Liker et al. (1999) argue that the main issue in product development is 'reintegrating' what has been divided as organisations, supply chains, and their products become more complex. They also argue that the effort to reintegrate design and manufacturing processes is aimed at creating a single 'mind' for the supply chain to be able to think as one entity. Construction is familiar with the debate of integrating design and construction. Such arguments are likely to encourage researchers to cross into other sectors of industry and benefit from their experience in the management of innovation. This issue will be explored further later in this chapter.

The conditions that are critical to promoting and facilitating innovation have attracted the attention of researchers for several decades. Dulaimi's (1995) research into the challenges facing innovation in construction has drawn attention to the importance of a psychological safety net and management willingness to support innovative champions in motivating individuals to innovate. The diversity of the concepts related to innovation has resulted in the emergence of many theories to formally relate innovation to its factors and determinants.

ALIGNMENT AND INTEGRATION

Innovation is a complex concept that encompasses several sub-concepts and various activities. It is an organisational phenomenon that involves teamwork and occurs in and between organisations. However, organisations do not innovate in isolation, as innovation activities involve extensive interactions with the organisation's environment (Chesbrough and Bogers, 2014). This interaction is described using the models of 'networks' or 'systems' that explain the flow of knowledge across the boundaries of several organisations. Powell and Grodal (2005) argued that a network can represent the relationships of the firm with external parties, such as partners, suppliers and clients. The increased interest in the concept of 'open innovation' (Chesbrough, 2003) is an indication of increased recognition by researchers and practitioners that working across organisational boundaries creates greater opportunities for innovation and enhances project performance.

It has been argued that organisations achieve competitive advantages through innovation as they differentiate their products, processes and services rather than compete on cost (Porter, 1990; Slaughter, 1998). Furthermore, innovation becomes essential for project success by enabling and empowering the supply chain to understand, develop and deliver effective solutions to the clients' needs. The nature of the construction process is that it must deal with daily problems, especially on-site, for which it has to develop new ideas to resolve problems encountered during the execution of the project. However, innovation can also be a process in which organisations or individuals are driven by a desire to improve performance. The construction industry is known for conservatism; professionals embrace accepted industry practices and norms in fulfilling clients' needs; changes are seen to be risky and could even be taken as a threat; and slack resources are rarely permitted (Nam and Tatum, 1997). Further, construction projects also have significant coordination and integration problems due to the extreme specialisation of functions and/or involvement of various professions (Nam and Tatum, 1992).

The construction industry has seen the process of developing innovative products restricted mostly to the early stages of the construction process, namely the design and planning stages. This may be explained by the general view in the industry that the construction industry's lack of innovation is due to the focus on meeting the 'client requirements'. The dynamic of the process is such that after the approval of the design by the client and the signing of contracts, the commercial positions accepted by the different parties are set. Hence, there is little incentive to convince the different parties to accept changes to the design and execution of the project that could alter the agreed commercial positions. This situation seems to undermine any effort to propose new ideas that could deliver innovation.

The issue of integration of the design and manufacturing/construction is evidently fundamental to the realisation of a successful innovation. The construction industry, fragmented by significant subcontracting practices, is challenged to bring together different groups and organisations, creating a new 'virtual organisation' to work and interact as one organisational unit. One of the main challenges critical to the success of innovation is whether this network of organisations will facilitate or hinder the development of more innovative solutions. In examining the innovation process, the alignment of the motivation of the different participants in the project, groups or individuals will need to bridge the gaps between competing paradigms across this diverse network. The emergence of the champion of the innovation will create an opportunity for aligning the supply chain by identifying the benefits, short term and long term, that the different stakeholders in the project would expect as an outcome of their investment in the innovation. This idea will be further explored later in this book.

ENABLING INNOVATION

Early research in construction innovation has raised the issue of organisational environment, demonstrating awareness that to foster successful innovation, organisations need to provide a positive climate for innovation that supports and encourages the development and delivery of new ideas (Tatum, 1989). Further, research has argued the importance of organisations having 'innovation champions' who are enthusiastic and committed to the innovation process and who can drive the successful development of innovation (e.g., Roberts and Fusfeld, 1981; Nam and Tatum, 1992, 1997; Winch, 1998). However, there needs to be careful consideration and understanding of the role champions of innovation can play and how their role is manifested in a construction project dynamic. Early research into construction innovation criticised the lack of attention to the importance of the role of champions who can exert great influence in the process of construction innovation (Nam and Tatum, 1997).

Tatum (1989) explained how the role of the innovation champion in a project-based process such as construction can be different from that in a corporate, closed environment. Despite several innovation-related studies in the past, only a few of them have examined innovation at a project level, and there is still a lack of empirical research in this area. Park et al. (2004) argued that the role of the project manager (PM) in a construction project is essentially that of a champion driving innovation on-site and, hence, improving project performance. Several individual and situational factors may also significantly affect the PM's championing role and its effectiveness, and influence directly and/or indirectly the level of innovation and project performance. Park et al. defined project managers' championing behaviour as their observable actions directed towards seeking, stimulating, supporting, carrying and promoting innovation in the project.

Research in construction innovation has also targeted the issue of climate for innovation (Dulaimi, 2022). Earlier, Nam and Tatum (1992) argued that it is not the availability of ideas that hinders construction innovation but the decision to use them or the environment that influences them. The environment basically refers to the organisational climate for innovation that is often described in terms of psychological climate. The psychological climate for innovation is a multi-dimensional construct that can be conceptualised and operationalised at the individual level (Koys and DeCotiis, 1991). It represents the cognitive interpretation of an organisational situation perceived by individuals and the signals they receive concerning organisational expectations for behaviour and potential outcomes of the behaviour (James and Sells, 1981; Scott and Bruce, 1994). The perceived corporate priorities and sensitivities about acceptable levels of uncertainties and risks that the organisation would tolerate may play a crucial role in motivating or inhibiting innovation. The conduct by which organisations signal an expectation for innovation is by committing and providing resources and support for innovation (Kanter, 1988; Amabile, 1997). Park et al. (2004) explained that the 'supportive organizational climate in construction may include acknowledgement of and reward for creativity; tolerance of risk, failure and mistakes; commitment of necessary resources; innovative culture that values innovation and change and clear strategic vision of the company, among others'. It is therefore argued that the perception of the project environment in terms of support for innovation and resource supply may encourage PMs to engage in a host of championing activities; such perception would also play a vital role in fostering innovative practices. The PM can influence construction innovation through providing leadership and, hence, provide clarity to the project team in terms of the decisions they need to make in order to deliver a successful project. Thus, PMs should understand the project environment and context, ability and willingness of the project team, including the supply chain, and then choose the leadership style that motivates and promotes teamwork to deliver innovation. The PMs can play a key role in convincing and selling innovative ideas to potential allies and obtain necessary support and approval from them; they can also coordinate different entities such as subcontractors, designers and other approval agencies and facilitate the implementation of internally generated ideas in the project. Another major role that PMs can play is an important leadership role to build a consensus within the project team to rally around one idea that the team can support. The PM can also play the role of a gatekeeper who facilitates and provides access to knowledge and information, across the supply chain, from various sources, and promotes the generation of new ideas by motivating and inspiring team members to work together to deliver successful innovation. Furthermore, PMs, to some extent, can use their influence, even authority, to support a particular direction for innovation based on their vision of the project. This would arise when the team members are finding it difficult to develop a consensus in reconciling their different ideas. Providing clarity of corporate priorities, defined through their understanding of client and market needs, should pave the way for the new ideas needed to address project challenges. Finally, PMs need to exhibit commitment in the innovation process by expending their energy, taking responsibility and a reasonable amount of risk. Because innovation would bring changes or even certain risks or uncertainties, the PM's conviction and confidence can overcome inertia and resistance and provide impetus to those who are involved in the innovation. Dulaimi et al.'s (2005) investigation showed that the PM's education is positively related to championing, indicating the extent to which PMs rely on their personal knowledge to become an effective innovation champion.

This book is organised into four parts, each with several chapters addressing key issues that researchers have tackled at global, national, industry and project levels.

1. INNOVATION POLICY AND INDUSTRY STRATEGIES

The calls from governments for the construction industry to improve its performance go back several decades. The response from the construction industry to such calls was dominated by the argument that the industry works to client requirements, which was interpreted as dismissing the calls for radical change and embracing an innovation agenda. De Valence (Chapter 2) argues the importance of innovation for the development of a modern industry and points to the construction industry's low research and development (R&D) expenditure being the lowest in OECD countries. In his chapter, de Valence challenges the narrow view of construction activities as merely contractors' and subcontractors' activities, including related industries in manufacturing, materials and professional services. The uniqueness of every sector of industry, especially construction, means that the path they take in innovation and development will be different. The nature of the fourth industrial revolution

(Industry 4.0), and indeed Industry 5.0, driven by digital technological innovations, is challenging even to the structure of the construction industry and creating opportunities for new entrants and allowing smaller firms, small and medium enterprises (SMEs), to challenge the dominance of larger firms.

The importance of the construction industry in any society and economy is undeniable. This is why governments have intervened to influence change in the construction industry. The innovation agenda is one area where governments sought to play a key role in influencing the performance of the construction industry in terms of its impact on the economy, society and the environment. Serpell et al., in Chapter 4, examine innovation at the national level in Chile, where the attitude and involvement in R&D activities compare negatively with that of other sectors. The chapter examines how the government in Chile aimed to play an effective role in enhancing innovation levels in construction. The analysis maps the role of seven different institutions that had indirect/strategic influence, such as the Ministry of Education, and those that played an active role in the innovation agenda, such as the National Agency for Research and Development as well as the Undersecretariat of Regional and Administrative Development. The fact that the construction industry is dominated by small companies means that such companies would not have the resources or opportunity to be involved in innovation. Any effort to examine the statistics related to construction industry performance and innovation efforts should consider the unique nature of the industry, its stakeholders, project organisation and practices.

Serpell et al. further reported that the construction industry in Chile has a poor record in innovation due to its lack of initiative to deliver effective innovation that has a 'lasting impact on [construction] companies'. They argue that it is the failure of such companies to provide an environment conducive to innovation that motivates and encourages the development and adoption of new ideas. The result is that the construction industry is involved in the development of new ideas but not at the pace expected by its stakeholders. Therefore, there is a need to accelerate the pace of innovation if we are to bridge the gap between what the industry offers and the expectations and aspirations of its stakeholders and society.

The belief that innovation can drive social and economic growth and development led Ofori (Chapter 3) to argue that developing countries may need innovation more than developed countries; otherwise, developing countries would risk being left behind. The higher backlog of buildings, facilities and infrastructure that developing countries need puts greater pressure to enhance innovation capabilities and capacity. Innovation not only delivers much-needed built environments but helps the country reduce its reliance on the import of material, technology and skills. There is a significant role the government can play to support and motivate innovation through changes to policy and regulations, as well as competency-building through investing in training and skills development. It is important for developing countries to learn from innovations in developed countries. However, the dynamics of innovation and the environment that supports such efforts should reflect the unique needs and culture of developing nations. It is argued that Construction 5, which takes the human element into account, should be the main consideration for developing countries now.

Ofori (Chapter 3) states that 'in developing countries it is necessary to look for leap-frogging opportunities and to seek synergies', which should put innovation at the top of the agenda of such industries.

2. IMPLEMENTING INNOVATIONS IN ORGANISATIONS AND PROJECTS

The decision to invest in innovation is complex in nature due to the intricacy of the factors and issues that influence such a decision and the process of making such a decision. The construction industry is still accused of being resistant to change and cautious in investing in new ideas. Abdelmegid et al. (Chapter 5) argue that innovation in the construction industry can be analysed using systems thinking, which recognises the complexity of the industry and enables the deployment of hard and soft skills critical to understanding the different dimensions of innovation. A careful analysis of the challenging nature of the external environment, balanced with the internal strengths and resources of the organisation, is critical to understanding the dynamics of innovation in construction. Therefore, the analysis of the construction firm's internal dynamics that produce innovation, technical and organisational, needs to be linked to the interplay between the different external forces that influence the adoption and diffusion of that innovation. The external forces extend beyond those within the industry itself.

Abdelmegid et al. explain how systems thinking can be the 'catalyst for holistic and creative thinking' in addressing the challenges and problems facing a construction organisation. It can be deduced that systems thinking facilitates, or could be said forces, the adoption of a more holistic perspective and the consideration of a more complex picture of the problems and their context. This tool could also require key stakeholders to view problems 'as interconnected components within a larger system' and enable a more creative problem-solving approach. Critical to the success of such an exercise will be the development of comprehensive metrics for measuring the tangible and intangible outcomes of innovation effort.

The collaborative approach of a multidisciplinary team and the collaborative network of stakeholders would create a transparent environment where participants can develop a mutual understanding of the problem. The proposed use of systems thinking would influence the creation of a collaborative culture, facilitate the development of a shared view of the problem/challenge and enable the development of a definition of the innovation that will be supported by all stakeholders.

Memon et al. examine in their chapter (Chapter 6) the impact of business models on the diffusion of innovation in construction. This chapter provides an important analysis of a topic that has recently attracted attention from researchers and practitioners. The need for the construction industry to address long-term goals such as sustainable development is viewed through the examination of traditional business models adopted in the industry. Memon et al.'s research examines how the digital revolution is transforming industries and driving innovation. The accelerated pace of innovation has rendered many of the existing organisational structures, practices and processes obsolete. Therefore, traditional cultures and leadership will not be able to deliver the innovative solutions where digital technologies have changed organisational dynamics and improved communications and relationships with key stakeholders. Memon et al. explore how business model innovations (BMIs) influence and support the effective diffusion of innovation in the construction industry. The authors argue for the need to shift from traditional to new business models to better create, deliver and capture value.

Bajracharya and Dulaimi examine, in Chapters 7 and 8, the dynamics of corporate relationships in the construction industry and how such dynamics influence innovation at the project level. They argue that construction firms and related organisations hardly interact with each other unless they are engaged in a project. The nature of such relationships would limit collective learning and cooperative effort to bring in innovation and improvement in the industry, other than that for time-limited projects. Therefore, the authors present an analysis of the decision-making processes at both the corporate and project level to reflect the hierarchical nature of such activities. The chapters develop a framework and then use a case study to explain how the alignment of motivations of the different entities involved in the project is critical to generating the commitment and momentum to invest effort and resources for the success of innovation.

3. INNOVATIONS IN CONSTRUCTION PROJECT PROCUREMENT

A predominantly project-based industry such as construction is expected to be more concerned about how projects are procured and delivered to client satisfaction. It is not surprising that the interest in innovation in construction has focused research efforts on project procurement. The increased complexity of the buildings and the organisational setup that delivers them has attracted researchers to investigate the strategies, processes and practices and the organisational structures that are believed to influence the effective delivery of projects. It is not surprising that researchers investigated how procurement can support innovation efforts to enable the construction industry to deliver important sustainability goals. The sustainability agenda has presented the construction industry with significant challenges and, at the same time, with significant opportunities for change, development and innovation. Sustainability can be credited for driving innovation in the construction industry, which has faced changes in regulations as well as changes in stakeholders' attitudes towards its products and services. Concerns for the environment, in the main, have motivated governments to issue new laws and set new regulations and standards that forced changes to design, material and component specifications as well as construction practices and facilities management. It is not surprising, then, to find that government policies and regulations have been one of the main drivers for change and innovation in construction.

Social procurement policies have affected change in contractors' behaviour to deliver social innovation. Denny-Smith et al. (Chapter 9) examine social procurement in Australia, where they investigated how social policy related to the indigenous people of Australia has influenced a change in behaviour of construction contractors. Indigenous procurement policy (IPP) is an example of how change in government policy and regulations can motivate or force the construction industry to develop and adopt new practices and change behaviour to meet these new requirements. This change will create demand in the market for clients wanting IPP-compliant projects. Hence, it would justify, for the management of construction organisations, the investment of new resources to facilitate the development of new ways of working capable of delivering a new set of targets and practices. Therefore, clients who can create a significant pipeline of projects could motivate the construction industry to engage such clients and develop a greater understanding of the needs and priorities that would inform innovation efforts. Such innovation and change of behaviour will also deliver social needs and satisfy the priorities that are key elements of sustainable development. The pressure to recruit, employ and engage indigenous workers and businesses has facilitated a better understanding of the potential of such capabilities and, therefore, generated further economic activities. Of course, this will increase uncertainty as contractors explore changing their practices. Denny-Smith et al. conclude that IPP, as a radical innovation, has changed how contractors think, organise and plan their supply chain to engage and collaborate with indigenous subcontractors. This engagement and collaboration effort is expected to create familiarity with indigenous contractors and workers and, hence, generate confidence in their capabilities and that efforts to adopt innovation are likely to generate benefits.

Researchers and practitioners have always argued that project procurement frameworks create challenges and impede efforts to improve construction project performance. Research has investigated the impact and influence of procurement frameworks and strategies on the ability and motivation of the construction team to develop and adopt innovation in construction projects. The fact is that the adopted procurement framework defines the organisational structure, communication channels and power structure of the project organisation, which would shape the innovation ecosystem. Over the last few decades, the construction industry experienced changes in procurement strategies that have created opportunities to explore different working dynamics. In Chapter 11, Walker and Love examine the emergence of integrated project delivery (IPD) and its potential to enhance collaboration and enable clients to effectively engage the supply chain. The alliance created by the IPD and the prospect of a continued pipeline of work packages, such as a 15-year programme, have motivated greater innovation and encouraged collaborative working. The long-term prospect of opportunities to work together seems to have created the justification for the businesses involved in the alliance to invest greater resources and effort to learn from past experiences so that the 'corporate knowledge' can be retained for the benefit of the alliance. Members of the alliance would also see opportunities for greater benefits of innovation beyond the project that triggered that innovation. The long-term engagement of the alliance and their client created the drive for innovation beyond the parameters of a single project, which has been one of the key challenges to motivating innovation in construction (Dulaimi, 1995). The innovation ecosystem that spans several organisations, members of the alliance, developed a more strategic dimension to learning that encouraged collaboration across organisational boundaries. Changes to organisational structures, leadership, practices, processes, attitudes and behaviour are, therefore, essential.

Doloi and Kumaraswamy (Chapter 12) examine how procurement strategies have evolved and how innovation in this area is enabling clients to address emerging challenges. The chapter focuses on developing countries where there is apprehension as to how innovative procurement systems, although arguably capable of delivering better performance, may increase risk for clients, especially public organisations. Collaboration may signal more efficient and effective working arrangements; however, clients in emerging economies are seen to be concerned about collusion in corrupt practices. Further, Doloi and Kumaraswamy present the arguments that test the challenges for public clients using their buying power to generate social value while simultaneously seeking better value for money. This raises the issue of whether we need to monetise social goals to be able to better judge whether we are offering what is of value to our society. Doloi and Kumaraswamy argue that this 'requires applying innovation towards poverty alleviation, improving jobs for marginalised groups, and fostering policies that aim to eliminate poverty'. Procurement frameworks adopted for a particular project will define the communication patterns and relationships, thus influencing organisational learning across the supply chain. The development of innovative procurement frameworks and systems has the potential, thus, to boost the innovative potential of the project organisation.

Eriksson et al., in their chapter (Chapter 10), extend the discussion and analysis that link procurement frameworks and innovation to explore the impact of stakeholders and their perspectives in public procurement on innovation levels. The chapter emphasises the role of the client and other stakeholders in setting the agenda and driving innovation within the supply chain. The authors argue that public procurement is an important policy tool that can drive innovation to deliver social value, in particular sustainable development. One of the main challenges identified here is the lack of resources and capabilities of public organisations needed to be able to articulate their needs and effectively influence and contribute to the procurement decision-making process. Stakeholders should be clear about the tension between short-term and long-term goals and benefits of the proposed projects to be able to make informed decisions and justify the investment. Eriksson et al. argue, based on the reviewed literature, that the different perspectives of internal and external stakeholders would require public clients to have the skills needed to align differing goals and expectations.

4. INNOVATIVE TECHNOLOGICAL SYSTEMS AND ORGANISATIONS

Mahammedi et al. start their chapter (Chapter 15) by examining the state of health and safety (H&S) in construction and highlighting its impact on not only the wellbeing of workers but also how it affects performance and productivity. They discuss the adoption of digital technologies and their role in improving communication and encouraging greater collaboration that would create opportunities to improve H&S in construction projects. Mahammedi et al. describe that 'Health and Safety in Construction Industry 4.0 encompasses a range of technologies and practices that aim to improve safety standards on construction sites'. The increased use of wearable devices, automation and the deployment of drones enable vast amounts of data to be collected and utilised to prevent accidents and improve the wellbeing of workers. Such advanced technologies also provide an alternative to human deployment in high-risk tasks. This chapter provides a review of the various technologies that can be incorporated throughout the lifecycle of a project, including the deconstruction stage, to enhance H&S. The authors further analyse the challenges facing the industry in responding to this opportunity. The challenges cover the substantial cost of investment, the lack of needed skills, as well as the challenges of adopting new technologies by organisations that are based on traditional structures, practices, cultures and mindsets.

Fasaghandis et al. (Chapter 13) propose a tool that can be used in construction to measure corporate progress in terms of developing the environment, abilities and leadership for innovation. The research identifies five categories that cover external and internal aspects that would influence the ability and potential of construction organisations to influence innovation efforts. Several indicators were developed to measure the potential of organisations to deliver successful innovation. The tool is used to guide deeper analysis and facilitate discussions within the relevant organisations to help them understand the opportunities to improve innovation performance.

Saad et al. examine (Chapter 14) how innovation transpires in complex social systems in construction. Their chapter examines the case of modern methods of construction (MMC), where traditional site activities are transferred off-site, with the expectation of reducing waste, enhancing quality, reducing cost and speeding up delivery. The industry uptake of such technologies remains low despite the fact that such technologies have been available for decades. The chapter presents an analysis of public clients, as a major force in the construction industry, and how their perspective and approach to construction procurement have shaped and influenced their decision regarding adopting MMC. One of the arguments presented is the extent to which the prevalent business model adopted in construction can recognise the value of MMC.

AlHarmoodi and Dulaimi (Chapter 16) investigate the impact of stakeholder integration on innovation to deliver sustainability. The concept of open innovation, where the innovation activities do not occur within one organisation (closed innovation) but across a network of collaborating firms, challenges the construction industry to collaborate and address the challenges it faces. Open innovation, a setup with which the construction industry is familiar through its projects, requires the supply chain to collaborate effectively. The integration of supply chain expertise and resources has the potential to strengthen the knowledge base for innovation. The harmonisation, alignment and synergising of such a network of stakeholders will be impacted by the adopted procurement strategies and frameworks, as well as by project management leadership. AlHarmoodi and Dulaimi conclude that 'leadership for innovation highly influences the effectiveness of innovation through providing vision and encouraging innovation, building interested teams with strong team-level and individual-level identity toward the innovation'. Leadership has an important role in 'integrating stakeholders by developing a knowledge base about them, facilitating interactions among them, and adapting to each other's behaviours'.

OVERVIEW AND WAYS FORWARD

The call for the construction industry to be more innovative and to adopt new technologies, materials and practices will continue as it has done for more than three decades. Innovation is critical if the construction industry is to address the challenges and meet the expectations of its stakeholders, in particular the challenge of providing more sustainable facilities and infrastructure. Sustainability is one of the main challenges that will continue to pressure the industry to be more innovative due to the significant impact construction activities have on society and the global environment. The more recent agenda to develop a smart built environment, fuelled by the digital revolution experienced across many industries, has created both the means and a sense of urgency to deliver innovation at a faster pace in construction. The efforts to transform the industry to a 'smart construction industry' could shift focus from the physical structure to the intelligent systems and tools that create, operate and control the built environment. Such a shift could create greater opportunities for interface with other industries, such as the manufacturing and information technology industries, where the pace of innovation is evidently faster than that in construction. It is less certain whether such pressures will motivate the construction industry to increase its investment in innovation.

The different chapters in this book will present a comprehensive and detailed review of research into innovation in construction. Such reviews will show repeated arguments that there is a significant gap between the technology and practices adopted in the construction industry and that of other sectors. A recurring theme of reports and research criticises the construction industry's investment record in innovation and R&D. The conservative nature of construction could explain why there are repeated calls for government intervention, as a client and policymaker, to drive and motivate the construction industry to develop and adopt new ideas and innovations.

The book will provide researchers with the opportunity to explore and investigate the forces that motivate and encourage organisations, teams and individuals to take the risk of developing and adopting new ideas. Construction projects require several business organisations, with variable capability, to work collaboratively to deliver products and services that satisfy several stakeholders while meeting their organisations' objectives to remain stable and successful. That is why research into innovation in construction is complex and multi-dimensional.

Research needs to focus on the impact of project procurement strategies on the ability, willingness and effectiveness of the supply chain to support a focused innovation effort. There is greater scope for research effort to critically examine our knowledge and understanding of the corporate 'mind' that influences and drives the decision-making process at corporate and project levels. The short-term mindset of the industry can be linked to how businesses in construction operate and function in order to remain stable and profitable. Several chapters in this book examine business models that enable corporations to function effectively and efficiently. National policies and regulations play an important role in shaping such mindsets and influencing the decision-making processes that define corporate roles in any innovation opportunity. Strong leadership at corporate and project levels could drive a strong agenda for collaborative working to motivate the development of knowledge and skills needed to deliver successful innovation.

This book will provide ample evidence and examples of the urgency and the pressure on the construction industry not only to be more innovative but also to accelerate the pace of such efforts to be able to, at least, accommodate the change in the way we live and work due to the digital revolution.

REFERENCES

Amabile, T.M. (1997) Motivating creativity in organizations: On doing what you love and loving what you do. *California Management Review*, 40(1), 39–58.

Chesbrough, H. (2003) *Open Innovation: The New Imperative for Creating and Profiting from Technology*. Boston, MA: Harvard Business School Press.

Chesbrough, H. and Bogers, M. (2014) Explicating open innovation: Clarifying an emerging paradigm for understanding innovation. In H. Chesbrough, W. Vanhaverbeke and J. West (eds), *New Frontiers in Open Innovation* (pp. 3–28). Oxford: Oxford University.

Drucker, P.E. (1985) The discipline of innovation. *Harvard Business Review*, May–June, 67–72.

Dulaimi, M.F. (1995) The challenge of innovation in construction. *Building Research and Information*, 23(2), 106–109. doi:10.1080/09613219508727437.

Dulaimi, M.F. (2022) The climate of innovation in the UAE and its construction industry. *Engineering, Construction and Architectural Management*, 29(1), 141–164. https://doi.org/10.1108/ECAM-07-2020-0492

Dulaimi, M.F., Baxendale, A.T. and Langford, D.A. (1996) The construction process as a process of innovation. Proceedings of the CIB W65 'The Organisation and Management of Construction: Shaping theory and practice', Glasgow, UK, 572–577.

Dulaimi, M.F., Nepal, M. and Park, M. (2005) A hierarchical structural model of assessing innovation and project performance. *Construction Management and Economics*, 23(6), 565–577.

Egan, J. (1998) *Rethinking Construction. The Report of the Construction Task Force*. London: Department of Environment, Transport and Regions.
Gann, D.M. and Salter, A.J. (2000) Innovation in project-based, service-enhanced firms: The construction of complex products and systems. *Research Policy*, 29(7–8), 955–972.
Guan, J. and Liu, N. (2016) Exploitative and exploratory innovations in knowledge network and collaboration network: A patent analysis in the technological field of nano-energy. *Research Policy*, 45(1), 97–112. https://doi.org/10.1016/j.respol.2015.08.002
Jacquetta, S., Igor, M. and Chuan, C. (2016) Implementation of innovative technologies in small-scale construction firms: Five Australian case studies. *Engineering, Construction and Architectural Management*, 23, 177–191.
James, L. and Sells, S. (1981) Psychological climate: Theoretical perspectives and empirical research. In D. Magnussen (ed.), *Towards a Psychology of Situations: An Interactional Perspective* (pp. 275–295). Hillsdale, NJ: Erlbaum.
Jelodar, M.B., Wilkinson, S. and Tooky, J. (2018) 'Construction Innovation: A system controlled by construction project dynamics', RICS Cobra, 23–24 April, London, UK.
Kanter, R.M. (1988) When a thousand flowers bloom: Structural, collective, and social conditions for innovation in organization. *Research in Organizational Behaviour*, 10, 169–211.
Koys, D.J. and DeCotiis, T.A. (1991) Inductive measures of psychological climate. *Human Relations*, 44(3), 265–285.
Latham, M. (1994) *Constructing the Team*. London: HMSO.
Liker, J.K., Collins, P.D. and Hull, F.M. (1999) Flexibility and standardisation: Test of a contingency model of product design-manufacturing integration. *Journal of Product Innovation Management*, 16, 248–267.
Lloyd-Walker, B.M., Mills, A.J. and Walker, D.H.T. (2014) Enabling construction innovation: The role of a no-blame culture as a collaboration behavioural driver in project alliances. *Construction Management and Economics*, 32, 229–245.
Nam, C.H. and Tatum, C.B. (1992) Strategies for technological push: Lessons from construction innovations. *Journal of Construction Engineering and Management*, 118(3), 507–524.
Nam, C.H. and Tatum, C.B. (1997) Leaders and champions for construction innovation. *Construction Management and Economics*, 15, 259–270.
Park, M., Nepal, M.P. and Dulaimi, M.F. (2004) Dynamic modeling for construction innovation. *Journal of Management in Engineering*, 20(4). https://doi.org/10.1061/(ASCE)0742-597X(2004)20:4(170)
Porter, M.E. (1990) The competitive advantage of nations. *Harvard Business Review*, March–April, 73–93.
Powell, W.W. and Grodal, S. (2005) Networks of innovators. In J. Fagerberg, D.C. Mowery and R.R. Nelson (eds), *The Oxford Book of Innovation* (pp. 56–85). Oxford: Oxford University Press.
Roberts, E.B. and Fusfeld, A.C. (1981) Critical function: Needed roles in the innovation process. In R. Katz (ed.), *Career Issues in Human Resource Management* (pp. 182–207). Hoboken, NJ: Prentice Hall.
Scott, S.G. and Bruce, R.A. (1994) Determinants of innovative behaviour: A path model of individual innovation in the workplace. *Academy of Management Journal*, 37(3), 580–607.
Slaughter, S. (1998) Models of construction innovation. *Journal of Construction Engineering and Management*, 124(3), 226–231.
Tatum, C.B. (1987) Process of innovation in construction firms. *Journal of Construction Engineering and Management*, 113(4), 648–663.
Tatum, C.B. (1989) Organising to increase innovation in construction firms. *ASCE, Journal of Construction and Engineering Management*, 115, 602–617.

Walker, D.H.T. (2016) Reflecting on 10 years of focus on innovation, organisational learning and knowledge management literature in a construction project management context. *Construction Innovation*, 16; 114–126.
Winch, G. (1998) Zephyrs of creative destruction: Understanding the management of innovation in construction. *Building Research and Information*, 26(4), 268–279.
Wittmann, C.M., Hunt, S.D. and Arnett D.B. (2009) Explaining alliance success: Competences, relational factors, and resource-advantage theory. *Industrial Marketing Management*, 38(7), 743–756.

2. Growth, productivity and structural change: economic arguments for investment in construction innovation

Gerard de Valence

1. INTRODUCTION

Research on economic growth has focused on understanding the causes of growth in Western Europe and the US during the first industrial revolution. Although economic activity is a complex phenomenon and a range of causal factors have been identified, they all have in common a lowering of barriers to innovation, allowing increased creation of new products and processes in agriculture, industry, and transport. With innovation and sustained technological progress, per-capita income rose as productivity increased and industrialisation led to the restructuring of economies around the emerging industries.

Early theories of economic growth focused on the role of investment and the accumulation of buildings, structures, machinery, and equipment that comprise the capital stock of an economy (Solow 1957). As national accounting developed and more data became available, the role of construction in long-run economic growth became a major research topic as the relationship between infrastructure investment, the capital stock, and GDP growth rates was established (Swan 1956). The focus then moved from the quantity of the capital stock to its quality and the effects of embodied technology on productivity when replacing old, worn-out machinery and equipment with new models and higher levels of performance, or in many cases, completely new capabilities (Kendrick 1973). Then, the explosive growth of the information technology (IT), computing, and telecommunications industries provided a new model of industry development driven by competition, research and development (R&D), and innovation (Katz and Shapiro 1985).

By the 1990s, innovation had become central to models of long-run economic growth. The development of new growth theory by Romer (1990) was based on technological opportunity and the appropriability of knowledge. The primary factor influencing the amount and type of R&D investment undertaken is the private return it generates. In these models, investment in R&D, innovation, human skills, and physical capital leads to spillover effects and positive externalities, resulting in increasing returns to scale at the macro level that drive long-run growth. There is now extensive research on the positive relationship between R&D and national economic growth rates, productivity levels, and per-capita income in countries and regions worldwide (OECD 2015). Therefore, many countries developed policies to promote innovation in some form (Warwick 2013, Bloom et al. 2019).

Innovation and technological change are closely related. Typically, the former is seen as leading to the latter, which is a broader concept that includes the externalities involved and the speed of diffusion of inventions and innovations throughout the economy. Innovation is regarded as the crucial element in the development of modern industry because it is responsible for the direction and rate of technological change, which varies both between industries and within industries as they go through periods of growth, competition, and restructuring as the technology they are based on goes through cycles of invention and innovation (Hughes 1989). This process pushes the technological frontier outward as firms that are successful adopters of new technology replace firms that fail, as high-R&D and high-productivity firms replace low-R&D and low-productivity firms (Dosi 1988).

Therefore, the role of new technology in economic growth is crucial because new investment does not merely duplicate old capital but embodies technical progress, shifting production to a new level. Investing in innovation and human and physical capital is important for generating the externalities that lead to growth and further innovation. An important factor is the industrial structure of an economy, as a small number of industries are responsible for a large share of R&D expenditure and patenting, such as the information and communication technology (ICT), electronics, biology, pharmaceuticals, aerospace, and automotive industries (Hausmann and Hidalgo 2011). Particularly in the US, the ICT-intensive industries between 1995 and 2005 that were high-productivity growth industries (Gordon and Sayed 2020) were also industries that heavily invested in R&D.

The industrial clusters and networks that support high-R&D industries have developed over many decades, and most of that R&D is done in the home countries of the major firms. The development of these high-tech industries in the second half of the twentieth century was extensively studied and documented, especially the computing hardware and software industries (Sutton 1999). These industries have a few large firms and a dominant technology that has gone through several generations over a few decades, making them ideal case studies of firm, product, and process lifecycles. Major manufacturing industries are similar, where detailed national data based on the System of Industrial Classification (SIC) allowed structural change in manufacturing to be researched, linking innovation to SIC codes, entry and exit of firms, and investment in R&D, patents, and copyrights. The Construction SIC has one of the lowest R&D expenditures in OECD countries, giving it the reputation as a low-innovation, low-productivity growth industry.

With a narrow view of construction innovation, only the contractors and subcontractors in the Construction SIC are included, but this chapter takes a broad view of construction innovation that includes related industries in manufacturing, materials, and professional services (de Valence 2019). The firms involved in construction are an example of a production network made up of industries where relevant data is spread across different SIC classifications. Construction involves many firms and organisations in the design, construction, and maintenance of the built environment, and this value chain and the way it is organised into a system of production is wide, deep, and complex (Hobday 1998). Firms from many different industries participate, together making up a large industrial sector linking mining, manufacturing, professional services, contractors, and subcontractors. One reason why innovation is hard to see clearly in construction is because so much happens in these related industries rather than onsite. Further, innovation is harder to see and study when production is geographically dispersed and project based (Gann 2003).

The remainder of the chapter is organised into seven sections. The following section is an overview of innovation as an investment that is neither riskless nor costless, but driven by competition and major new technologies, which take at least a couple of decades to be adopted by the industry. Section 3 looks at innovation in construction products and two distinctive characteristics. First, it is typically incremental, proceeding in discrete steps over long periods of time, with the development of concrete over more than 100 years given as an example. Second, many construction product and process innovations are hidden from consumers and users.

Section 4 is a discussion on the dynamics of innovation, diffusion, and productivity. Section 5 focuses on innovation and procurement. Construction is in a catch-22 situation, where neither contractors nor clients can rationally commit to significant, expensive investments in innovation for the vast majority of construction projects. Section 6 discusses funding for R&D from private equity and venture capital funds. This is a significant metric because investment in construction technology startups is a proxy for innovation and the development of IP. Section 7 looks at the role of government and two areas where governments have successfully promoted (and invested in) innovation through regulation: building standards and decarbonisation, and building information model (BIM) mandates. Section 8 discusses Construction 4.0 technologies and the organisation of construction. There is a short conclusion on technical, regulatory, and institutional aspects of innovation in construction.

2. INNOVATION AS INVESTMENT

Schumpeter (1942: 83) used 'perennial gales of creative destruction' to describe the long-term effects of innovation and technological change on industries and firms. Firms search out new goods and new production processes to increase profits, so competition between firms is not only based on costs and prices but also on the R&D that leads to new products, new methods of production, or new forms of organisation. For Schumpeter, this was technological progress. Because of the search for profits, another innovation comes along, allowing the development process to restart and the cycle to repeat. The clustering of innovations and their cascading effect on economic development was at the core of Schumpeter's business cycle theory, an idea since developed by Perez (2003), who used the clustering of innovations to explain why each technological revolution gives rise to an 'opportunity explosion' focused on specific industries, which leads to overinvestment and financial bubbles.

The development of new technology requires investment in R&D and intellectual property (IP). If firms cannot capture the benefits of innovation and IP for some reason, because of imitation, piracy, or secure supply of materials, for example, they will not invest in innovation. Technology and R&D have significant costs associated with them, and Schumpeter recognised that investment in R&D was a calculated risk for a firm but, if successful, would benefit society as a whole. The model is based on the rate of return in determining investment in innovation and the role of firms in deciding to invest in innovative activity.

Schumpeter made two key propositions: there is a relationship between the size of a firm and investment in innovation, with most innovation done by a relatively small number of large firms; and there is a relationship between innovative activity and a firm's market power, which is the extent of control a firm has over product and supplier prices. In Schumpeter's model of the economy, the main agents for change are entrepreneurs, who see that improvements and progress can be made, and he compared them with the capitalists who bear financial risks, inventors who have the ideas, and managers who are concerned with production and distribution. Fundamentally, innovation is neither costless nor riskless; it is an investment in the future.

The concepts of linkages and diffusion were used to develop a theory of economic growth through leading industrial sectors with strong linkages to other industries. Initial investment creates knowledge for the investor and others to make more informed decisions about future investments. The 'leading sectors' in Schumpeter's theory are those that generate the most externalities and where diffusion of technologies to other industries occurs most rapidly. Romer's (1990) 'new growth theory' brought technological progress inside these models of economic growth by allowing innovators to capture the benefits of their R&D.

In one way or another, research on innovation and technological change over the last 80 years has addressed some aspect of Schumpeter's model. These decades saw the rapid development of high-tech industries like aerospace, automobiles, pharmaceuticals, computing hardware, software, and telecommunications, so his influence on the way innovation and technological change is analysed has been profound. In those industries, the drivers, characteristics, and dynamics of technological change have been the subject of intense interest and refinement. Innovation is now understood to be radical, incremental, or disruptive; technological change can be embodied in physical capital or disembodied as intellectual property (Dosi et al. 1994). The diffusion of general purpose technologies (GPTs) like the internal combustion engine, electricity, and computers takes at least a couple of decades, and they are taken up at different rates in different industries, depending on where an industry is in its lifecycle (Helpman and Trajtenberg 1990).

Innovating in a large, complex, and long-established industrial sector like construction can be difficult. The institutional architecture can impose regulatory hurdles or other policy disadvantages on new technologies, and government expenditures often support existing technology. Lenders and insurers are risk-averse. There are subsidies and price structures that favour incumbents and ignore externalities like the environment and public health. Educational curricula, career paths, and professional standards are oriented to existing technology (Bonvillian and Weiss 2015). The dominance of existing technologies is reinforced by market imperfections such as network economies, lumpiness, split incentives, and the need for collective action (Mowery and Rosenberg 1998). New technologies have to significantly lower costs and improve performance to successfully compete with existing technologies and vested interests. These are all factors that inhibit innovation and long-run economic growth. However, failure to adopt new technology can have many reasons. It is embodied in expensive capital goods and requires scarce complementary factors such as infrastructure and skilled labour. Outright resistance by workers was widely seen historically (Mokyr 1994).

How firms utilise technological capabilities differentiates them within a diverse, location-based production system (Bresnen and Marshall 2001). It is widely recognised there are differences between industries in the way that technology is adopted, adapted, and applied, but differences within industries generally get less attention. Innovation within industries depends on the different individual characteristics of firms, which affect the rate and extent of adoption of new technologies, the effects of competition, and how the adoption of new technology by one firm influences the adoption of technology by other firms (Bloom and van Reenen 2010). For construction this is significant, because the many small and medium-sized construction firms are technology followers; they are not typically early adopters and cannot invest in R&D. Technological inertia can be the outcome of rational behaviour by utility-maximising individuals and profit-maximising firms (Mullett 2022).

However, a global contractor might have over 50,000 employees. The suppliers of basic materials and sophisticated components are large multinational or multilocational industrial firms, and there are large global consultants and project managers. Many of these firms are publicly listed with the management and financial resources required to invest in twenty-first-century technology, if and when they decide to do so. The ability of firms to capture knowledge externalities, innovate around new technologies, and adapt to the impacts of emerging technologies and their requirements is the crucial issue.

Modern construction is a technocratic system of production using industrialised materials like glass, steel, and concrete (Pfammatter 2008). It is based on accepted standards and codes, contractually defined professional roles, and there is a high degree of technological lock-in (David 1985) due to the culture, age, and complexity of the system (Davis 2006). The 'embeddedness' of construction technology is found inside professional institutes and organisations, trade and industry associations, government regulations and licensing, building standards and codes, insurance, and finance. Innovation is difficult but not absent because the economic basis for the organisation of construction also favours innovation that is slow and gradual, as the introduction of new products demonstrates.

3. INNOVATION IN CONSTRUCTION PRODUCTS

One of the methods used to measure innovation is the share of a company's revenue due to new products or new versions of existing products (Björk et al. 2023). This is not exact; a new product will typically require years of R&D, and then follows an adoption path to gain market share over periods ranging from a few years to several decades as it improves and becomes more widely used. Increasing adoption results in improved versions, wider use, and increased diffusion of the product. Perez (2003) called these the installation and deployment phases of the technology cycle.

Industry R&D statistics typically only include expenditure by research labs and institutions, but firms' investment in the equipment, software, and training needed for innovation is found in their capital expenditure. Innovation data from national statistical agencies does not capture organisational innovations like lean construction or production improvements that increase efficiency, such as the lifting capacity of tower cranes. These agencies also do not usually collect data on new products or services, so patent and copyright data is often used instead. A further complication arises from the fact that many new products are improved versions of existing ones. The focus here is on two distinctive characteristics of innovation in construction products: it is typically incremental, proceeding in discrete steps over long periods of time, and many product innovations are hidden and not apparent to users because they are inside the structure. This is followed by a short discussion of the effect these characteristics have on the measurement of real output construction productivity.

Incremental Innovation

Firms refine and improve products and processes over time. Although this form of innovation is incremental, it should not be dismissed as unimportant. An example is the steady increase in lifting capacity of tower cranes since their invention in 1949. Another example is the development of computer-aided design (CAD) software, which went on for two decades before Autodesk was started in 1982, one year after the first IBM PC. Over the decades, building information models (BIMs) have advanced through 2D and 3D versions to 4D (schedule) and 5D (cost) iterations. Now, software linked to cameras on helmets or drones can provide real-time augmented reality (AR) images from a building site linked to the BIM of the project.

These incremental innovations improve existing processes in construction, with many of the technical advances coming from the materials suppliers or component, plant, and equipment manufacturers who have been responsible for the introduction of new products such as excavators, cranes, facades, and lifts. These incumbent firms are part of a network, where some sub-markets are concentrated and oligopolistic (e.g., cement, lifts and elevators, glass facades), with a few large, incumbent firms exactly like those Schumpeter suggested would be most likely to engage in R&D, invention, and innovation. Across the construction supply chain, these incremental innovations and the firms that develop or exploit new technologies don't create new industrial networks but join the existing one.

Concrete is an example of how effective incremental innovation in construction materials can be. By the 1880s, the increasingly widespread use of concrete had changed its status from a hobby to a modern industry, as scientific investigation into its material properties revealed its shear and compressive characteristics. With the development of reinforced concrete, there was change in architectural concepts of structures and approaches to building with concrete (Peters 1996). The industrial standards of concrete technology influenced ways of thinking based on building systems and standardised building elements. These became identified with what was known as the Hennebique System, a simple-to-use system of building with reinforced concrete columns and beams patented in 1892. By 1905, the Hennebique System had spread across Europe and elsewhere, and his company employed 380 people in 50 offices with 10,000 workers onsite (Pfammatter 2008).

Concrete then set the agenda for the development of construction as a system of production over the next hundred years, driven by the modernist movement in architecture as it explored the possibilities of this material for increasing the height and scale of buildings, and modern construction materials and methods spread around the world (Cody 2003). Concrete shows how incremental innovation in materials played a significant role in the reorganisation of site production methods as mixers, pumps, and chemicals were refined and developed in a long process of interconnected innovations. One of the characteristics of a successful technology is these spillover effects, with advances in one industry leading to complementary developments in related industries. A similar story of long-term incremental improvement can be seen in other construction materials like glass and ceramic products, in building services such as air conditioning, lifts, and elevators, and in the performance of equipment such as cranes, excavators, and trucks.

For over a hundred years, since Hennebique, there has been ongoing refinement and development of the world's most widely used construction material, as shown in Table 2.1. Current developments in 3D concrete printing (3DCP) combine BIM, new concrete mixtures and chemicals, new pumps, and printing machines (Sanjayan et al. 2019). As was the case in the early 1900s, a combination of new materials and new machinery is required for a new technology to work onsite at scale.

The information needed to create a 3D blueprint is generated during design, and it is a relatively small step to move from a BIM model to instructions for a 3D printer. There are three methods for 3D printing: stereolithography, patented in 1986; fused deposition modelling, patented in 1989; and selective laser sintering, patented in 1992. In the mid-1990s, research into 3DCP was focused on developing the equipment needed and the performance of the materials used, often in collaboration with one of the multinational concrete producers. Twenty years later, there were over a dozen experimental prototypes built, documented in Sanjayan et al. (2019) with details on the materials science required to identify successful mixtures and admixtures.

By 2022, the commercialisation of 3DCP was under way, with two types of systems available: one using a robotic arm to move the print head over a small area, intended to produce structural elements and precast components, and the other a gantry system for printing large components, walls, and structures. In 2023, the Additive

Table 2.1 *Incremental innovation in concrete since 1800*

Date	Innovation
1800	Portland cement
1867	Monier reinforcement system
1892	Hennnebique System
1911	Concrete mixer
1913	Ready-mix concrete
1918	Duff Abram's water/cement ratio law
1925	Concrete pumps and vibrators
1928	Freyssint pre-stressed concrete
1938	Air entrained agents
1952	Silica fume
1955	Chemical admixtures, slipforming
1960	Fibre-reinforced concrete
1964	Superplasticers and high-performance concrete
1980	Roller-compacted concrete
1991	Fibre-reinforced polymers
2001	Porous concrete
2003	Self-compacting concrete
2011	New generation of sealants and additives

Source: Jahren (2011).

Manufacturing Marketplace had 34 concrete printing machines listed, ranging from desktop printers to large track-mounted gantry systems that can print three- or four-storey buildings. There are also companies offering 3DCP as a service at an hourly or daily rate.

Significantly, many 3DCP companies are new entrants, including the providers of printing mixes that supply the 3DCP industry, some with funding from industry majors. Other entrants have corporate backing (Black Buffalo by Hyundai, for example), but the great majority are recent startups funded by private investors. Many of these entrants can be expected to fail or be taken over by incumbents.

Hidden Innovation

Hidden innovation improves the performance of a product but is not visible to the user or customer (Dosi et al. 1994). Examples of hidden new technologies in automobiles are disk brakes and electronic ignition, neither of which would register with most car drivers. A great deal of construction innovation is hidden, either inside the structure or in process innovations in onsite work, such as powered hand tools or new lifting and hoisting equipment. Higher standards for building materials' fire, thermal, or seismic performance are not visible; improvements in building air quality or energy use are not obvious to users, and products like ducting, composite wiring, and plastic pipes are inside the structure. Hidden innovation is often an incremental upgrade of a current product or system, and process innovations that improve site efficiency do not benefit clients and users directly.

Piping and plumbing provide good examples of hidden innovation in construction of the built environment. Flushing toilets are one of the most successful inventions of all time and are themselves visible. However, the piping, drainage, and sewerage systems required, and the pumps and equipment needed to make the system work, are largely out of sight. Flushing toilets, electricity, central heating, and ducted air conditioning were new functions that became incorporated into housing, and by the mid-twentieth century, building design and materials were regulated by codes that set performance standards. The functionality of housing has not fundamentally changed since, and continuing production, product, and process innovations have been hidden from consumers and users.

Construction is a dispersed market with many customers requiring different functions spread out over different locations. Contractors are typically local firms, and their suppliers range from small workshops to giant multinational corporations, allowing opportunities for substitution between products that are inside the structure and hidden from users and consumers. Local markets also have characteristic preferences in design and materials, particularly for houses. With design decisions and specifications made from project to project and familiar products usually favoured, gaining market share in construction for a new product will be a slow process that requires commitment and resources to be successful. This characteristic also means many innovations have never achieved scale or significant market share.

Consider the market for external cladding for buildings. To the traditional materials of wood and brick have been added glass, concrete, metal, ceramic, vinyl, composite, and fibre cement alternatives, all of which have continuing shares of the total market. There are over a dozen major manufacturers who all have large market shares in some, but not all, regional markets (like Europe, East Asia, Central, South, and North America). A recent innovation is HardiePlank, a fibre cement panel invented in Australia in 1986 that took 20 years in the US to reach 20% of the housebuilding cladding market (Hardie 2022). Contractors are responsible for process improvements that increase onsite efficiency and rely on manufacturers for product innovation. The network of relatedness between products means manufacturers typically develop goods and services close to those currently produced.

Incremental and hidden innovation have affected the measurement of construction productivity through the price indexes used to estimate real output, which overstate price growth because they do not fully capture quality or product efficiency changes, and they do not capture changes in the efficiency of delivering services when new products are introduced (de Valence 2001). The resulting underestimation of real output reduces the level of productivity as output per person employed or hour worked. These general problems become acute during periods of major technological revolution because price indexes do not adjust for quality and performance improvements or the introduction of new products. In his study on lighting, Nordhaus (1997) divided product changes in the long run into three categories: 'run-of-the-mill changes', where price indexes miss relatively little of the quality change or impact of new goods; 'seismically active sectors', with major changes in the quality and provision of goods; and 'tectonic shifts', where the production process has changed radically, to such an extent price indexes do not attempt to capture the qualitative changes. In the twentieth century, tectonic shifts included household appliances, medical care, utilities, telecommunications, transportation, and electronic goods. He noted, 'many tectonic shifts are buried in run-of-the-mill or seismically active sectors. For example, the lowly toilet is classified as furniture but delivers "a service that would delight a medieval prince"' (Nordhaus 1997: 58).

4. INNOVATION, DIFFUSION, AND PRODUCTIVITY

There are major differences between industries in rates of technological progress and productivity growth, and those differences have been persistent for long periods of time. Efforts have been made to find both theoretical and empirical explanations for the factors that drive these differences, and one explanation is R&D intensity, as a measure of the extent of innovation in one industry compared with another (Sutton 1999). Research intensity is typically measured by R&D as a percentage of sales or income.

R&D intensity in an industry is determined by two key variables. One is technological opportunity, which determines the productivity of R&D and the potential returns available from innovation. The second is the ability of innovating firms to appropriate a significant share of the economic value created by an innovation, or to capture the externalities created through new knowledge (Nelson and Winter 1982). Although both influence R&D intensity, technological opportunity determines the rate of technological progress in an industry in the long run. Nelson and Wolff (1997) argue cross-industry differences in technological opportunities are due to R&D opportunities differing between industries. In turn, differences between industries in technological progress will be driven by differences in R&D intensity, appropriability, and available opportunities. Their theory is that industries with high R&D intensity and technological opportunity must have a high rate of new technological opportunities to make up for those that are being exhausted.

Diffusion is the time taken for an innovation and its use to become widespread across firms and industries. Because industries develop along specific technological trajectories that typically have decreasing exploitable technological opportunities, the boost to industry productivity of a new technology or product depends on how easy it is for new ideas and knowledge to be transferred to other firms in the same industry (Helpman and Trajtenberg 1990). As a technology or product becomes widely used, the focus then turns to incremental improvements. The initial boost to productivity and efficiency runs out of steam and wears off, and productivity growth declines (Gordon and Sayed 2020).

How and why a new technology spreads through the economy and society is determined by many factors; however, studies of historical cases such as tractors, electricity, TV, the internet, and phones have given good examples of technology diffusion and the time taken for the parallel changes in forms of organisation, methods of production, and patterns of consumption that are required. These are not decisions firms and households make lightly, and historical experience shows it takes 15 to 30 years for a new technology to reach 90% of its potential market (Lipsey et al. 2005). However, while economic systems adapt over time, they do not necessarily produce optimal outcomes and can be subject to technological lock-in, where sub-optimal technologies become established and entrenched, as suggested by Arthur (1989) and David (1985).

Would technological lock-in be an explanation for the lack of productivity growth in construction? The use of industrialised materials has required modern methods for decades. There have been constant incremental improvements in materials and manufactured products and equipment, and regular updating of the relevant standards and codes. Research by Garcia and Molloy (2023) found, for the US from 1987 to 2019, a small increase in productivity has been absorbed by higher but unobserved (i.e., not in the data) quality. The result was no growth in construction industry productivity over the period, but also no evidence of lock-in or use of inferior technology.

This is the mystery of construction productivity: why has there been no increase in labour productivity since the first attempts to measure it in the 1960s? Despite the many efforts made by governments and industry organisations over the decades, and despite the significant improvement in human and physical capital, the rate of growth of construction productivity has been poor since the 1960s (Cassimatis 1969, McKinsey 2017). Measurement issues leading to underestimation of output are widely believed to be the main problem; however, this may not be the case. Goolsbee and Syverson (2022) found some underestimation but concluded the lack of growth in construction productivity is real.

Could construction be a highly developed system of production with a high level of technical efficiency? Technical efficiency is defined as the ability of firms and industries to produce as much output as possible, given the inputs of labour and capital used and the level of technology available. At maximum efficiency, increasing output requires adding another input to the system of production, such as an extra worker or another machine. Research using data envelopment analysis (DEA) measures the efficiency of firms and industries, defined as the ability of firms and industries to produce as much output as possible, given the inputs of labour and capital used and the level of technology available.

There is a relationship between technical efficiency and productivity. The same inputs of labour and capital are used, but efficiency is the quantity of output given the inputs, while productivity is the ratio of output to those inputs. Labour productivity, for example, is the number or value of units produced divided by the number of hours worked or the number of people employed. The level of technical efficiency is determined by the use of the capital stock available to workers. As the level of capital per worker (machines, software, etc.) increases so does output per worker. However, as the level of output per worker increases, it approaches the limits of the technology currently used in production and, at a high level of technical efficiency, productivity growth is no longer possible.

There have been DEA papers on construction in Spain (Fernández-López and Coto-Millán 2015), Sweden (Edvardsen 2004), Europe (Nazarko and Chodakowska 2017), Hong Kong (Wang and Chau 2001), China (Hu and Liu 2018), New Zealand (Abbott and Carson 2013), and Australia (Chancellor 2015). The DEA studies find that the overall level of technical efficiency in construction is high, and for the best firms, very high, and is strongly pro-cyclical, rising and falling as the volume of work increases and falls. The industry in all the countries where construction has been analysed with DEA is efficient, based on data on the volume of work, industry value added, capital stock, and employment.

This can go a long way in explaining the productivity problem. When the level of work is high and increasing, productivity improves until the industry is approaching the efficiency frontier, where more workers are needed to increase output. Therefore, productivity stops growing. As the volume of work falls during the contraction phase of the building cycle and firms retain workers in the expectation of future work, the level of industry productivity falls, ending up where it started.

If productivity growth is no longer possible with the technology currently used in the system of production, which in construction has been developing for well over a hundred years, the industry will focus on efficiency and getting the most out of the technology, labour, and capital available. Efficiency trumps productivity in construction with current technology.

Therefore, to move the industry to a higher level of productivity, new technology will be required beyond the technologies currently used for prefabrication and offsite manufacturing (OSM). Technologies that are commercially available include 3D printing, drones, exoskeletons, video and laser scanning, Internet of Things (IoT) linked sensors, cloud computing, simulation, and augmented reality. Technologies that are under development include actuators, robots, AI, big data and data analytics, reality capture, and blockchain applications. Collectively, these are often called modern methods of construction (MCM), or Construction 4.0.

As these technologies become available to the industry, many after decades of development, their use will lead to further innovations in site organisation and management, the mechanisation and automation of tasks, and the production of elements and components. For example, online platforms link design libraries with production by prequalified fabricators for OSM for automated production of building elements and components (Wei et al. 2020). These digital platforms provide designs and specifications for manufacturing, often through a standardised kit-of-parts. Platform-based industrialised building is inherently more flexible than a factory-based, capital-intensive system of production. By aggregating buyers of building components, a platform solves the fundamental problem of OSM, which is maintaining a sufficient volume of production at a factory's break-even level or more.

There are fast-growing firms that are technology leaders, some of whom are incumbents, but many are entrants. These 'frontier firms' are pushing the technological frontier through innovation, experimentation, and development (Andrews et al. 2015). Frontier firms bring with them radical new production technologies and new organisational methods. Some of these firms are new entrants (e.g., 3DCP startups, Procore, Built Robotics, FBR), but incumbents are also on the frontier. Examples are the partnership between Trimble and Microsoft, companies like Autodesk, Skanska and ARUP, and remote-controlled excavators from Caterpillar and Komatsu. High-productivity firms at the technological frontier invest in R&D and innovate to take advantage of technological opportunities, whether incumbents or new entrants.

5. CONSTRUCTION INNOVATION AND PROCUREMENT

The method used for procurement has a significant effect on technological opportunity and innovation. The effects of appropriability of IP, substitutability between suppliers, and risks associated with innovation for clients are a set of mutually reinforcing factors that have worked against innovation in construction. Because the traditional design–bid–build method does not allow capture of IP and knowledge externalities by contractors in their tenders, there is a perverse disincentive to innovate. Tendering rules or codes have been developed to maintain the integrity of the bidding process, not to encourage innovation, and a successful tenderer's scope to be innovative is limited. There is an opportunity to maximise profits within the tender price by novel ways of organising work or driving down subcontract prices, but bidders are not asked to put forward design suggestions, there are no criteria for evaluation of novel proposals, and tenderers cannot be treated equally if one is preferred on an alternative tender that is non-conforming in terms of the original invitation (Craig 1997).

With non-traditional procurement methods such as design and construct (D&C), build, own, operate (BOO), or build and maintain (B&M), this disincentive is reduced. Firms can appropriate the benefits of innovation and R&D efforts through improved performance. Contractors can make novel proposals to owners, and society benefits from such innovation. The answer proposed by Craig and others is that the best way to increase innovation lies in changing the methods and systems used to procure building and construction projects.

It may not be that simple. If all firms have access to the same technology and can survive by continual, but gradual, improvement, they are subject to a 'ratchet effect' (Weitzman 1980). First identified in the 1930s by sociologists studying workers subject to performance pay, they found that workers choose to restrict their output because they rationally anticipate that employers will respond to higher output by raising output requirements or by cutting piecework pay or worker incentives within firms (Ickes and Samuelson 1987). It was also an unintended consequence of Soviet planning (Weitzman 1976). If a factory met or exceeded its planned target, the target for subsequent years was increased, thus reducing incentives and effort for the factory manager.

In construction, the ratchet effect can be seen in bidding for projects, where tenderers will typically not deviate far from a client's expected cost for the project (Park and Chapin 1992), and all tenderers have access to the same information (Kagel and Levin 1986). Because of the ratchet effect, a firm avoids revealing a significant cost advantage on one project that might jeopardise margins on future projects. Importantly, it allows for innovations that improve productivity and efficiency, which are neither disruptive nor expensive to contractors but will deliver a windfall gain if a project comes in well under budget. These gains will be hidden from the client and competitors as much as possible. This suggests that there will be cost-reducing innovations available to contractors if they decide to invest, but the pressure to find them will be affected by client demands, upfront costs, market conditions, and a competitor's likelihood of using them.[1]

If clients avoid the risk associated with innovation on their projects and do not include it in their budgets, they act as a significant barrier to innovation because they are concerned about both construction costs and operating costs, and do not think they individually will benefit significantly from a successful innovation. Ivory (2005) suggests clients carry a 'disproportionate' share of innovation risk and, as a result, 'should not be routinely expected to take on the risks (in terms of cost and time overruns or poor building performance) and costs associated with innovation' (Ivory 2005: 868–9). This risk minimisation objective also applies to financiers and insurers of construction projects.

While Ivory's argument might be generally true, exceptions prove the rule. An example is the Heathrow Terminal 5 (T5) project. This project demonstrates the effect a determined client with a clear strategy to encourage innovation to improve performance can have. Innovation on the project was documented by Davies et al. (2009), and the alliance relationship by Gil (2009). In its role as the client, BAA took on all the risk for the £4.3 billion project under the unique T5 Agreement that the 60 first-tier contractors signed. In total, 15,000 suppliers were involved. The overall project was divided into 147 sub-projects, with an integrated team led by BAA responsible for each one. Unlike the majority of megaprojects, T5 was delivered on time and on budget.

The key relevant point about T5 was that innovations were actively sought out and rewarded (Wolstenholme et al. 2008). These included product innovations in offsite fabrication such as the roof structure, technological innovations such as the tunnelling process and equipment, process innovations such as the two logistics centres (Potts 2009), and management innovations such as the insurance provisions and incentives built into the T5 Agreement. Deakin and Koukiadaki (2009) detailed the industrial relations aspects of the project.

Caldwell et al. (2009) suggested the risk associated with these large, complex projects can provide the motivation for clients to pursue and reward innovation by the major contractors and suppliers, who on T5 demonstrated a capability for innovation that is left unused under traditional tendering and procurement methods. However, most construction projects are less complex, many are standardised and repetitive, and clients have no reason to support innovations that might marginally affect their project's delivery or performance but increase the risk of cost overruns. Construction is in a catch-22 situation, where neither industry incumbents nor its clients can rationally commit to significant, expensive investments in innovation for the vast majority of construction projects.

6. VENTURE CAPITAL AND INDUSTRY INVESTMENT IN INNOVATION

Investment in construction innovation R&D is now coming from private equity and venture capital funds such as Brick & Mortar, WND Ventures, Ironspring, Building Ventures, and Foundamental. This is a new development; before 2017 there was little interest in new construction technology from investors. Then, a Californian startup called Katerra reached a $1 billion valuation in 2017 and got $865 million in 2018 from SoftBank. The company's goal was complete vertical integration of design and construction, building factories to manufacture cross-laminated timber panels and then delivering and assembling the buildings onsite. Over five years, Katerra went through four different business models as they sought to achieve sufficient scale to keep their factories busy, but in June 2021 filed for Chapter 11 bankruptcy, and the factories were sold (de Valence 2022: 94–96).

Despite the failure of Katerra and other firms attempting to make the economics of manufactured housing work, there has been a rapid increase in the quantity of venture capital going into construction tech. Venture capital funding is a significant metric because investment in construction technology startups is a proxy for innovation and the development of IP and other forms of intangible capital (Chaoran 2020). In 2018 the amount of venture capital going into construction technology startups significantly increased in the EU and US. Based on a dataset for a widely defined Architecture Engineering and Construction industry from Traxn, Foundamental (2024) found nearly $30 billion was invested in construction technology startups between 2014 and 2023. The estimate for 2021 was a high of $8.7 billion, followed by $4.9 billion in 2022, and falling to $1.3 billion in 2023.

A wide range of technologies advancing rapidly is unusual. In the past, there would be one or two new technologies developing at any one time, but now there is a broad front of simultaneous advances that mutually reinforce each other and underpin the fourth industrial revolution. The problem with the concept of Industry 4.0 in general, and Construction 4.0 in particular, is the large number of technologies included (Ghaffar et al. 2022), all of which have their own development paths and many of which are enabling technologies, like IoT and virtual reality (VR). In the same way that the concept of an industrial revolution by itself is not an insight into the inventions and innovations involved, Industry 4.0 is not an explanation of what is happening to industries and firms. Some Industry 4.0 technologies like BIM and platforms are in use, some like 3D printed concrete are moving out of the experimental stage to commercialisation, and others like smart contracts are still conceptual (Sawhney et al. 2020).

It was at the first public demonstration of VR headsets in 1990 that William Gibson made his now famous observation: 'The future is already here, it's just not very evenly distributed' (Kelly 2016: 215). Those first nausea-inducing systems were clunky and expensive, but after a couple of decades of development, the costs of the key components, particularly small high-resolution screens and sensors, had fallen to the point where consumer products were possible. Released in 2016 were VR headsets from Oculus, Microsoft, and Samsung, and in 2017, everyone from architects to zookeepers started thinking and talking about how this 'new' technology could be used. Smart helmets arrived, and the Microsoft HoloLens 2 was released in February 2019, connected to a BIM running on Trimble software to provide an AR view of the project. In another five years, these may be commonplace on construction sites.

7. GOVERNMENT AND INNOVATION: STANDARDS, CODES, AND BIM MANDATES

As well as general industry policies (Warwick 2013) targeting innovation, training, or business investment, construction is also subject to many other government regulations, legislation, and policies (Gruneberg 2020). The number of different government departments and agencies involved as both repeat clients and in regulating the built environment can be a barrier to innovation because coordination is difficult and there are opportunities for incumbents to delay or derail progress when reforms are proposed.

Although the public sector is collectively the largest client for construction, expenditure is spread over departments like health, education, transport, and defence, and there is unrelenting pressure from public sector clients for the lowest possible cost of work. Over the years, inquiries and reports in many countries have recommended leveraging public procurement of buildings and structures to increase innovation, but in practice, there are significant institutional constraints on government buying power.

While governments can have major impacts through tax, innovation, and R&D policies, their effect is often uneven and hard to discern. For example, large firms in capital-intensive industries like cement respond to industry policies differently than large contractors, as do professional service SMEs compared with construction trade SMEs. There have, however, been two areas where governments have successfully promoted (and invested) in industry development and innovation through regulation: building standards and decarbonisation, and BIM mandates.

Building Standards and Decarbonisation

The regular revision and upgrading of building codes and product standards is a policy area where governments, usually through regulatory agencies, invest in industry

R&D. This is a more complex proposition than the use of BIM mandates because there is often no specific policy or set of policies that would qualify as an industrial strategy. Nevertheless, the use of building standards and codes to influence industry development has a long history with some notable successes, because buildings are designed and delivered in conformance with those regulations.

Standards and codes establish allowable tolerances and how much variation is allowed for products and processes. They underpin quality control and are the basis of inspections to verify the work being done, so a standard is a document structured around requirements for conformity and measures that certify meeting those requirements. During the late nineteenth century, governments and insurers began raising the standards they set in building codes for access, light, safety, amenity, and appearance, significantly improving the design and construction of buildings (Pfammatter 2008).

The first standard was agreed upon in Paris for the International System of Electrical and Magnetic Units in 1881, and the International Electrotechnical Commission was established in 1906 to develop and distribute standards for the units of measurement used today. The British Standards Institution was founded in 1901, as were French and German institutes. In the US, the Underwriters Laboratory was founded in 1894 by William Merrill, an electrical engineer, to provide testing of building materials for insurers, and the 1897 National Electrical Code on electrical wiring and equipment installation was the first US modern code. Insurers led the way in developing standards and methods for fireproofing the steel-framed buildings that were becoming common, issuing a model building code in 1905 to reduce fire risk. Also in the US, the American Society for Testing Materials goes back to 1898 with their standard for the steel used to fabricate railway tracks. In 1902, it became the American Section of the International Association for Testing Materials, which eventually became the International Organization for Standardization (ISO) in 1947. The American National Standards Institute was formed in 1918 (Gustafsson 2020).

The ISO now has more than 22,000 different standards covering every aspect of organisation management and production control. National testing and standards institutes are members of the ISO. They meet annually to review programs, and countries fund it in proportion to their trade and GDP. There is a six-stage process for getting a standard published, typically based on research from the member institutes, and each standard has a guide for developing and maintaining it. Multiple standards are being combined to make them easier to manage (Rich and Malik 2019). Although agreeing new standards is a lengthy process, they are universally accepted and applied because of the rigorous scientific and engineering research they are based on. Therefore, an important element in a strategy to increase innovation in construction is to increase funding for testing laboratories.

Periodic revisions to building standards and codes allow innovations and new products to be introduced in an incremental but typically slow process. While that reduces risk for designers and contractors, it also affects the rate of built environment product innovation and improved building performance. Revisions can be opposed or delayed, for example, by residential builders worried about increased costs in a price-sensitive market or by product manufacturers protecting market share. Nevertheless, a regular review and update process, like the US three-year cycle for building codes, keeps them relevant and focused on the key issues of building quality, energy use, and embodied carbon emissions (Vollebergh and van der Werf 2014).

The role of building standards and codes in decarbonisation, reducing energy use, and cutting greenhouse gas (GHG) emissions is well known (Popp et al. 2010). A carbon budget for both the construction and operation of the built environment is required. The UN produces an annual *Global Status Report for Buildings and Construction* that says: 'Cutting building-related emissions by improving energy efficiency is a crucial aspect of meeting net zero by 2050 climate change goals. Building energy codes provide a tool for governments to mandate the construction and maintenance of low-energy buildings' (UN 2021: 59). Countries with energy codes are moving towards the electrification of building operations, particularly for heating and cooking. This is a necessary requirement to reach net zero by 2050 because residential energy use accounts for around 40% of total emissions (Popp 2019).

BIM Mandates

Government policies that promote digital construction are important, and countries have taken different approaches to promoting BIM (Lee and Bormann 2020). Broadly, Scandinavian and Western European countries, Singapore, and the UK followed a government-driven approach, while Australia and the United States (US) took a more industry-driven approach. BIM policies have roadmaps with stages for BIM adoption, using both level of use and size of project as targets, and are intended to allow time for the industry to adjust.

From its origins in 1970s CAD programs, BIM has taken four decades of development to reach widespread, but far from universal, use in construction. While capable, these systems are expensive and complex, require significant skill levels, and do not integrate easily with each other or with systems used elsewhere, such as manufacturing and logistics. Resolving that complexity and managing the design process may well be a key role that AI plays in future construction. At present, BIM is required for construction-specific technologies like cloud-based digital twins, 3D concrete printing, and online platforms that coordinate the design, manufacture, and delivery of building components.

A few countries have implemented national BIM mandates (Lee and Bormann 2020):

- 2004: Singapore – public construction projects
- 2007: Finland – all public projects over €1 million
- 2007: US General Service Administration and the Army Corps of Engineers
- 2010: South Korea – public construction over KRW500 million from 2016
- 2011: UK – public building
- 2018: Spain – public construction
- 2019: Abu Dhabi – all major projects
- 2020: Germany – federal infrastructure projects

Many other countries have published roadmaps, standards, and guidelines without following up with a mandate, including Austria, Australia, France, Switzerland, and Japan at this stage. In every case, the underlying assumption is that BIM will become business as usual over the decade of the 2020s. However, at the beginning of the decade, countries that were early movers, like the US, Singapore, Finland, and the UK, clearly have the highest use of BIM. There are also state mandates in the US and Australia: Wisconsin required BIM for projects over $5 million in 2010, and Queensland for public projects in 2018. By the early 2020s, most major projects for both public and private clients worldwide were being done with BIM.

The UK provides an example of the policy approach being followed by governments to increase the use of BIM. A 2011 Construction Strategy introduced a BIM mandate which required increasing levels of use over time. With the mandate, the government provided clauses covering contentious issues in construction contracts, such as intellectual property and data ownership, that worked with rather than against industry practice and culture. In 2015, standards BS 8541-5 and 8541-6 on offsite construction and modular buildings were released. The Construction Strategy was extended to 2016–2020, with a single shared building model to be held in a centralised repository for the operation of assets over their lifecycle. The 2018 BIM Framework, based on ISO 19650, provided a roadmap for firms and clients, and the development of the necessary standards provided a toolkit.

A decade after 2011, the UK had a high level of use compared with other countries. A survey by the UK BIM Alliance (2021) with over 1,100 respondents found 65% were implementing BIM and used it on around half their projects, while 30% were using ISO 19650. However, over half of the subcontractors and cost consultants, and over 40% of project managers and facility managers, were not implementing BIM. Nevertheless, 56% of respondents thought BIM would become business as usual in three to five years, and the other 44% thought it would take longer. Architects, engineers, and large contractors have adopted BIM faster than services engineers, facilities managers, and smaller contractors employing fewer than 50 people.

The introduction of international standards for BIM and digital twins in 2019 was a milestone in the transition to a digitised system of production (ISO 2019). The ISO 19650 standards provide a framework for creating, managing, and sharing data on built assets, an essential step in a technocratic system like construction because standards play a key role in the diffusion of new technology and establishing consensus on what is to be done and how. Although the transition may take decades as Construction 4.0 technologies diffuse and improve, as the history of technological change suggests, that history also suggests the industry will restructure around these new technologies.

8. INNOVATION, CONSTRUCTION 4.0, AND STRUCTURAL CHANGE

Industries go through periods of growth, competition, and restructuring as the technology they are based on goes through cycles of invention and innovation. As a result, technological change is not constant but proceeds in fits and starts, with bursts of innovation leading to new products and processes when new technologies arrive, followed by periods of refinement and incremental improvement. This process separates firms that are successful adopters of new technology from firms that are laggards:

> [T]he innovation that has shaped recent economic growth is not an autonomous event that falls like manna from heaven. Nor is it a result of R&D and ICT investments alone. Instead, a surge of new ideas (technological or otherwise) is linked to output growth through a complex process of investments in technological expertise, product design, market development, and organizational capability. (Corrado and Hulten 2010: 101)

The long-term effects of GPTs give a 'technological shock' to an existing system of production, which leads to a transition period where incumbent firms have to adjust to the new business environment, and new entrants appear to take advantage of the new technology, leading to growing sunrise industries and declining sunset industries in the economy. The creative part is the invention and innovation that lead to new products and services; the destructive part is the demise of less successful firms and the decline of industries as they become less relevant.

The arrival of major new technologies that lead to periods of rapid change is not common (Lipsey et al. 2005). They are rare but powerful because they are used by many industries in many different ways. Steam power was such a general purpose technology (GPT) and is one of the three GPTs originally identified by David (1985) as the drivers of the first, second, and third industrial revolutions: steam, electricity, and information technology (IT). It is widely believed that AI is a new GPT (Acemoglu and Restrepo 2019). For construction, AI may be as disruptive as steam power was in the nineteenth century, when contractors replaced master builders. As Figure 2.1 shows, such restructurings have happened to construction three times, as new GPTs were adopted. The industrial structure of construction evolved through three stages: first, from master builders and craft guilds to contractors and tradesmen, then to the modern project manager (PM) and subcontractor structure.

Technological developments are combining intelligent machines with engineered materials, deep learning capabilities, human augmentation, and new organisational concepts, and are pushing against established custom and practice in construction. Because the system is so well established the effect of new technology and the changes it brings could spread slowly and unevenly across the industry's many small and medium-sized firms, as was the case with computers. On the other hand, a period of rapid innovation and disruptive change led to a restructuring during the second half of the nineteenth century as a new system of production eventually resulted

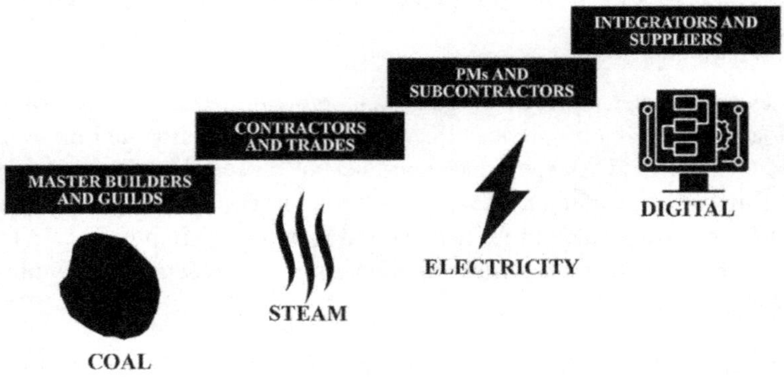

Figure 2.1 Four industrial revolutions since 1800

in a new form of industry organisation led by contractors instead of architects and engineers. That disruptive transition took several decades, as industrial materials replaced craft ones and site work was mechanised and reorganised. Then, over the twentieth century, contractors evolved into project managers and the traditional trades became subcontractors.

At this point, how quickly and how far AI, 3D printing, and other Construction 4.0 technologies will spread through construction over the next decade is unknowable. There are many technical, regulatory, and institutional hurdles to be overcome, so diffusion might be slow for many reasons. On the other hand, if continuing innovation in big data, analytics, and AI can improve project processes, and digital fabrication increases onsite productivity, they could be adopted quickly by companies that have the systems and resources to take advantage of them. AI in construction is already in use for generative design, drone monitoring, and managing equipment maintenance.

The effects of new technology on a large, long-established industrial sector take time and will be unevenly spread between firms and countries (Ørstavik et al. 2015). Nevertheless, as Table 2.2 shows, Construction 4.0 technologies have profound economic implications for the organisation of construction. It is possible, or probable, that AI will lead to a reorganisation of construction from one centred on project managers to one based on AI-powered platforms providing cheap, outsourced business processes and integrators that combine digitised methods of onsite and offsite production.[2] As that happens, the organisation and structure of the industry could change from one centred on PMs and contractors to one based on project integrators that combine site preparation with a mixture of OSM, onsite or nearsite production using 3D printing in modular fabs, and mass-produced items.

Productivity growth delivers improved products and new ways of making them; however, reorganisation of industry and reallocation of capital and labour towards

Table 2.2 Construction and the fourth industrial revolution

Activity	The fourth industrial revolution
Production of components and materials	Platforms integrate design and production with voice-controlled 3D models of buildings and components. Generative design used for projects and planning. AI-assisted selection of components and modules from online design libraries. AI monitors compliance with building standards and codes. Developments in digital fabrication, design software, materials, and equipment allow a range of new production technologies.
Mechanisation and automation of tasks	Site workers have exoskeletons and smart helmets available. Many onsite tasks can be done by teams of robots or by humanoid robots. Heavy machinery and equipment are operated remotely, with autonomy for repetitive tasks.
Organisation of projects	Cloud-based platforms integrate delivery of the physical project with its digital twin. Real-time data and monitoring of activities and tasks. Standardised, outsourced cloud-based business processes used. As-built digital twins required by clients on completion and handover.

more productive uses are necessary requirements. The history of the evolution of the corporate form over the twentieth century by Langlois (2023) details the technological trajectory of US large-scale corporations and the associated legal and regulatory institutions. The focus is on how these new institutions interacted with innovation and technological change to shape the organisational form and managerial structure of firms in various industries. Bloom et al. (2016) described management as a technology, arguing that some forms of management practices are a technology because they raise productivity, and they found differences in management practices account for about 30% of total factor productivity differences between countries and across firms. Lipsey et al. (2005) list two dozen GPTs since 9000 BCE and include two organisational forms: mass production and the factory system; and lean production and the Toyota system. Lepinoy et al. (2022) discuss innovation, production systems, and business models in construction.

An example of the interaction of productivity, technology, and organisational form was the Suez Canal, the largest project ever undertaken when started in 1862. The first phase began with a French contractor and used up to 40,000 Egyptian serf labourers to dig 75 kilometres, but in 1863 the workers were withdrawn by the Egyptian Government, the chief contractor resigned, and the project was reorganised. By 1866,

French industry had produced the steam-powered excavators, conveyors, cranes, tractors, railways, dredges, and barges the new (English) contractor required, and by 1867, the site was fully mechanised. Between 1866 and 1869, another 85 kilometres were dug, 50 million cubic metres of earth moved, and a novel form of site organisation based on 'planned processes' began to emerge. In the year before opening in 1869, there were 7,000 labourers and 6,900 European workers on the project (Peters 1996).

Fifteen years later, construction of the Panama Canal started, and Peters argues that the modern form of site organisation and project management emerged when the US military took over in 1904, after the first attempt at the project failed. He concludes, 'at Suez, the most innovative thinking had gone into design and manufacture of machinery. At Panama, Goethals and his team emphasised the design of integrated systems and processes' (Peters 1996: 244). The driver of these changes was steam-powered machinery, a new GPT that fundamentally changed existing industry practices and delivered a shock to the existing system of production.

9. CONCLUSION

Innovation is not constant but proceeds in fits and starts, with bursts of new products and processes when new GPTs arrive, followed by periods of refinement and incremental improvement. While construction will be affected by fourth industrial revolution technologies, like other industries, the path taken will be distinct and different from the path taken in those other industries. This path dependence varies not just from industry to industry, but from firm to firm as well. Machine learning, AI, digital fabrication, automation, and robotics are an interconnected set of technologies that are evolving quickly, enabled by expanding connectivity and the massively scalable hardware available today.[3] There is potential for new entrants, new business models, and disruption to the current organisation of production.

The effect of R&D is most pronounced where technological opportunity is high and firms have the size and economies of scale to invest in R&D. Therefore, innovation in construction has often come from the suppliers of materials, components, and equipment, who have been responsible for the introduction of many new products and machinery that have affected site processes, such as plastics, drywall, glass facades, lifts, and hoists. A narrow view of construction innovation would include only contractors and subcontractors, but because construction requires inputs from materials, manufacturing, and professional service firms, this chapter has taken a broad view of construction innovation that includes these related industries in manufacturing, materials, and professional services in the discussion.

Innovation requires both invention and diffusion. A new product or production method must attract investment, workers, and customers before its full impact becomes apparent, in a process that typically takes two to three decades for a GPT. The introduction of new construction products is affected by two of the primary characteristics of the industry. The first is fragmentation due to the large number of small firms, which slows the diffusion of new technology and the adoption of new products because small firms have limited capacity for investment in innovation and new capabilities and do not do R&D. Unlike manufacturing, construction is not concentrated in a few firms or a few places. Second, products have to conform to the relevant standards and codes. Although agreeing new standards is a lengthy process, they are universally accepted and applied because of the rigorous scientific and engineering research they are based on. Therefore, an important element in a strategy to increase innovation in construction is to increase funding for testing laboratories.

The procurement and contracting system typically used in construction is for specific projects and is focused on cost and risk management, not the innovation and R&D that increases productivity. By the end of the twentieth century, construction had a deep, diverse, and specialised value chain that resists integration because it is resilient, flexible, and adapted to economic variability. Therefore, one possibility is that AI and the fourth industrial revolution will be used to improve construction productivity and efficiency but will not fundamentally alter the current industry structure with PMs and large contractors at the core. The impact of a new GPT on a developed and embedded system of production like construction might be gradual, changing industry practices over time. There are good reasons to think this may be wrong, however, because of advances in machine learning and automation, and the broadening potential of 3D printing and digital fabrication.

For both industry majors and smaller firms, a cycle of innovation appears to have begun with new entrants attracting investment from outside the industry, offering many possibilities, opportunities, and threats to incumbents. This new cycle is starting more than 200 years after the emergence of modern construction in the first industrial revolution, and after 20 years or more of development of fundamental fourth industrial revolution technologies like BIM, digital twins, and virtual reality.

Nevertheless, in the various forms that AI, automation, 3D printing, design for manufacturing and assembly (DfMA), and procurement platforms take on their way to the construction site, they will become central to many of the tasks and activities involved as they improve over time. Because construction as a system of production is so wide and deep, the fourth industrial revolution will affect a large number of people, and through them, the wider economy and society. This is an opportunity for government and industry to work together to develop policies and roadmaps for firms and people employed in construction. The future is not determined; although innovation reshapes and restructures industry and the economy, decisions made today create the future.

NOTES

1. Given a variety of locations with different relative prices, there will be a best location for the supply of the most productive factor. Therefore, firms can raise productivity by moving to a site with a larger supply and lower relative price of the most productive factor, so for any one location, there will be a better

technology in use somewhere else (but with different relative prices). However, firms face search and switching costs when looking for new technology, and sunk costs in adopting one.

2. Arthur (2009) argues the three key characteristics of advanced technologies such as AI are combination, recursiveness, and phenomena. In combination, technologies allow AI-enabled platforms to generate value. The value for users increases as predictions improve with increasing processing and networking power, which creates recursiveness because AI consists of technologies that are made up of other technologies. The phenomenon is data-driven learning.
3. As Agrawal et al. (2023) point out, if AI-assisted prediction increases the expected value of innovation and reduces costs, it will greatly increase innovation productivity.

REFERENCES

Abbott, M. and Carson, C. 2013. A comparative approach to determining the growth of productivity of the New Zealand construction industry, *Australasian Journal of Construction Economics and Building*, 13(4): 65–77.

Acemoglu, D. and Restrepo, P. 2019. Artificial intelligence, automation, and work, in Agrawal, A., Gans, J. and Goldfarb, A. (eds) *The Economics of Artificial Intelligence: An Agenda*, London: The University of Chicago Press, 197–236.

Agrawal, A., McHale, J. and Oetti, A. 2023. Superhuman science: How artificial intelligence may impact innovation, *Journal of Evolutionary Economics*, 33(2): 1–45.

Andrews, D., Criscuolo, C. and Gal, P.N. 2015. *Frontier Firms, Technology Diffusion and Public Policy: Micro Evidence from OECD Countries*, OECD Productivity Working Papers, 2015-02.

Arthur, W.B. 2009. *The Nature of Technology: What it is and How it Evolves*, London: Penguin.

Arthur, W.B. 1989, Competing technologies, increasing returns and 'lock-in' by small historical events, *Economic Journal*, March. 116–131.

Björk, J., Frishammar, J. and Sundström, L. 2023. Measuring innovation effectively: Nine critical lessons, *Research-Technology Management*, 66(2): 17–27.

Bloom, N., Sadun, R. and van Reenen, J. 2016. *Management as a Technology*, Working Paper 22327, National Bureau of Economic Research.

Bloom, N. and van Reenen, J. 2010. Why do management practices differ across firms and countries?, *Journal of Economic Perspectives*, 24(1): 203–224.

Bloom, N., van Reenen, J. and Williams, H. 2019. A toolkit of policies to promote innovation, *Journal of Economic Perspectives*, 33(3): 163–184.

Bonvillian, W.B. and Weiss, C. 2015. *Technological Innovation in Legacy Sectors*, New York: Oxford University Press USA.

Bresnen, M. and Marshall, N. 2001. Understanding the diffusion and application of new management ideas in construction, *Engineering Construction and Architectural Management*, 8(6): 335–345.

Caldwell, N.D., Roehrich, J.K. and Davies, A.C. 2009. Procuring complex performance in construction: London Heathrow Terminal 5 and a Private Finance Initiative Hospital, *Journal of Purchasing & Supply Management*, 15(3): 178–186.

Cassimatis, P.J. 1969. *Economics of the Construction Industry*, New York: National Industrial Conference Board.

Chancellor, W. 2015. Drivers of productivity: A case study of the Australian construction industry, *Construction Economics and Building*, 15(3): 85–97.
Chaoran, C. 2020. Technology adoption, capital deepening and international productivity differences, *Journal of Development Economics*, 143: 1023–1088.
Cody, J. 2003. *Exporting American Architecture 1870–2000*. London: Routledge.
Corrado, C. and Hulten, J. 2010. Measuring intangible capital: How do you measure a 'technological revolution'?, *American Economic Review*, 100: 99–104.
Craig, R. 1997. The 'Tendering Contract': Fairness, equality and innovation, in Davidson, C. and Mequid, T. (eds) *Procurement – A Key to Innovation*, CIB W92 Publication 203, 91–100.
David, P.A. 1985. Clio and the Economics of QWERTY, *American Economic Review*, 75(2): 332–337.
Davies, A., Gann, D. and Douglas, T. 2009. Innovation in megaprojects: Systems integration at Heathrow Terminal 5, *California Management Review*, 51(2): 101–125.
Davis, H. 2006. *The Culture of Building*, Oxford: Oxford University Press.
Deakin, S. and Koukiadaki, A. 2009. Governance processes, labour–management partnership and employee voice in the construction of Heathrow T5, *Industrial Law Journal*, 38(4): 365–389.
de Valence, G. 2022. *Creative Destruction and Constructing the Built Environment: From the First Industrial Revolution to the Fourth*, Sydney: Construction Economics Research.
de Valence, G. 2019. Reframing construction within the built environment sector, *Engineering, Construction and Architectural Management*, 26(5): 740–745.
de Valence, G. 2001. Construction deflators and measurement of output, *ARCOM 2001 Conference*, 693–702.
Dosi, G. 1988. Sources, procedures, and microeconomic effects of innovation, *Journal of Economic Literature*, 26(3): 1120–1171.
Dosi, G., Freeman, C. and Fabiani, S. 1994. The process of economic development: Introducing some stylized facts and theories on technologies, firms and institutions, *Industrial and Corporate Change*, 3(1): 1–45.
Edvardsen, D.F. 2004. *Four Essays on the Measurement of Productive Efficiency*, Goteborg University.
Fernández-López, X.L and Coto-Millán, P. 2015. From the boom to the collapse: A technical efficiency analysis of the Spanish construction industry during the Financial Crisis, *Construction Economics and Building*, 15(1): 104–117.
Foundamental, 2024. https://foundamental.com/aecvc/the-state-of-constru-tech-in-q4-2023/
Gann, D.M. 2003. Innovation in the built environment, *Construction Management & Economics*, 21: 553–555.
Garcia, D. and Molloy, R. 2023. *Can Measurement Error Explain Slow Productivity Growth in Construction?* Finance and Economics Discussion Series 2023-052, Washington: Board of Governors of the Federal Reserve System.
Ghaffar, S.H., Mullett, P., Pei, P. and Roberts, J. 2022. What is Construction 4.0, in Ghaffar, S.H., Mullett, P., Pei, P. and Roberts, J. (eds) *Innovation in Construction: A Practical Guide to Transforming the Construction Industry*, Cham: Springer Nature Switzerland, 355–408.
Gil, N. 2009. Developing cooperative project client–supplier relationships: How much to expect from relational contracts, *California Management Review*, Winter: 144–169.
Goolsbee, A. and Syverson, C. 2022. *The Strange and Awful Path of Productivity in the U.S. Construction Sector*, Working Paper 30845, National Bureau of Economic Research.
Gordon, R.J and Sayed, H. 2020. *Transatlantic Technologies: The Role of ICT in the Evolution of U.S. and European Productivity Growth*, Working Paper 27425, National Bureau of Economic Research.
Gruneberg, S. 2020. *A Strategic Approach to the UK Construction Industry*, Abingdon, UK: Taylor & Francis.

Gustafsson, I. 2020. *How Standards Rule the World: The Construction of a Global Control Regime*, Cheltenham, UK and Northampton, MA, USA: Edward Elgar Publishing.
Hardie. 2022. *Annual Report*, James Hardie Pty Ltd.
Hausmann, R. and Hidalgo, C.A. 2011. The network structure of economic output, *Journal of Economic Growth*, 16: 309–342.
Helpman, E. and Trajtenberg, J. 1990. Diffusion of general purpose technologies, in Helpman, E. (ed.) *General Purpose Technologies and Economic Growth*, Cambridge, MA: MIT Press, 85–119.
Hobday, M. 1998. Product complexity, innovation and industrial organisation, *Research Policy*, 26: 689–710.
Hu, Xiancun and Liu, Chunlu. 2018. Measuring efficiency, effectiveness and overall performance in the Chinese construction industry, *Engineering, Construction and Architectural Management*, 25(6): 780–797.
Hughes, T.P. 1989. *American Genesis: A Century of Invention and Technological Enthusiasm 1870–1970*, Chicago: University of Chicago Press.
Ickes, B.W. and Samuelson, L. 1987. Job transfers and incentives in complex organizations: Thwarting the ratchet effect. *RAND Journal of Economics*, 18(2)(Summer): 275–286.
ISO 2019. International Standard 19650. https://www.iso.org/obp/ui/#iso:std:iso:19650:-1:ed-1:v1:en
Ivory, C. 2005. The cult of customer responsiveness: Is design innovation the price of a client-focused construction industry?, *Construction Management and Economics*, 23: 861–870.
Jahren, P. 2011. *Concrete: History and Accounts*, Trondheim: Tapir Academic Press.
Kabri Construction Research 2022. https://constructionphysics.substack.com/p/a-brief-history-of-construction-startups?s=r
Kagel, J.H. and Levin, D. 1986. The winner's curse and public information in common value auctions, *American Economic Review*, 76(5): 894–920.
Katz, M.L. and Shapiro, C. 1985. Network externalities, competition, and compatibility, *American Economic Review*, 75(3): 424–440.
Kelly, K. 2016. *The Inevitable: Understanding the 12 Technological Forces That Will Shape Our Future*, New York: Random House.
Kendrick, J.W. 1973. *Postwar Productivity Trends in the United States 1948–1966*, New York: National Bureau of Economic Research.
Langlois, R. 2023. *The Corporation and the Twentieth Century: The History of American Business Enterprise*, Princeton, NJ: Princeton University Press.
Lee, G. and Bormann, A. 2020. BIM policy and management, *Construction Management and Economics*, 38(5): 413–419.
Lepinoy, O., van der Heide, G. and Moore, C. 2022. Some changes are invisible to the eye: Transformation of business models, organizations and culture, in Ghaffar, S.H., Mullett, P., Pei, P. and Roberts, J. (eds) *Innovation in Construction: A Practical Guide to Transforming the Construction Industry*, Cham: Springer Nature Switzerland, 355–408.
Lipsey, R.G., Carlaw, K.I. and Bekar, C.T. 2005. *Economic Transformations: General Purpose Technologies and Long-term Economic Growth*, Oxford: Oxford University Press.
McKinsey. 2017. *Reinventing Construction Through a Productivity Revolution*, McKinsey Global Institute.
Mokyr, J. 1994. Progress and inertia in technological change, in James, J. and Thomas, M. (eds) *Capitalism in Context: Essays in Honor of R.M. Hartwell*, Chicago: University of Chicago Press, 230–254.
Mowery, D.C. and Rosenberg, N. 1998. *Paths of Innovation: Technological Change in 20th Century America*, Cambridge: Cambridge University Press.
Mullett, P. 2022. Challenges to innovation in construction, in Ghaffar, S.H., Mullett, P., Pei, P. and Roberts, J. (eds) *Innovation in Construction: A Practical Guide to Transforming the Construction Industry*, Cham: Springer Nature Switzerland, 89–116.

Nazarko, J. and Chodakowska, E. 2017. Labour efficiency in construction industry in Europe based on frontier methods: Data envelopment analysis and stochastic frontier analysis, *Journal of Civil Engineering and Management*, 23(6): 787–795.

Nelson, R. and Winter, S. 1982. *An Evolutionary Theory of Economic Change*, Cambridge, MA: Harvard University Press.

Nelson, R. and Wolff, E. 1997. Factors behind cross-industry differences in technical progress, *Structural Change and Economic Dynamics*, 8(2): 205–220.

Nordhaus, W.D. 1997. Do real-output and real-wage measures capture reality? The history of lighting suggests not, in Bresnahan, T. and Gordon, G. (eds) *The Economics of New Goods*, Chicago: University of Chicago Press, 27–70.

OECD. 2015. *The Future of Productivity*. Paris: Organisation for Economic Co-operation and Development.

Ørstavik, F., Dainty, A. and Abbott, C. (eds). 2015. *Construction Innovation*. Hoboken, NJ: John Wiley & Sons Inc.

Park, W. and Chapin, W. 1992. *Construction Bidding: Strategic Pricing for Profit*, 2nd ed., New York: Wiley.

Perez, C. 2003. *Technological Revolutions and Financial Capital: The Dynamics of Bubbles and Golden Ages*, Cheltenham, UK and Northampton, MA, USA: Edward Elgar Publishing.

Peters, T.F. 1996. *Building the Nineteenth Century*, Cambridge, MA: MIT Press.

Pfammatter, U. 2008. *Building the Future: Building Technology and Cultural History from the Industrial Revolution until Today*, Munich: Prestel Verlag.

Popp, D. 2019. *Environmental Policy and Innovation: A Decade of Research*, CESifo Working Paper No. 7544.

Popp, D., Newell, R. and Jaffe, A.B. 2010. Energy, the environment, and technological change, in Hall, B. and Rosenberg, N. (eds) *Handbook of the Economics of Innovation: Vol. 2*, Oxford: Academic Press/Elsevier, 873–937.

Potts, K. 2009. From Heathrow Express to Heathrow Terminal 5: BAA's development of supply chain management, in Pryke, S. (ed.) *Construction Supply Chain Management*, Oxford: Blackwell.

Rich, N. and Malik, N. 2019. *International Standards for Design and Manufacturing: Quality Management and International Best Practice*, London: Kogan Page.

Rodrik, D. and Stantcheva, S. 2021. Fixing capitalism's good jobs problem, *Oxford Review of Economic Policy*, 37: 824–837.

Romer, P.M. 1990. Endogenous technological change, *Journal of Political Economy*, 98: 71–102.

Sanjayan, J.G., Nazari, A. and Nematollahi, B. 2019. *3D Concrete Printing Technology – Construction and Building Applications*, London: Elsevier.

Sawhney, A., Riley, M. and Irizarry, J. (eds). 2020. *Construction 4.0: An Innovation Platform for the Built Environment*, Abingdon: Routledge.

Schumpeter, J. 1942. *Capitalism, Socialism and Democracy*, New York: Harper.

Solow, R.M. 1957. Technical change and the aggregate production function, *Review of Economics and Statistics*, 39(3): 312–320.

Sutton, J. 1999. *Technology and Market Structure*, Cambridge, MA: MIT Press.

Swan, T.W. 1956. Economic growth & capital accumulation, *Economic Record*, 32(2): 334–361.

UK BIM Alliance. 2021. https://www.ukbimalliance.org/wp-content/uploads/2021/06/UKBIMA-State-of-the-Nation-Survey-Report-2021.pdf

UK Cabinet Office. 2015. *Government Construction Strategy 2016–2020*, London.

UN. 2021. United Nations Environment Programme, *Global Status Report for Buildings and Construction: Towards a Zero-emission, Efficient and Resilient Buildings and Construction Sector*, Nairobi. https://globalabc.org/sites/default/files/2021-10/GABC_Buildings-GSR-2021_BOOK.pdf

Vollebergh, H.R.J. and van der Werf, E. 2014. The role of standards in eco-innovation: Lessons for policymakers, *Review of Environmental Economics and Policy*, 8: 230–248.
Wang, Y.S. and Chau, K.W. 2001. An assessment of the technical efficiency of construction firms in Hong Kong, *International Journal of Construction Management*, 1(1): 21–29.
Warwick, K. 2013. *Beyond Industrial Policy: Emerging Issues and New Trends*, OECD Science, Technology and Industry Policy Papers, No. 2.
Wei, W., Zhou, F. and Liang, P.F. 2020. Product platform architecture for cloud manufacturing, *Advances in Manufacturing*, 8: 331–343.
Weitzman, M. 1980. The 'ratchet principle' and performance incentives, *Bell Journal of Economics*, 11: 302–308.
Weitzman, M. 1976. The new Soviet incentive model, *Bell Journal of Economics*, 7: 251–257.
Wolstenholme, A., Fugeman, I. and Hammond, F. 2008. Heathrow Terminal 5: Delivery strategy, *Proceedings of the ICE – Civil Engineering*, 161(5).

3. Innovation in construction: the perspective of developing countries

George Ofori

INTRODUCTION

Background: The Innovation Imperative

It has long been observed that innovation is at the root of national economic growth and development (Schumpeter, 1934; Solow, 1956) and especially so in developing countries (Crespi and Zuniga, 2011). It is seen as the basic foundation for productivity, efficiency, and corporate competitiveness and growth, as well as a basis of organisational change and improvement (Pérez-Luño et al., 2007). UNCTAD (2019) noted that science, technology, and innovation have been seen as a vital route to structural change, economic diversification, productivity growth, jobs, and national competitiveness. In the World Bank report, *A Practitioner's Guide to Innovation Policy*, Cirera et al. (2020) note that the central role business innovation plays in economic growth is almost uncontested; it is established by a large body of evidence. Therefore, it is important to understand the nature of innovation, its attributes, prerequisites, determinants, and key players. This is especially pertinent as authors note that innovation is a complex and nuanced process (Cirera et al., 2020; Mohnen and Hall, 2013); it does not occur in any particular order; and context matters.

More recently, UNCTAD (2021a) noted that the world is undergoing a technological transformation that is reshaping how and where goods and services are produced. Referred to as the Fourth Industrial Revolution (Industry 4.0), it is an era where production is characterised by the integration of cyber-physical systems such as robotics, 3D printing, artificial intelligence, and machine learning. The concept has moved on. Leng et al. (2022) noted that Industry 5.0 aims to place the well-being of humans at the centre of manufacturing systems, thereby achieving social goals beyond employment and growth to attain sustainable development. Thus, Cirera et al. (2020) observed that improving the effectiveness of policies aimed at fostering innovation is even more urgent because the risks of a further increase in the technological divide between industrialised and developing countries are high, and there are opportunities for technological catch-up.

As is suggested for the whole economy, Blayse and Manley (2004) are among authors who have noted that increasing the level of innovation in the construction industry will enhance the contribution of the industry to economic growth. Innovation will enable the many types of organisations which undertake the varied tasks in construction to initiate, develop, and distribute new products, and develop technical capabilities and improvements in their operations (Bossink, 2004; Gambatese and Hallowell, 2011). However, as Ofori (2022) and other authors (see, for example, Barrett, 2007; Ribeirinho et al., 2020) suggest, studies in most parts of the world have found that the performance of the construction industry in terms of efficiency, productivity, quality, and health and safety on its projects, and the operating effectiveness of its products, compares poorly with those of other industries. This has been attributed to a number of factors, including a low rate of innovation in construction (World Economic Forum (WEF) and Boston Consulting Group, 2016; Buehler et al., 2018).

In the developing countries, it is important for the capacity, capability, and performance of the construction industries to be improved owing to the critical role construction plays in the national economies and in long-term socio-economic development (Giang and Low, 2010; Ofori, 2023). Thus, it is necessary to explore all the elements and factors which will contribute to the improvement of the performance of the industries, innovation being among the most important. CIB (1999) defines construction industry development as a deliberate process to improve the capacity and effectiveness of the construction industry in order to meet the demand for building and civil engineering products and to support sustained national economic and social development objectives. Fostering innovation is a key element of this process.

Aim and Objectives

The aim of this chapter is to explore how the construction industries in developing countries can take advantage of innovation to improve their performance as they undertake the projects needed for the development of the nations.

The objectives of the chapter are:

- to explain innovation and its importance at the levels of the nation, the industry, the company, and the project
- to discuss the challenges and constraints encountered in seeking to attain innovation and consider the process of innovation in the area of construction, and in the context of developing countries
- to consider what is currently being done, and what else needs to be done to attain innovation in construction in developing countries, including exploring the prerequisites
- to propose a research agenda for studies on innovation in construction in developing countries.

INNOVATION

Defining Innovation

Innovation is often viewed as involving the development of new products, adoption of new methods or processes, or the introduction and utilisation of new material inputs into a process (see, for example, Roger, 2004). It can also be the adoption of new forms of organisation at the corporate, project, or team level, and the entry into new markets. UNCTAD (2019) defined innovation as 'a process of discovery of better ways to arrange productive resources in order to address individual or social needs' (p. 5). Earlier, Schumpeter (1934, p. 65) had referred to innovation as the 'carrying out of new combinations'. Cirera et al. (2020) define innovation as 'the development, introduction, and utilization of upgraded or new products, processes, technologies, marketing strategies, organizational structures, and business models' (p. 7). This challenges the long-standing view that innovation is linear and primarily about science, research and development (R&D), and inventions, a conceptualisation which focuses on more radical innovations. Indeed, authors point out that most organisations modify and imitate what others have created and adapt external knowledge sources to develop their innovation outcomes (Pérez-Luño et al., 2011). In the broad conceptualisation of innovation, the ability to create new products and processes is less important than the capacity to choose, assimilate, change, use, and adapt existing technologies to enable competitive levels of costs and quality (Lall and Pietrobelli, 2002).

Thus, the OECD (2009) describes innovation as a broad concept encompassing a wide range of activities and processes: markets, entrepreneurship, networks and competition, but also skills and organisations, creativity, and knowledge transfers. UNCTAD (2019) presented these types of innovation: product and service innovation; process innovation; organisational (institutional) innovation; marketing innovation; pro-poor and inclusive innovation – types of innovation designed to address the needs of poorer, marginalised groups; grassroots innovation – involves grassroots actors (non-government organisations, communities) in the process of applying knowledge to local development challenges; frugal innovation – products designed or redesigned to reduce their cost and complexity; and social innovation – new collaborative arrangements with social and environmental benefits. At the national level, the objective of innovation relates mostly to the diffusion of technology within the particular sector and beyond.

Many factors affecting the success of innovation in business organisations have been identified. They include (see a review in Son, 2019): innovation efforts, resources, and strategic choices. The process of innovation has been well researched, but there is no consensus on it. The conceptualisation of innovation proposed by Crépon, Duguet, and Mairesse (1998) (CDM model) is often applied. The model suggests that innovation is a process that begins with the decision of whether to invest in research or other activities leading to innovation, and the degree to which a firm invests (**innovation intensity:** such as R&D, technology acquisition, and informal innovation activities), which in turn creates innovation outputs (**innovation output:** process innovation, product innovation, and decision to invest in innovation activities), leading to the **innovation process.**

Innovation in Developing Countries

'A developing country'

What is a developing country? The debate on this question has evolved over the years, and the terms and categories have changed. Casadella and Uzunidis (2017) suggested that a developing country fits these main conditions: instability and political crises; unemployment, impoverishment, pronounced social inequalities; defective markets, informal practices; incipient financial institutions and regulation; inadequate and inefficient collective infrastructure; an economy subject to the vagaries of the international environment; and fragility due to unpredictable capital movements. However, they noted that these characteristics mainly apply to a special group, the 'least developed countries', which have very low income per inhabitant; low health, nutrition, and education indicators; and face economic vulnerability owing, among other factors, to the weakness of their productive systems. (There were 46 such countries in 2023 (UNDP, 2023).)

The World Bank definition of a 'developing country' is most commonly used. That economic indicator is based on gross national income (GNI) per capita as determined by the Bank and adopts this classification of countries (World Bank, 2024):

a. Low-income countries – per capita GNI of US$1,135 and below in July 2022
b. Lower-middle-income countries – per capita GNI of between US$1,136 and US$4,465
c. Upper-middle-income countries – per capita GNI of between US$4,466 and US$13,845
d. High-income countries – per capita GNI of US$13,846 and above.

The 'low-income' and 'lower-middle-income' nations are referred to as 'developing countries'.

It is widely acknowledged that it is inappropriate to use one economic indicator (GNI per capita) to 'measure' 'development' (Fantom and Serajuddin, 2016). The World Bank admits that it does not directly measure the level of development of a country nor indicate the welfare of its residents but noted that it is a useful indicator that is often closely correlated with other measures of the quality of life, such as life expectancy at birth, mortality rates of children, and enrolment rates in school. There is a large and growing volume of literature on the subject of 'measuring the development of nations' (see, for a review, Obeng-Odoom, 2023). There are various indicators of 'development', such as the UN Human Development Index (HDI) and the Global Multidimensional Poverty Index (GMPI). The HDI 'was created to emphasize that people and their capabilities should be the ultimate criteria for assessing the development of a country, not economic growth alone' (UNDP, 2024a). It 'is a summary measure of average achievement in key dimensions of human development: a long and healthy life, being knowledgeable and having a decent standard of living. HDI is the geometric mean of normalised indices for each of the three dimensions' (UNDP, 2024a). Countries are ranked and grouped under: low human development; medium human development; high human development; and very high human development (UNDP, 2024b). The other indicator mentioned here, GMPI, complements traditional monetary poverty measures by capturing the acute deprivations in health, education, and living standards that a person faces simultaneously (see, for example, Oxford Poverty and Human Development Initiative and UNDP, 2023).

State of innovation in developing countries

What is the state of innovation in developing countries? Watkins et al. (2015) noted that in developing countries, where governance capacities are often inadequate, the resulting ineffective innovation systems lead to stagnant economies and low national development. UNCTAD (2021b) considers Industry 4.0 as comprising artificial intelligence, robotics, the Internet of Things, and other new technologies designed to sense, predict, and interact with the physical world and support production; it is transforming the way things are produced. Industry 4.0 in manufacturing is expected to result in higher productivity accompanied by lower environmental impact. However, 'too many firms in developing countries are still too far from using these new technologies in their production processes'.

In developing countries, external sources of technology and knowledge are more significant than R&D. The innovation efforts there include investments in in-house R&D, technology acquisition, and other informal innovation activities. Moreover, activities undertaken in routine work in the organisation might lead to the accumulation of internal sources of knowledge. Thus, innovation inputs can include in-house knowledge sources (such as through R&D); transfer or imitation of external technology and knowledge sources (such as in the acquisition of machinery or equipment); and informal innovation activities (Son, 2019). It is pertinent to note that, in technology transfers, it is necessary to seek appropriate technologies that are suitable for application and adaptation by the target transferee in their local context, rather than copying technologies that have worked in developed countries in the past. Adaptation may include substitution of some imported materials or components with locally available resources that may, in turn, be also value-enhanced, as well as add value with further innovations to both the processes and products. It should also be noted that it has been found that rather than technology being transferred in one direction, from industrialised countries to developing countries, there is, indeed, a two-way flow, with both parties benefiting in a 'technology exchange' (Kumaraswamy and Shrestha, 2002).

Adebowale et al. (2014) noted that, in relation to poor countries, it is fundamental to understand innovation in a broader socio-economic perspective with a focus on competence-building in all parts of the economy. They observed that recent work has suggested that there is a potential for innovating for poor segments of the world population. They refer to work on 'bottom of the pyramid innovation', which proposes that multinational firms addressing low-income households in developing countries may realise attractive rates of return (Prahalad, 2005). Work on 'below the radar innovation' (Clark and Chataway, 2009) suggested that firms in developing countries may address the needs of the poor regions in their home market that are neglected by large foreign companies and use this to build world-class production capacity (Kaplinsky, 2011). Son (2019) developed a mechanism to capture all innovation efforts and adapted the CDM model to formulate a structural model of the innovation process in developing countries, and used it to determine the determinants of product, process, and joint product-and-process innovation in small and medium-sized enterprises (SMEs) in Vietnam.

The innovation challenge in developing countries

The challenges in developing countries with respect to innovation include: market and system failures that hamper innovation; inadequate research volume, quality, and appropriateness to context; a business environment imposing high costs on firms; inadequate infrastructure for testing and prototyping new ideas; insufficient innovation finance; and inadequate competencies in the institutions established to design and deliver innovation programmes (Cirera et al., 2020). Again, Cirera and Maloney (2017) identified these barriers to innovation in developing countries: (1) weak firm capabilities to undertake innovation; (2) absence of critical innovation complementarities (including appropriate skills, access to foreign technologies, competition, an enabling environment, and good managerial and organisational practices); and (3) weak government capabilities to manage the complexity and breadth of the innovation policies implied by the previous two barriers.

What is the state of knowledge of innovation in developing countries? Van Uden et al. (2014) suggested that studies on innovation in developing countries are needed because, as noted above, an increasing body of literature provides evidence that innovation and productivity are key factors to grow out of poverty (Crespi and Zuniga, 2011). For example, introducing a special issue of the *African Journal of Science, Technology, Innovation and Development* on: 'Innovation Research and Economic Development in Africa', Adebowale et al. (2014) noted that the volume of the literature on innovation policy studies in Africa is relatively low, although the region confirmed its understanding of science and technology being a force for development with the launch, in the early 1980s, of the Lagos Plan of Action for the Economic Development of Africa, 1980–2000, which was followed by the current New Partnership for Africa's Development (NEPAD) (Secretariat of NEPAD, undated). The authors attributed this in large part to the weak scientific infrastructure in most African countries. Jauhiainen and Mugabe (2024) analysed 2,869 papers on innovation in Africa across science, social science, and humanities journals from 2015 to 2021 and found that 46% of them lacked contributions from African organisations or scholars; another 25% involved collaborations between African and non-African authors or institutions. Underlining the breadth of the concept of innovation, the themes of the papers included the knowledge economy, business innovation, agricultural advancements, and the role of mobile phones in knowledge dissemination and marketing.

Innovation and the new technologies

UNCTAD (2021a) reviewed these frontier technologies: artificial intelligence; Internet of Things; big data; blockchain; 5G networks; 3D printing; robotics; drones; gene editing; nanotechnology; and solar photovoltaic (PV). Subsequently, UNCTAD (2023) examined 17 frontier technologies at the leading edge of green innovation: artificial intelligence; green hydrogen; biofuels; concentrated solar power; blockchain; nanotechnology; big data; 5G; electric vehicles; gene editing; robotics; drone technology; 3D printing; wind energy; biogas and biomass; and solar PV. It highlighted their potential economic benefits, assessed country capabilities to use, adopt, and adapt them, and proposed ways forward for all countries. It stated: 'Developing countries now have opportunities to catch up, reduce poverty, and help tackle climate change and set the world on a more sustainable course' (p. 1).

UNCTAD (2021a) noted that experience shows that, over time, new technologies are likely to permeate various sectors of the economy and social activities. Therefore, it suggested that countries at all stages of development should promote the use, adoption, and adaptation of frontier technologies and prepare their people and firms for the tasks. Thus, developing countries should adapt and use frontier technologies to increase productivity, promote economic diversification, and create jobs. These countries have to address challenges that include: (a) *demography* – developing countries typically have expanding and younger populations; this increases the supply of labour and reduces the incentives for automation; (b) *technological and innovation capabilities* – these countries have fewer skilled people; (c) *slow diversification* – these countries typically innovate by acquiring and adapting external technologies, but this process is slowest in the poorest countries; (d) *financing mechanisms* – most developing countries have increased their R&D expenditures, but these are still relatively low, there is little private funding of industrial technologies; and (e) *intellectual property rights and technology transfer* – stringent intellectual property protection will restrict the use of frontier technologies.

UNCTAD (2021a) noted that technological progress can perpetuate inequalities or create new ones, for example, by limiting access to privileged groups and affluent countries. Thus, governments should act to ensure universal access, create and shape the market for inclusive and sustainable innovations, and invest in the necessary human and physical resources. They should establish ethical frameworks, particularly for the use of AI. Developing countries face these challenges in promoting access to the benefits of frontier technologies: (a) *income poverty* – many in developing countries cannot afford some of the goods or services; (b) *digital divide* – many developing countries lack adequate digital infrastructure, and Internet costs are high; and (c) *shortage of skills* – including digital skills.

INNOVATION IN CONSTRUCTION

Defining Innovation in Construction

Following the broad definition of innovation above, innovation in the construction industry may be defined as the process of development, distribution, and application of technology, a new or improved product, process or service, and knowledge with the purpose of improving productivity and other relevant performance parameters to suit the requirements of clients and users. A commonly cited definition is that of Slaughter (1998), who defined it as the 'actual use of a non-trivial change and improvement in a process, product, or system that is novel to the institution developing the change' (p. 226). Toole (1998) defined the innovation process as the application of technology that is new to an organisation and that significantly improves the design and construction by decreasing the cost, increasing the performance, and improving the business process. The completed item can also be both the venue and a subject of innovation, in managing it, operating it, and modifying it, or in how it is used. Davis et al. (2016) observed that there are problems in conceptualising, defining, articulating, and assessing innovation in construction. Wang et al. (2023) found that the scope of construction innovation covers multiple dimensions and can be perceived from different levels.

The construction industry does not have a good reputation for innovation (see, for example, Dulaimi, 1995; Winch, 2003; Glass et al., 2022). The industry has generally been considered to be slow to adopt new technologies and to lag behind other sectors in innovation (Demirkesen and Tezel, 2022). The World Economic Forum and Boston Consulting Group (2016) noted that in the last few decades, the pace of innovation, with improved communication and increased productivity following the digital revolution, has been remarkable in many sectors, but the construction industry has been slower to adopt and adapt to new technologies than other sectors. The types of innovations in construction are discussed next, followed by factors contributing to innovation in the industry.

Types of Innovation in Construction

Different categories of innovation are identified in construction. For example, Slaughter (1998) presented these groups from which firms can select and implement innovations: incremental; modular; architectural; system; and radical. Ewart (2019) noted that innovation studies in construction predominantly focus on major corporations, complex projects, national interests, and a desire to increase economics and efficiency on a large scale. In a study in rural Borneo, Ewart (2019) demonstrated that an accumulation of minor innovations can have significant consequences. Again, while the focus in innovation studies is put on large organisations, the rest of the long construction project value chain plays a role. Sexton and Barrett (2003) observed that innovation-led performance improvement in the construction industry is influenced by innovation by small firms, depending on organisational factors, forms of interaction among the companies and projects participating in the projects, and the operating environment. As observed above for firms in general (Lall and Pietrobelli, 2002), Sexton and Barrett (2003) noted that the innovation potential of construction firms depends on their ability to acquire existing knowledge, create new knowledge, and to make use of it for the realisation of new construction solutions.

Innovation scholars have found that different skills, resources, and cultures are needed to encourage each type of innovation (Pérez-Luño et al., 2011). Thus, appropriate strategies and implementation systems are vital to enable the organisation to implement and succeed with their innovation (Popadiuk and Choo, 2006). In construction, Steele and Murray (2004) explored how to build a culture of innovation in organisations and demonstrated the importance of creativity, the promotion of a culture for innovation, and the development of intellectual capital. Yusof et al. (2014) noted that different skills are required for innovation creation or innovation adoption, and different cultures are needed to encourage each of them. Ewart (2019) showed that the innovations in construction he identified were inherently social rather than purely economic, as there were changes to the technologies used for roofing and foundations, entangled with new social structures. Thus, Ewart (2019) suggested that researchers should think more broadly about the forms and effects of innovation in construction. Oti-Sarpong et al. (2022) suggested that the adoption and use of advanced technologies will reconfigure existing arrangements in the construction industry, creating a problem of institutional complexity for the firms in the industry, especially contractors.

Factors Influencing Innovation in Construction

There is a large body of literature on determinants and influences of innovation in construction. Blayse and Manley (2004) outlined these primary influences: clients and manufacturers; the structure of production; relationships between individuals and firms in the industry and between the industry and external parties; procurement systems; regulations and standards; and the nature and quality of the resources of the organisation. Seaden and Manseau (2001) developed a conceptual model for analysing innovation in construction, describing the linkages between the business environment, business strategy, innovative practices, and business outcomes. It is evident that both market forces and company resources are important determinants of the innovative capability of a company. Pries and Janszen (1995) analysed the innovative and strategic behaviour of the construction industry and its companies. They showed a relationship between innovation and the environment of the industry.

It is recognised that the construction industry, project, process, and product have several features that influence innovation (Xue et al., 2014). First, firms in the industry have little control over the demand for their works and services in terms of what they are required to produce, when they produce these items, and where the items are to be built (De Valence, 2011; Ofori, 2022). Moreover, construction firms tend to have low bargaining power compared with their clients (WEF and Boston Consulting Group, 2016). The firms face the effects of seasons on their activities and, given the structure of demand, rapid changes in volumes of activity. The resulting risks and uncertainties in the project-based nature of construction make firms reluctant to invest in new technologies. Uusitalo and Lavikka (2021) note that construction companies must develop competence in handling the short-term variation in building projects by seeking to establish broad resource bases to be able to adapt their operations to the continuously shifting market situation. Seeking a design and production approach that would enable construction companies to serve the entire market has been a long-standing objective (Lessing and Brege, 2015).

Second, construction is highly regulated, owing to the impact of its inputs, processes, and outputs on health and safety, and, as is recognised more recently, the well-being of participants and occupants or users (Hillebrandt, 2000; Di Leva, 2021). Due to technical regulations and standards, and, increasingly, laws incorporating social requirements such as the UK's Public Services (Social Value) Act 2012 (HM Government, 2012), many aspects and performance parameters are strictly determined and controlled by agencies and players outside the industry. As noted by several authors, the regulations limit the opportunities for innovation, such as in terms of differentiation of product and services (Blayse and Manley, 2004; Pries and Janszen, 1995).

Third, the construction project involves a large number of organisations with different economic, technical, and social motivations, interests, and aspirations (Miozzo and Dewick, 2004; Bygballe and Jahre, 2009). The project is described as a temporary organisation (Turner and Müller, 2003). At the corporate level, organisations work mostly on one-off projects, and there are often poor linkages between project and business processes, and among strategic and tactical considerations. Also, the lack of long-term relationships among the many project participants in the fragmented construction industry (Dubois and Gadde, 2002) and lack of integration in the often extensive and complex supply chain (Akintoye et al., 2000) are considered to inhibit innovations. Sadeh et al. (2021) noted that over the decades, few construction companies have achieved technological transformation. They sought to explain this heterogeneity and the determining factors by comparing the characteristics of firms classified as 'haves' and 'have-nots' regarding digital transformation. The relatively low level of innovation in construction is also attributed to the reluctance within the industry to take up new approaches and changes that will affect established, familiar ways of organising work within and between firms. These are exacerbated by the adversarial culture among different participants in construction projects, which has been identified in many countries (Gadde and Dubois, 2010; Morrell, 2015; Yang et al., 2023).

Davis et al. (2016) suggested that the criticism of the slow pace of innovation in construction may be unwarranted, considering the structure of the industry and the nature of the construction business. Also, the distinction between innovations on projects and in the firm innovation makes it difficult to extract innovations. They found that internal organisational and process innovation account for the majority of improvements identified in Australia. Wang et al. (2023) noted that the characteristics of the construction industry, such as site-based operations, project-based business, and temporary supply chains, make it difficult to measure innovation in construction using the criteria applied for other industries. Ninan et al. (2022) investigated the role of narratives in shaping innovation. They studied 133 innovations across different construction projects in the UK, such as 'innovations that enable project completion', 'innovations that improve productivity', 'innovations in health and safety', and 'innovations in sustainability'. They argued that while factors such as incentives and rewards can motivate innovation, narratives shape or guide innovation in particular directions and areas.

Some factors can be motivators as well as inhibitors of innovation in construction (Slaughter, 2000). For example, those spurring competition also hinder long-term investment. The industry presents low barriers to entry, which fosters high levels of competition. At the same time, there are options for new ways of practice. Indeed, the fact that construction involves discontinuous projects means that each one presents opportunities for innovation in terms of inputs, processes, and practices. New methods and equipment can be introduced. The many project participants can be brought together in different ways, at different stages.

Over time, the changing needs of types of projects, location, conditions in which they are to be constructed, developments in basic technology including the availability of new materials, changes in the requirements of the clients and other stakeholders, and so on, provide incentives for innovation in terms of new designs, construction technology, approaches to project management, and so on. In the era of globalisation, it became evident that innovation was key to overcoming the disadvantages of entry and operating outside the home country, given the 'local' nature of construction (Ofori, 2003).

It is pertinent to note that construction enables innovation in other sectors through the items it builds. Zeng et al. (2023) noted that innovation is an important driving force for high-quality regional economic development. Noting that cities are the gathering places of various innovation elements and resources, they examined the impact of smart city construction on regional innovation in China and found that the construction of a smart city has improved the level of innovation in the region, and investment in science and technology and human capital are key vehicles for realising this.

Construction 4.0 (and 5.0?)

In recent years, many new, advanced technologies have been used in construction. These include many digital technologies, manufacturing-based approaches, 3D printing, analytics and big data, digital twins, sensors, and AI/machine learning. These developments are usually considered to be drawn from the idea of Industry 4.0, leading to the term Construction 4.0 (Sawhney et al., 2020). Das et al. (2023) noted that Industry 4.0, characterised by the exponential pace of technological innovations, is compelling organisations to transform or be displaced. They proposed a conceptual maturity model for construction enterprises for business scenarios leading to Industry 4.0.

It is pertinent to consider the comprehensive review of Construction 4.0 by van der Heijden (2023). He observed that Construction 4.0 has the potential to transform the way constructed items and the built environment are designed, built, and operated. Van der Heijden (2023) assessed its implications from both:

1. a narrow perspective considering technology adaptation, referring to digital technologies, data analytics, and automation to improve efficiency and productivity (Statsenko et al., 2023), including building information modelling (BIM), the Internet of Things, robotics and automation, prefabrication and off-site construction, and virtual reality (VR) and augmented reality (AR); and
2. a broader perspective considering its implications on society, environment, governance, and technology itself (Wang and Guo, 2022).

The analysis by van der Heijden (2023) revealed: (i) a gap between theoretical proposals and the practical implementation of Construction 4.0 technologies, processes, and strategies; (ii) a variety of barriers hindering the effective adoption of Construction 4.0, including high initial costs, a shortage of the necessary skilled personnel, inadequate regulations, hesitancy among construction leaders, and an aversion to change in the industry; and (iii) a lack of understanding among policymakers and scholars of the impact of Construction 4.0 on the dimensions of the broader perspective outlined above. Van der Heijden (2023) suggested that by applying Construction 4.0 appropriately, the construction industry may be able to leapfrog the '4.0' revolution and embrace a '5.0' approach by incorporating a human-centric perspective and focusing on how technology and automation can help address the challenges of the twenty-first century, leading the way towards achieving Society 5.0.

De Soto et al. (2022) also studied the implications of Construction 4.0 for the workforce and organisational structures. Aghimien et al. (2021) found that in the era of Industry 4.0, construction organisations need transformation capabilities relating to technology governance, knowledge management, and the creation of new resources and processes to gain a competitive advantage and transform their service delivery. Most recently, Omotayo et al. (2024) explored current and future developments in the construction industry considering Construction Industry 4.0 and 5.0, and the planning, design, construction processes, as well the behaviour of construction professionals, research in the built environment, and new interactions of built environment practitioners with professionals from fields relating to construction.

There has been caution regarding new technologies. Sherratt (2021) considered the ethical and social challenges of implementing Construction 4.0. Earlier, du Plessis and Sherratt (2020) observed that while it is promised that the use of technologies will improve productivity in construction and also enhance the operation and maintenance of the future built environment, by being able to both automate and optimise their operations, Construction 4.0 could be directing the industry towards a 'smart dystopia', exposing built items to vulnerabilities such as that to cyberattacks, energy loss, and concerns about privacy and exclusion.

INNOVATION IN CONSTRUCTION IN DEVELOPING COUNTRIES

The Need and the Situation

It is generally agreed in the literature that the volume and rate of innovation in the construction industry in developing countries is inadequate (van Egmond-de Wilde de Ligny, 2011; Ofori, 2016; Sayyed et al., 2023). However, innovation in construction is even more key in the context of developing countries, for many reasons. First, there is a higher volume of a backlog of built items and infrastructure which these nations require for their development, and the citizens need to meet the basic standards of quality of life. Second, there is a shortage of resources, both financial and material, considering the methods being applied currently. It is necessary to reduce the reliance on imports and build local capacity. In turn, this would enhance the multiplier effect of construction activity in the local economy. Third, there are opportunities to foster innovation. As Ofori (2016) suggests, it is also worth noting that innovation is even more complex in the context of developing countries owing to the usual absence of a policy framework, incentive programmes, and the generally unconducive operating environment of businesses. Thus, a greater infusion of ingenuity is required to realise innovation. Table 3.1 presents an analysis of the imperatives, opportunities, threats, inhibitors, weaknesses, and terminators.

As noted above, organisations tend to imitate and adopt innovations that others have created. Yusof et al. (2014) found that construction participants in Malaysia, both contractors and consulting firms, are innovation adopters. They suggested that this knowledge could be used by organisations, government, and professional institutions to formulate strategies, policies, and programmes to undertake, engender, or support innovation. Sayyed et al. (2023) examined the relationships between a construction project's level of innovation, collaborative procurement management, and project performance, focusing on developing countries. They found that a project's level of innovation has a strong influence on project performance.

Rust and Koen (2016) noted that innovation in the construction industry should be stimulated in order to develop the uniquely South African technological solutions required to provide and maintain the country's economic and social infrastructure. They argued that investment in construction innovation should be based on knowledge of the current state of construction technology and the industry and the drivers and trends that will shape the future industry landscape (see below). Their construction industry technology foresight study identified these focus areas of technological development in South Africa: advanced materials and products; ICT and robotics; alternative energy systems for the built environment, especially solar power; housing and alternative technologies for municipal service delivery; clean processes and technologies; and logistics and supply chain optimisation. Kedir et al. (2022) undertook a scenario analysis of the factors influencing the adoption of industrialised construction in three developing countries that have high housing demand: Ethiopia, Kenya, and South Africa. They identified 19 factors, including product performance,

Table 3.1a *Innovation in construction: imperatives, opportunities, threats*

Imperatives	Opportunities	Threats
• Satisfying the SDGs and improving the quality of life. • Providing the foundation for economic and social development; and enhancing the multiplier effects from construction. • Improving the performance of the construction industry. • Need to understand innovation in general, and in construction in developing countries.	• Possibility of innovation to improve the performance of the construction industry for the benefit of the industry and the nation. • High level of need for innovation because of backlog of need and demand for all forms of infrastructure. • Technology and knowledge of new ways at the broader levels, and in construction, now available. • Market opportunities for truly new products, services, and approaches. • Awareness and agreement that 'something should be done', including by governments and the general citizenry, facing housing shortages, high costs of housing, and poor infrastructure services. • Possibility to reduce imports of materials, equipment and companies. • Greater understanding of innovation and its requirements, especially in the context of developing countries.	• Poor access to good infrastructure services, including transport, hindering economic growth. • Environmental, climate change, and sustainability considerations posing new pressures. • Health and quality of life issues associated with poor urban provision and management. • Increasing levels of urbanisation with implications for effective management. • Prevailing obstacles to innovation in business, including policy and administration.

Table 3.1b *Innovation in construction: inhibitors, weaknesses, and terminators*

Inhibitors	Weaknesses	Terminators
• Low finance, little assistance, low access to funding. • Narrow housing market. Lack of access to mortgages. • Willingness to accept the unknown. • Regulations hindering new technology … old laws. • Intellectual property protection.	• Levels of technology and knowledge … education based on metropolitan bodies of knowledge. • Low government leadership and support. Strategies, policies, and procedures. Lack of linkage between national development plans and construction industry development proposals. • Weak administrative systems. • Low linkages between construction and other sectors leading to low multiplier.	• Entry and dominance of foreign companies and products. • Narrow view and treatment of innovation.

and policies and regulations, and found that the key accelerators are governments' commitment to invest in future construction and incentivise companies to adopt industrial construction methods, and for those companies to produce competitive products. The policies to be adopted depend on the specific situation.

As mentioned above, at the national level, a key objective in innovation is the diffusion of the new technology, processes, or product. Van Egmond-de Wilde de Ligny and Erkelens (2008) noted that the diffusion of innovative technologies in the market is a difficult and complex process with varying degrees of success. Moreover, innovative technologies, although they promise superior performance compared with incumbent technologies, are not diffused in a balanced manner. Applying the concepts of the technological regime, actor network, and technology sets to analyse technology diffusion in the dwelling segment of the construction industry in Costa Rica, they found that bottlenecks in the innovation system curtailed the diffusion of innovative construction technology.

Works on Advanced Technology Applications in Developing Countries

How have developing countries taken to frontier construction technologies? Mansour et al. (2021) proposed a multi-criteria decision-making model to measure the strategic readiness and levels of maturity of construction firms in Malaysia with respect to Industry 4.0 implementation. They found that the human capital component (the intellectual agility, knowledge, skills, and competencies of professionals) is the most critical factor affecting the success of Industry 4.0 implementation. Arowoiya et al. (2021) evaluated the critical success factors affecting the implementation of AR in construction. They found that there is full awareness of information and communication technology (ICT) among respondents, while around half have used mobile AR computing. The most critical success factors were: the need for innovation in executing construction projects; advancement in computer design interface and hardware devices; improvement of accuracy in AR technology; developments in mobile computing; and AR collaboration with ICT software. They recommended that government policies should support, enforce, and implement the use of new technologies in construction projects to enhance efficiency and productivity.

Adepoju and Aigbavboa (2020) examined the opportunities and threats of Construction 4.0 for the workforce in Nigeria. They found 'a high level of implementation of Prefabrication and BIM'; opportunities such as increased collaboration, creativity, workers' efficiency; and a major threat, which is the retrenchment of workers, especially those with manual jobs. They recommended more sensitisation of the workforce to Construction 4.0, reskilling of the workforce, and development of government policies to protect the workforce against retrenchment. Olugboyega and Windapo (2021) sought to identify indicators for tracking the reduction in BIM adoption barriers in South Africa. The results revealed that BIM adoption barriers, such as the lack of BIM resources, and non-readiness of employees to accept change, have been significantly reduced. They concluded that the construction industry in South Africa is a BIM-frontier industry.

Dauda et al. (2024) noted that widespread adoption of digitalisation in the construction industry in India faces impediments. They found a willingness in the industry to embrace digitalisation, citing benefits such as enhanced productivity and other beneficial impacts on processes. Larger organisations were more proactive, while smaller businesses had lower adoption rates owing to challenges in resources and knowledge. They identified constraints under these themes: finance/resource, culture/organisation, regional disparities, data security/privacy concerns, and awareness/capacity-building. Oke et al. (2023) examined the level of awareness and usage of digital technologies in the construction industry in Lagos State, Nigeria, and found that while the professionals had very good knowledge of some digital technologies, the implementation of the tools in the industry is low. They proposed a digital transformation roadmap for developing countries involving the inter-related efforts of higher education institutions, financial institutions, government, and professional institutions. Acheng et al. (2022) observed that the adoption of BIM remains low in developing nations. Studying the situation in Uganda, they assessed relationships between profession, work experience, role in the supply chain, and involvement in academia against the awareness, BIM tools used, BIM functions, perceived benefits of BIM, and barriers to BIM adoption. Enhanced productivity was the highest benefit of BIM, along with complete project information and improved design quality. The top barriers to adoption were acquisition and training costs.

Aiyetan and Das (2023) studied the factors that influence the use of drones in the construction industries of South Africa and India. They found that the barriers and factors could be categorised under economic, operational, socio-legal and environmental, and construction-related aspects. The strategic measures that can enhance the use of drones in the industry in developing countries include appropriate government policy, regulations and legal provisions; enabling competency-building through training and piloting licences; permitting airspace for the use of drones for construction; reckoning the cost of drones and their operation in the project cost; and creating an organisational culture for adopting technological change.

POSSIBLE ACTION

Possible Approaches

Many possible approaches to enhancing the incidence, volume, and pace of innovation in construction in developing countries have been suggested. The earlier works are reviewed in Rwelamila and Ogunlana (2015), and some of the more recent ones are considered above. In this section, possible areas of research are discussed. First, some of the drivers are considered. Then, a research agenda is outlined.

The Tableau

What are the drivers of innovation in construction in developing countries? The World Economic Forum (2024) identified these top ten short-term risks globally: AI-generated misinformation and disinformation; extreme weather events; societal and/or political polarisation; cyber insecurity; inter-state armed conflict; lack of economic opportunity; inflation; involuntary migration; economic downturn; and pollution. Over the next 10 years, it outlined these additional risks among the top six: critical change to earth systems; biodiversity loss and ecosystem collapse; natural resource shortages; and adverse outcomes of AI technologies. These risks underlie drivers of action in sectors to address global challenges.

In construction, Rust and Koen (2016) identified these trends in South Africa: globalisation; conjuncture of the economy (cyclical nature of the economy); innovation economy (that boosts productivity and creates new industries); regulation and legislation (including corporate liability, shareholder pressure, and consumer activism); sustaining humanity (the need to improve human well-being and quality of life for a growing population through poverty reduction, protection of basic human rights, improved health and food security, and delivery of basic services); infectious diseases (and pandemics); environment and energy impact (sustainability of ecosystem services that provide food, water, materials, energy, and ability of the environment to absorb waste products); urbanisation; housing and services (addressing backlogs which lead to slums lacking adequate access to water, sanitation, security of tenure); skills for today and tomorrow (a shift towards knowledge workers); job creation and job quality; and advances in technology (frontier technologies and technology trends that influence the construction industry).

Considering developing countries in general, these trends can be added: infrastructure development; climate change and its impact; economic stress at individual and national levels owing to the post-pandemic and geopolitical situation; frequency and impact of natural disasters. These drivers should influence the urgency, nature, and pace of innovation in construction.

Research Agenda

As noted above, innovation is even more important in developing countries, and more so in the construction industries. Blayse and Manley (2004) suggested that further research be undertaken to explore the relationships among the innovation influences, and between those influences and other aspects of business strategy and operating environment, in the context of broader societal considerations. Some proposals for action are presented here in the form of a research agenda. Researchers should consider examining the proposals and studying how to realise them.

A research agenda for innovation in construction in developing countries should consider innovation in these areas: materials; technologies (techniques and tools); processes; management and organisational approaches at different levels (team, project, company) (including new 'participants' such as community involvement and traditional structure). It should focus on sustainability. It should also seek synergies among measures and actions, and look for leapfrogging opportunities. The research should go beyond analysing the literature, no matter how advanced the method. It should seek to realise innovations with impact.

The overall banner for innovation in construction in developing countries could be: 'Towards attaining the Sustainable Development Goals (SDGs) through innovation in construction – Providing the basis for national development and quality of life with continuous innovation.'

From the above discussion, research in construction innovation should be based on the broad definition and conceptualisation of the process. It should go beyond R&D, technology, and products. It could focus on addressing the problems of the construction industry that hinder improvement in its performance, its products and the services provided in the products. In developing countries, the research topics could be in two categories: conventional subjects and contemporary subjects, such as application of frontier technologies.

At the level of individuals and tasks

This would involve innovation in technology, materials, skills, and processes:

- In technologies, the frontier approaches such as Construction 4.0 and 5.0 should be explored in the context of developing countries. The segments include effective digitalisation including BIM and all other aspects of construction; modern methods of construction; and appropriate equipment.

- In materials, there should be an exploration of modern materials, including areas like nanotechnology. The approach should encompass materials design in addition to material science. The human–technology interface should be explored in the context of developing countries.
- In terms of skills, the concern should be attracting talent into careers in construction; fostering diversity in the industry; and ensuring the well-being and safeguarding of the workforce. Preparing the construction workforce and society for change.

In teams

There should be a focus on the formation of appropriate teams for particular types of projects and contexts. The development and maturity of teams as part of the project management process also need to be studied. Leadership is key in developing countries, and leadership development is an important subject.

At the level of projects

- Appropriate project performance parameters are required in developing countries to guide companies and practitioners. These should go beyond time, cost, quality, safety and health, and environmental considerations. These factors could also be studied: the broader concept of sustainability and client, user, or stakeholder satisfaction.
- The procurement arrangements used should also be appropriate and aim to provide conditions for effectiveness in project management. Some incentives that can be offered in this regard to foster innovation include: (a) innovation competitions (such as those related to design with specific requirements such as affordability); (b) making provision in tenders for the submission of alternative designs with alternative prices, in addition to the price for the base design that forms the basis of the tender; (c) provisions for value engineering exercises that encourage innovations (that may lead to cost, time, and/or energy savings, and other forms of performance improvement).
- On project implementation and management, the topics worth exploring include: governance (addressing corruption and lack of professionalism); avoiding short-termism in projects from the companies' perspectives, such as in attending to team selection and development; and post-mortemism, in effective on-project monitoring to draw lessons from the particular ongoing projects.
- Operating environment – effective regulation, standards, and benchmarking; importance of context; and project–location interface to enhance linkage effects.

In companies

- Project–company linkages, and how each project contributes to corporate growth.
- Strategic planning for corporate growth and development, and corporate social responsibility.

- Construction SME development, strategic alliances, and joint venture formation; public–private enterprises and ventures.
- Managing the construction value chain and exploring synergies among firms in various related sectors in forming commercial relationships.
- Benefiting from frontier technology applications, appropriate planning, learning, and application; benefiting from government policies and programmes. Gatekeeping and horizon-scanning periodically at the corporate level.
- Building a culture for innovation in the organisation; investment in human resource development and other capacity-building.

At national (industry) level

- Reducing fragmentation in the construction industry through appropriate strategies by multiple professional institutions, building a culture for cross-discipline collaboration.
- Forming and continuously strengthening agencies for construction industry development, reducing duplication, and plugging gaps in the administration of aspects of development and construction.
- Gatekeeping and technology horizon-scanning periodically occur at the industry level.
- Stakeholder, community, and market pressure on the construction industry to improve its performance and reputation.

At national (government) level

- Policies, regulations, and government administrative systems are responsible for development and building control. National-level, all-sector technology development, and innovation programmes.
- Appropriate coverage of construction in the national development plan, and in the national science, technology, innovation, and digitalisation strategies; construction industry policies; and collaboration at the regional level towards attaining regional synergies in finding solutions to common problems.
- Gatekeeping and horizon-scanning periodically – broadly, and at the national level.
- Building and fostering a culture for innovation in the economy as a whole, and particularly in construction.
- Industry and government collaboration; how to make it effective.

CONCLUDING REMARKS

Innovation is a critical factor in historical accounts and empirical analyses of how nations achieve economic development and improvements in the standard of living. Developing countries are in danger of being left behind in technology in general and in construction specifically. It is suggested that the approach to all aspects of innovation in developing countries should differ from what has been agreed to be the norm in industrialised countries owing to the differences in context, operating environment, and resources. Innovation in developing countries is even more critical because of the need for them to build the infrastructure, systems, and capabilities to take advantage of advances in key areas of knowledge. Innovation in construction faces challenges owing to the inherent characteristics of the industry, its projects, processes, and products. Again, in developing countries, innovation is even more needed because the existing backlogs of work hold back their socio-economic advancement and entrench the citizenry in a poor quality of life.

Developing countries should seek to exploit the frontier technologies. The implications of Construction 5.0 should be the main consideration for developing countries now. It is more comprehensive in coverage and takes the human element into account. This segment is key as culture and society are relevant in innovations: their development and implementation. It is necessary to leave no company behind. This is a major task owing to the variety and multiplicity of firms in the construction industries of those countries, most of which are small. Effective diffusion of new technologies and ideas should be a national objective. To realise this, it is necessary to explore the role of the local context: the societal, economic, technological, political, and administrative milieu of innovations. In developing countries, it is necessary to look for leapfrogging opportunities and to seek synergies. It would also be appropriate to consider how construction innovations can engender development in other areas of the economy.

REFERENCES

Acheng, P.O., Kibwami, N., Mukasa, T.J., Odongkara, B.B., Birungi, R., Semanda, J. and Manga, M. (2022) Building information modelling adoption in Uganda's construction industry. *International Journal of Construction Management*, *23*(13), 2185–2208. https://doi.org/10.1080/15623599.2022.2047278

Adebowale, B.A., Diyamett, B., Lema, R. and Oyelaran-Oyeyinka, O. (2014) Introduction. *African Journal of Science, Technology, Innovation and Development*, *6*(5), v–xi. https://doi.org/10.1080/20421338.2015.1010774

Adepoju, O. and Aigbavboa, C. (2020) Implementation of construction 4.0 in Nigeria: evaluating the opportunities and threats on the workforce. *Academic Journal of Interdisciplinary Studies*, *9*(5), 254–264. 10.36941/ajis-2020-0102.

Aghimien, D., Aigbavboa, C. and Matabane, K. (2021) Dynamic capabilities for construction organizations in the fourth industrial revolution era. *International Journal of Construction Management*, *23*(5), 855–864. https://doi.org/10.1080/15623599.2021.1940745

Aiyetan, A.O. and Das, D.K. (2023) Use of drones for construction in developing countries: barriers and strategic interventions. *International Journal of Construction Management*, *23*(16), 2888–2897. doi:10.1080/15623599.2022.2108026.

Akintoye, A., McIntosh, G. and Fitzgerald, E. (2000) A survey of supply chain collaboration and management in the UK construction industry. *European Journal of Purchasing and Supply Management*, *6*, 159–168.

Arowoiya, V.A., Oke, A.E., Akanni, P.O., Kwofie, T.E. and Enih, P.I. (2021) Augmented reality for construction revolution – analysis of critical success factors. *International Journal of*

Construction Management, *23*(11), 1867–1874. https://doi.org/10.1080/15623599.2021.2017542
Barrett, P. (2007) *Revaluing Construction*. Wiley-Blackwell, Chichester.
Blayse, A.M. and Manley, K. (2004) Key influences on construction innovation. *Construction Innovation*, *4*(3), 143–154. https://doi.org/10.1108/14714170410815060
Bossink, B.A.G. (2004) Managing drivers of innovation in construction networks. *Journal of Construction Engineering and Management*, *130*(3), 337–345.
Buehler, M., Buffet, P. and Castagnino, S. (2018) The Fourth Industrial Revolution is about to hit the construction industry. Here's how it can thrive. Proceedings of the World Economic Forum Agenda, Davos, Switzerland, January, pp. 22–26.
Bygballe, L. and Jahre, M. (2009) Balancing value creating logics in construction. *Construction Management and Economics*, *27*, 695–704.
Casadella, V. and Uzunidis, D. (2017) National innovation systems of the south, innovation and economic development policies: a multidimensional approach. *Journal of Innovation Economics & Management*, *23*, 137–157. doi:10.3917/jie.pr1.0007.
CIB (International Council for Research and Innovation in Building and Construction) (1999) Managing Construction Industry Development in Developing Countries: Report on the First Meeting of the CIB Task Group 29. Arusha, Tanzania, 21–23 September. Rotterdam.
Cirera, X. and Maloney, W.F. (2017) *The Innovation Paradox: Developing-country capabilities and the unrealized promise of technological catch-up*. World Bank, Washington, DC.
Cirera, X., Frías, J., Hill, J. and Li, Y. (2020) *A Practitioner's Guide to Innovation Policy: Instruments to build firm capabilities and accelerate technological catch-up in developing countries*. World Bank, Washington, DC.
Clark, N. and Chataway, J. (2009) Below the radar: a fresh approach to innovation and development policy. *International Journal of Technology Management & Sustainable Development*, *8*(3), 171–175. doi:10.1386/ijtm.8.3.171/2.
Crépon, B., Duguet, E. and Mairesse, J. (1998) Research, innovation and productivity: an econometric analysis at the firm level. *Economics of Innovation and New Technology*, *7*(2), 115–158.
Crespi, G. and Zuniga, P. (2011) Innovation and productivity: evidence from six Latin American countries. *World Development*, *40*(2), 273–290. doi:10.1016/j.worlddev.2011.07.010.
Das, P., Perera, S., Senaratne, S. and Osei-Kyei, R. (2023) Paving the way for industry 4.0 maturity of construction enterprises: a state of the art review. *Engineering, Construction and Architectural Management*, *30*(10), 4665–4694. https://doi.org/10.1108/ECAM-11-2021-1001
Dauda, J.A., Chavan, N.N., Saka, A.B., Ajayi, S.O. and Oyegoke, A.S. (2024) An appraisal of barriers to digitalisation of construction industry in developing countries: perspective from India. *International Journal of Construction Management*, 1–13. https://doi.org/10.1080/15623599.2024.2362014
Davis, P., Gajendran, T., Vaughan, J. and Owi, T. (2016) Assessing construction innovation: theoretical and practical perspectives. *Construction Economics and Building*, *16*(3), 104–115. https://doi.org/10.5130/AJCEB.v16i3.5178
Demirkesen, S. and Tezel, A. (2022) Investigating major challenges for industry 4.0 adoption among construction companies. *Engineering, Construction and Architectural Management*, *29*(3), 1470–1503.
de Soto, B.G., Agusti-Juan, I., Joss, S. and Hunhevicz, J. (2022) Implications of Construction 4.0 to the workforce and organizational structures. *International Journal of Construction Management*, *22*(2), 205–217. doi:10.1080/15623599.2019.1616414.
De Valence, G. (2011) *Modern Construction Economics: Theory and applications*. Spon, London.
Di Leva, E. (2021) The challenge and promise at the intersection of environmental and social policies: how the World Bank established a policy framework that fully integrates

environmental and social concerns. *Global Social Policy*, *21*(2), 344–348. https://doi.org/10.1177/14680181211019170

Dubois, A. and Gadde, L.E. (2002) The construction industry as a loosely coupled system: implications for productivity and innovation. *Construction Management and Economics*, *20*, 621–632.

Dulaimi, M. (1995) The challenge of innovation in construction. *Building Research and Information*, *23*(2), 106–109.

du Plessis, C. and Sherratt, F. (2020) Construction 4.0 and built assets in-use: creating an e-topia or dystopia? *Proceedings of the Institution of Civil Engineers-Management Procurement and Law*, *173*(4), 182–189.

Ewart, I.J. (2019) The social consequences of minor innovations in construction. *Construction Management and Economics*, *37*(9), 537–549. https://doi.org/10.1080/01446193.2019.1614202.

Fantom, N.J. and Serajuddin, U. (2016) The World Bank's Classification of Countries by Income, Policy Research Working Paper, No. WPS 7528. World Bank Group, Washington, DC. http://documents.worldbank.org/curated/en/408581467988942234/The-World-Banks-classification-of-countries-by-income

Gadde, L.-E. and Dubois, A. (2010) Partnering in the construction industry: problems and opportunities. *Journal of Purchasing and Supply Management*, *16*(4), 254–263. https://doi.org/10.1016/j.pursup.2010.09.002

Gambatese, J.A. and Hallowell, M. (2011) Enabling and measuring innovation in the construction industry. *Construction Management and Economics*, *29*, 553–567.

Giang, D.T.H. and Low, S.P. (2010) Role of construction in economic development: review of key concepts in the past 40 years. *Habitat International*, *35*(1), 118–125. https://doi.org/10.1016/j.habitatint.2010.06.003

Glass, J., Bygballe, L.E. and Hall, D. (2022) Transforming construction: the multi-scale challenges of changing and innovating in construction. *Construction Management and Economics*, *40*(11–12), 855–864. https://doi.org/10.1080/01446193.2022.2141432

Hillebrandt, P.M. (2000) *Economic Theory and the Construction Industry*, 3rd ed. Macmillan Press, Basingstoke.

HM Government (2012) Public Services (Social Value) Act 2012. https://www.legislation.gov.uk/ukpga/2012/3/enacted

Jauhiainen, J.S. and Mugabe, J.O. (2024) Innovation research in and on Africa: a literature analysis in 2015–2021. *African Journal of Science, Technology, Innovation and Development*, *16*(3), 425–438. https://doi.org/10.1080/20421338.2024.2329017

Kaplinsky, R. (2011) Schumacher meets Schumpeter: appropriate technology below the radar. *Research Policy*, *40*(2), 193–203. doi:10.1016/j.respol.2010.10.003.

Kedir, F., Chen, Q., Hall, D.M., Adey, B.T. and Boyd, R. (2022) Formative scenario analysis of the factors influencing the adoption of industrialised construction in countries with high housing demand – the cases of Ethiopia, Kenya, and South Africa. *Construction Management and Economics*, *40*(9), 690–710. https://doi.org/10.1080/01446193.2022.2098508

Kumaraswamy, M.M. and Shrestha, G.B. (2002) Targeting 'technology exchange' for faster organizational and industry development. *Building Research & Information*, *30*(3), 183–195. https://doi.org/10.1080/09613210110115216

Lall, S. and Pietrobelli, C. (2002) *Falling to Compete: Technology development and technology systems in Africa*. Edward Elgar Publishing, Cheltenham, UK and Northampton, MA, USA.

Leng, J., Sha, W., Wang, B., Zheng, P., Zhuang, C., Liu, Q., Wuest, T., Mourtzis, D. and Wang, L. (2022) Industry 5.0: prospect and retrospect, *Journal of Manufacturing Systems*, *65*, 279–295. https://doi.org/10.1016/j.jmsy.2022.09.017

Lessing, J. and Brege, S. (2015) Business models for product-oriented housebuilding companies: experience from two Swedish case studies. *Construction Innovation*, *15*(4), 449–472.

Mansour, H., Aminudin, E. and Mansour, T. (2021) Implementing industry 4.0 in the construction industry: strategic readiness perspective. *International Journal of Construction Management*, *23*(9), 1457–1470. https://doi.org/10.1080/15623599.2021.1975351

Miozzo, M. and Dewick, P. (2004) Building competitive advantage: innovation and corporate governance in European construction. *Research Policy*, *31*, 989–1008.

Mohnen, P. and Hall, B.H. (2013) Innovation and productivity: an update. *Eurasian Business Review*, *3*(1), 47–65.

Morrell, P. (2015) *Collaboration for Change: The Edge Commission Report on the Future of Professionalism*. Ove Arup Foundation, London.

Ninan, J., Sergeeva, N. and Winch, G. (2022) Narrative shapes innovation: a study on multiple innovations in the UK construction industry. *Construction Management and Economics*, *40*(11–12), 884–902. https://doi.org/10.1080/01446193.2022.2037144

Obeng-Odoom, F. (ed.) (2023) Broken mirrors: towards holistic political economies of development, its discontents and alternatives. In *Handbook on Alternative Global Development*. Edward Elgar Publishing, Cheltenham, UK and Northampton, MA, USA, pp. 2–11.

OECD (2009) *Innovation in Firms: A microeconomic perspective*. https://www.oecd.org/berlin/44120491.pdf

Ofori, G. (2003) Frameworks for analysing international construction. *Construction Management and Economics*, *21*(4), 379–391. https://doi.org/10.1080/0144619032000049746

Ofori, G. (2016) Construction in developing countries: current imperatives and potential. In Kahkonen, K. and Keinanen, M. (eds) *CIB World Building Congress 2016 on Intelligent Built Environment for Life*, Tampere, 30 May to 3 July, Tampere Department of Civil Engineering, Construction Management and Economics, Tampere University of Technology.

Ofori, G. (2022) Defining and describing construction in developing countries. In R. Best and J. Meikle (eds) *Describing Construction: Industries, projects and firms*. Routledge, Abingdon, pp. 171–196.

Ofori, G. (2023) Get construction project performance parameters right to attain Sustainable Development Goals. *Sustainability*, *15*(18), 13360. https://doi.org/10.3390/su151813360

Oke, A.E., Aliu, J., Jamir Singh, P.S., Onajite, S.A., Shaharudin Samsurijan, M. and Azura Ramli, R. (2023) Appraisal of awareness and usage of digital technologies for sustainable wellbeing among construction workers in a developing economy. *International Journal of Construction Management*, *24*(5), 521–529. https://doi.org/10.1080/15623599.2023.2179628

Olugboyega, O. and Windapo, A. (2021) Modelling the indicators of a reduction in BIM adoption barriers in a developing country. *International Journal of Construction Management*, *23*(9), 1581–1591. https://doi.org/10.1080/15623599.2021.1988196

Omotayo, T., Egbelakin, T., Ogunmakinde, O. and Sojobi, A. (eds) (2024) *Innovations, Disruptions and Future Trends in the Global Construction Industry*. Routledge, Abingdon.

Oti-Sarpong, K., Pärn, E.A., Burgess, G. and Zaki, M. (2022) Transforming the construction sector: an institutional complexity perspective. *Construction Innovation*, *22*(2), 361–387. https://doi.org/10.1108/CI-04-2021-0071

Oxford Poverty and Human Development Initiative and United Nations Development Programme (2023) *Unstacking Global Poverty: Data for high-impact action*. United Nations Development Programme and Oxford Poverty and Human Development Initiative. University of Oxford, Oxford.

Pérez-Luño, A., Cabrera, R.V. and Wiklund, J. (2007) Innovation and imitation as sources of sustainable competitive advantage. *The Journal of the Iberoamerican Academy of Management*, *5*(2), 71–82.

Pérez-Luño, A., Wiklund, J. and Cabrera, R.V. (2011) The dual nature of innovative activity: how entrepreneurial orientation influences innovation generation and adoption. *Journal of Business Venturing*, *26*, 555–571.

Popadiuk, S. and Choo, C.W. (2006) Innovation and knowledge creation: how are these concepts related? *International Journal of Information Management*, *26*, 302–312.

Prahalad, C.K. (2005) *The Fortune at the Bottom of the Pyramid*. Wharton School Publishing, Upper Saddle River, NJ.

Pries, F. and Janszen, F. (1995) Innovation in the construction industry: the dominant role of the environment. *Construction Management and Economics*, *13*, 43–51.

Ribeirinho, M.J., Mischke, J., Strube, G., Sjödin, E., Luis Blanco, J.L., Palter, R., Biörck, J., Rockhill, D. and Andersson, T. (2020) *The Next Normal in Construction: How disruption is reshaping the world's largest ecosystem*. McKinsey & Company.

Roger, S. (2004) Information technology and organisational structure: vindicating theories from the past. *Management Decision*, *42*, 316–329.

Rust, F.C. and Koen, R. (2016) Positioning technology development in the South African construction industry: a technology foresight study. *Journal of the South African Institution of Civil Engineering*, *53(*1), Paper 759, 2–8.

Rwelamila, P.D. and Ogunlana, S. (2015) *W107 – Construction in Developing Countries Research Roadmap – Report for Consultation*. International Council on Research and Innovation in Building and Construction (CIB), Rotterdam.

Sadeh, H., Mirarchi, C. and Pavan, A. (2021) Technological transformation of the construction sector: a conceptual approach. *International Journal of Construction Management*, *23*(10), 1704–1714. https://doi.org/10.1080/15623599.2021.2006400

Sawhney, A., Riley, M. and Irizarry, J. (2020) Construction 4.0: introduction and overview. In A. Sawhney, M. Riley and J. Irizarry (eds) *Construction 4.0: An innovation platform for the built environment*, Routledge, Abingdon, pp. 1–21.

Sayyed, Y., Hatamleh, M.T. and Alaya, M. (2023) Investigating the influence of procurement management in construction projects on the innovation level and the overall project performance in developing countries. *International Journal of Construction Management*, *23*(3), 462–471. doi:10.1080/15623599.2021.1889088.

Schumpeter, J.A. (1934) *The Theory of Economic Development: An inquiry into profits, capital, credit, interest, and the business cycle* (vol. 55). Harvard University Press, Cambridge, MA.

Seaden, G. and Manseau, A. (2001) Public policy and construction innovation. *Building Research and Information*, *29*(3), 182–196.

Secretariat of the New Partnership for Africa's Development (undated) NEPAD in Brief. Midrand, South Africa. https://www.nepad.org/publication/nepad-brief

Sexton, M. and Barrett, P. (2003) Appropriate innovation in small construction firms. *Construction Management and Economics*, *21*(6), 623–633. https://doi.org/10.1080/0144619032000134156

Sherratt, F. (2021) The ethical and social challenges of implementing Construction 4.0. *Proceedings of the Institution of Civil Engineers-Civil Engineering*, *174*(2). doi:10.1680/jcien.2021.174.2.55.

Slaughter, E.S. (1998) Models of construction innovation. *Journal of Construction Engineering and Management*, *124*, 226–231.

Slaughter, E.S. (2000) Implementation of construction innovations. *Building Research & Information*, *28*(1), 2–17. https://doi.org/10.1080/096132100369055

Solow, R. (1956) A contribution to the theory of economic growth. *Quarterly Journal of Economics*, *70*(1), 65–94.

Son, T.K.L. (2019) Innovation Efforts in Developing Countries: The case of Vietnamese small and medium-sized enterprises, WIDER Working Paper 2019/7. United Nations World Institute for Development Economics Research, Helsinki. https://doi.org/10.35188/UNU-WIDER/2019/641-8

Statsenko, L., Samaraweera, A., Bakhshi, J. and Chileshe, N. (2023) Construction 4.0 technologies and applications: a systematic literature review of trends and potential areas for development. *Construction Innovation*, *23*(5), 961–993. https://doi.org/10.1108/CI-07-2021-0135

Steele, J. and Murray, M. (2004) Creating, supporting and sustaining a culture of innovation. *Engineering, Construction and Architectural Management*, *11*(5), 316–322.

Toole, T.M. (1998) Uncertainty and home builders' adoption of technological innovations. *Journal of Construction Engineering Management*, *124*(4), 323–332.

Turner, J.R. and Müller, R. (2003) On the nature of the project as a temporary organization. *International Journal of Project Management*, *21*(1), 1–8. https://doi.org/10.1016/S0263-7863(02)00020-0

United Nations Conference on Trade and Development (UNCTAD) (2019) *A Framework for Science, Technology and Innovation Policy Reviews: Harnessing innovation for sustainable development*. United Nations, New York.

UNCTAD (2021a) *Technology and Innovation Report 2021: Catching technological waves – Innovation with equity*. United Nations, New York.

UNCTAD (2021b) Why global cooperation on science, technology and innovation is more crucial than ever, 13 December. https://unctad.org/news/why-global-cooperation-science-technology-and-innovation-more-crucial-ever

UNCTAD (2023) *Technology and Innovation Report 2023: Opening green windows – Technological opportunities for a low-carbon world – Overview*. United Nations, Geneva.

United Nations Development Programme (UNDP) (2023) *The Least Developed Countries Report 2023: Crisis-resilient development finance*. United Nations, Geneva.

UNDP (2024a) Human Development Index (HDI). https://hdr.undp.org/data-center/human-development-index#/indicies/HDI

UNDP (2024b) *Human Development Report 2023–24: Breaking the gridlock – Reimagining cooperation in a polarized world*. United Nations, New York.

Uusitalo, P. and Lavikka, R. (2021) Technology transfer in the construction industry. *The Journal of Technology Transfer*, *46*, 1291–1320. https://doi.org/10.1007/s10961-020-09820-7

van der Heijden, J. (2023) Construction 4.0 in a narrow and broad sense: a systematic and comprehensive literature review. *Building and Environment*, *244*, 110788. https://doi.org/10.1016/j.buildenv.2023.110788

van Egmond-de Wilde de Ligny, E. (2011) Construction technology development and innovation. In G. Ofori (ed.) *New Perspectives on Construction in Developing Dountries* (pp. 185–228). CIB, Spon.

van Egmond-de Wilde de Ligny, E. and Erkelens, P. (2008) Construction technology diffusion in developing countries: limitations of prevailing innovation systems. *Journal of Construction in Developing Countries*, *13*(2), 43–63.

van Uden, A., Knoben, J. and Vermeulen, P. (2014) *Human Capital and Innovation in Developing Countries: a firm level study*. Institute of Management Research, Radboud University, Nijmegen.

Wang, K.Y. and Guo, F.Y. (2022) Towards sustainable development through the perspective of construction 4.0: systematic literature review and bibliometric analysis. *Buildings*, *12*, 1–33.

Wang, N., Xu, Z. and Liu, Z. (2023) Innovation in the construction sector: bibliometric analysis and research agenda. *Journal of Engineering and Technology Management*, *68*, 101747. https://doi.org/10.1016/j.jengtecman.2023.101747

Watkins, A., Papaioannou, T., Mugwagwa, J. and Kale, D. (2015) National innovation systems and the intermediary role of industry associations in building institutional capacities for innovation in developing countries: a critical review of the literature. *Research Policy*, *44*(8), 1407–1418. https://doi.org/10.1016/j.respol.2015.05.004

Winch, G.M. (2003) How innovative is construction? Comparing aggregated data on construction innovation and other sectors – a case of apples and pears. *Construction Management and Economics*, *21*, 651–654.

World Bank (2024) Data – World Bank country and lending groups. https://datahelpdesk.worldbank.org/knowledgebase/articles/906519-world-bank-country-and-lending-groups#:~:text=For%20the%20current%202024%20fiscal,those%20with%20a%20GNI%20per

World Economic Forum (2024) *The Global Risks Report 2024*, 19th ed. Geneva.

World Economic Forum and Boston Consulting Group (2016) *Shaping the Future of Construction: A breakthrough in mindset and technology*. Geneva.

Xue, X., Zhang, R., Yang, R. and Dai, J. (2014) Innovation in construction: a critical review and future research. *International Journal of Innovation Science*, *6*(2), 111–126. https://doi.org/10.1260/1757-2223.6.2.111

Yang, Y., Wei, Z. and Zhang, Z. (2023) Stakeholder relationship in construction projects: a mixed methods review. *Buildings*, *13*(12), 3122. https://doi.org/10.3390/buildings13123122

Yusof, N., Kamal, E.M., Lai, K-S. and Iranmanesh, M. (2014) Are innovations being created or adopted in the construction industry? Exploring innovation in the construction industry. *SAGE Open*, July–September, 1–9. doi:10.1177/2158244014552424.

Zeng, Y., Zhang, Z., Ye, Z. and Li, L (2023) Regional innovation effect of smart city construction in China. *PLoS ONE*, *18*(2), e0281862. https://doi.org/10.1371/journal.pone.0281862

4. Innovation in the construction industry: experiences from Chile

Alfredo Serpell, Claudia Valderrama-Ulloa and Ximena Ferrada

INTRODUCTION

Innovation is a highly relevant issue for developing a country and its industrial sectors. In Chile, several efforts have been made to promote innovation at the country level by creating institutions that support these processes, as well as by the publication of an official document: *Política Nacional de Innovación para la Competitividad, Orientaciones y Plan de Acción* (National Policy of Innovation for Competitivity, Orientations and Action Plan) 2009–2010 (Ministerio de Economía, 2010). At country level, the implementation of innovation policies and programmes in Chile has been defined by the Committee of Ministers of Innovation (CMI) and is carried out through the coordinated work of six public institutions or agencies: the Millennium Science Initiative, the National Agency for Research and Development (ANID), the Ministry of Education (MINEDUC), the Undersecretariat of Regional and Administrative Development (SUBDERE), the Foundation for Agrarian Innovation (FIA), and the Production Development Corporation (CORFO). These agencies have different objectives, from promoting scientific and technological research (ANID) to supporting innovation in entrepreneurship (CORFO). Likewise, the Ministry of Education and the Undersecretariat of Regional and Administrative Development implemented their programmes focused on developing innovation within the educational system and implementing regional innovation policies. The institutions, universities, research institutes, private research and development agencies, and others work together to put into practice the guidelines and fundamental objectives of the National Innovation Policy for Competitiveness through specific programmes.

In 2023, Chile reached 52nd out of 132 nations in the Global Innovation Index 2023 of the World Intellectual Property Organization (WIPO) (INAPI, 2023). According to this report, the country stands out for its institutional strength concerning its quality in regulatory matters, and also for the increase in the growth rate of labour productivity and the creation of new businesses, which are related to the increase in the use of the intellectual property system and trademark applications of residents, per capita (Invest Chile, 2023).

Even so, it should be noted that investment in R&D in Chile has not shown significant growth in the last 10 years (2010–2020), remaining at 0.34% of GDP. This means that the percentage of innovation in companies in Chile is only 15.1% (Ministerio de Ciencia, Tecnología, Conocimiento e Innovación, Chile, 2022).

Innovation supported by the state has been reflected since 2008 through Law 20,241 on tax incentives for developing innovation in Chile (Ministerio de Hacienda, Chile, 2008). Its objective is to contribute to improving the competitive capacity of Chilean companies by establishing a tax incentive for investment in research and development (R&D) that allows these entities to rebate, via taxes, 35% of the resources allocated to R&D activities. Although since its creation, R&D spending has remained without a significant increase, the change in the construction sector stands out, where it doubled its spending from 2,542 million in 2019 to 5,193 million Chilean pesos in 2020 (Construye 2025, 2021).

There is also the creation and promotion of different research funds whose objective is to promote R&D in areas such as the strengthening of human capital (national and foreign scholarships for graduate studies), research projects (such as FONDECYT – National Fund for Scientific and Technological Development), associative research centres, development of applied research (such as FONDEF – Scientific and Technological Development Promotion Fund or FONIS – Fund for Health Research and Development), and the creation of knowledge networks and strategies, all managed under the aegis of the National Agency for Research and Development (ANID) (ANID, 2024).

In the development of R+D+I (research, development, and innovation), the Production Development Corporation of the Ministry of Economy, Development, and Tourism (CORFO) stands out, supporting entrepreneurship, innovation, and competitiveness in companies in different areas. Its primary areas of work are entrepreneurship, networks and territories, creative industries, technological capabilities, gender strategy in companies, investment and financing, emerging companies, and innovation. In this last area of innovation, initiatives such as Súmate a Innovar (Join us to Innovate), Crea y Valida (Create and Validate), and Innova Región (Innovate Region), among others, stand out (CORFO, 2024).

The construction industry has been considered to lack innovation compared with other industries. Despite this, there has been a growing interest in innovation in the industry over the last 30 years, and different initiatives have been launched to promote innovation, both at the institutional level and the level of companies in the sector. However, there is still a large space for innovation development in the sector that needs to be fostered, and many companies are not prepared or do not have the competencies to carry out effective innovation activities. As a result, these companies maintain an inadequate state of technological development.

However, in the last 10 years, the conditions of the Chilean construction sector have changed. Clients demand more in executing their projects regarding quality, costs, technology, and complexity. This situation has led construction companies to experience reduced earnings margins. In addition, the presence and participation of many foreign construction companies with more international experience, better technology, and financial backing have become a real threat to local companies.

This reality is forcing local construction companies to look for ways to improve their performance and obtain a competitive advantage through innovation. In this sense, the capacity of a company to change, adapt to new requirements, learn, acquire experience, and assume a position of competitive leadership inside the market in which it participates is fundamental for success in the current world (Serpell, 1997).

This chapter addresses the issue of innovation in the Chilean construction industry to evaluate the state of development of the industry in this aspect and does so from two perspectives. First, an institutional perspective encompasses different organisations, both private and public, that promote innovation in the sector. The second perspective is a vision at the level of construction companies, highlighting the studies carried out at this level on the actions they have been taking in terms of innovation. Additionally, a methodology that has been developed as part of the research is presented, which allows for reviewing the degree of readiness of each construction company to carry out innovation, based on a set of innovation drivers obtained from an exhaustive study of the literature on the subject.

In what follows in this chapter, the research problem addressed in this work is presented, including the contribution of this work and the scope of the research conducted. The next section describes the research methodology used to study innovation from two perspectives, which is followed by the results obtained from the research. Finally, a section discussing the results is presented, followed by the main conclusions of the chapter.

THE RESEARCH

The research is descriptive and based on a literature review on the topics described above. The objective of this work is to carry out an exhaustive survey of the development of innovation in the construction industry, both at the institutional level and at the level of each construction company, during the last 30 years. The main research questions that were addressed are:

1. What have been the main drivers of innovation in construction at the institutional level?
2. What efforts have been made to understand the state of innovation in construction companies?
3. What actions have been taken to promote innovation at the level of construction companies?

The research carried out and the results obtained contribute to providing a global knowledge of the innovation process that has been conducted in the Chilean construction industry, highlighting the different activities undertaken and the institutions involved. It also highlights innovations carried out by construction companies in the last decade.

METHOD

This descriptive research collects detailed, accurate data to describe a phenomenon without manipulating variables or establishing causal relationships. The key is to provide an objective and complete representation of the observed facts. The phenomenon being investigated revolves around initiatives undertaken by the Chilean government and private entities, such as trade associations, construction companies, and other organisations, to foster innovation within the construction industry.

To achieve the above, various sources of information were reviewed, such as webpages, Chilean newspaper articles, professional journals, and reports from different Chilean construction institutions and construction companies. The terms 'innovation' and 'construction sector' were used in the search. The decision to utilise these sources was based on their provision of historical and current information regarding the state of innovation in the construction sector. With the results obtained, a thematic analysis was performed, similar to the one carried out by Serpell and Rubio (2023). The collected information was processed and organised, resulting in a well-structured classification conducive to analysis and presentation.

RESULTS

Below are detailed various groups that have promoted innovation in Chile over the last 30 years, such as CORFO and ANID, founded in 1939 and 1957, respectively. Also included are the Cámara Chilena de la Construcción (Chilean Construction Chamber), founded in 1951, and newer centres like CIPYCS and CTEC, established in 2016, among others.

Innovation in Construction at the Institutional Level

In the case of the construction industry in Chile, institutional innovation has been driven by the participation and interaction of various organisations, including financial support from the state, as well as initiatives at the level of the business association and the corporate level. In this regard, four innovation-driving groups stand out:

a. Regulatory and standardisation agencies (Instituto Nacional de Normalización, Corporación Nacional del Medio Ambiente, Mutual del Trabajador, and Asociación Chilena de Seguridad, among others): define the legal or regulatory framework within which companies and organisations in the construction sector can operate and provide financing for the study and adoption of innovative technologies in their fields of action.
b. Universities and their research centres: through individual and collaborative work between academic institutions, there are opportunities to develop or assimilate modern technologies and knowledge in different areas to increase development

and technology transfer in the sector, with the financial support of public and private organisations. The Center for Innovation of New Materials, Concrete and Wood of the Pontificia Universidad Católica de Chile, the Interdisciplinary Center for Productivity and Sustainable Construction (CIPYCS), and others stand out.

c. Institutes (Instituto de la Construcción, Instituto del Cemento y del Hormigón de Chile, and the Corporación de Desarrollo Tecnológico de la Cámara Chilena de la Construcción, among others): represent instances to promote and coordinate research, innovation, and technology transfer initiatives in companies and organisations of the construction sector, articulating public and private efforts, encouraging and proposing policies, standards, technical documents, and promoting best practices.

d. Development centres financed by state funds (CORFO and ANID) that have been created through the association of different companies, organisations, and institutions (national and foreign), whose objective is to develop Innovation in Productivity and Sustainability in the Construction sector. In 2016, CORFO launched the Construye 2025 programme, whose focus is on the transformation of the construction sector for productivity and sustainability. Today this programme is a strategic reference that unites the public, private, and academic sectors to generate innovative initiatives, boost innovation and digital transformation, increase innovation, improve human capital, promote industrialisation, improve and update the national regulatory framework, and strengthen the sustainability of the construction sector. Among its main results are the launching of the BIM and DOM 'Dirección de Obras Municipales' plan online (DOM is a state agency in municipalities that enforces the General Law of Urbanism and Constructions, inspecting and granting various building permits), the development of the R&D and Circular Economy roadmap, and the organisation of 'Edifica', which is a fair that, since 1990, has brought together stakeholders from the construction sector and its entire value chain to learn about the latest trends in solutions and technologies for construction (Construye 2025 and CORFO, 2016). Additionally, CORFO of the Ministry of Economy, Development, and Tourism aims to bolster entrepreneurship, innovation, and competitiveness, facilitating credit access for small to medium-sized enterprises through coverage from financial institutions. Concurrently, the National Agency for Research and Development (ANID) administers programmes to advance research, technological development, and scientific-technological innovation across all knowledge domains. Both centres have supported the creation of other innovation instances, such as:

  - The Technological Center for Innovation in Productivity and Sustainability in Construction (CTEC) is a technological centre for the dissemination of disruptive technologies and open innovation. It fosters synergies and collaboration in the construction ecosystem, promoting digitalisation, the creation of prototypes and pilots, R&D projects, and the formation of human capital to improve the productivity, sustainability, and competitiveness of the national industry.

- The Interdisciplinary Center for Productivity and Sustainable Construction (CIPYCS), whose mission is to promote development and innovation in the construction industry to make it more productive and sustainable through cutting-edge infrastructure and innovation and development services throughout Chile, employing R&D+i.
- Extension centres for small and medium-sized enterprises (SMEs) support these companies to help reduce the technological and human capital gap affecting the construction industry in Chile. These centres are created in universities or trade institutions. Although the extension centres were established in 2011, no official document has accounted for their effectiveness to date.

At the institutional level, in the private sector, two sources of innovation activities in the sector stand out:

a. Trade associations (Chilean Chamber of Construction (CChC) and professional associations): they represent instances of analysis of the problems faced by companies and organisations in the sector (associates), thus collaborating in the development and progress of the construction sector in general. The CChC's vision is to promote the development and promotion of construction as a fundamental lever for the country's development. In 2019, through the Technological Development Corporation of the CChC, the first innovative HUB in the industry was created to promote innovation, sustainability, productivity, and technology transfer among the actors of the construction industry. This was achieved through collaboration, connecting, and enhancing synergies between companies and entrepreneurs that can offer solutions to the various challenges of the sector (Innovación Chilena, 2019). The CChC recently published *Radiografía de Innovación en la Construcción* (Radiography of Innovation in Construction), which resulted from a study commissioned by this institution to a local university. The conclusions of this study are presented below (CChC, 2024).
b. Associative councils that bring together actors from the public and private sectors who can contribute their knowledge and experience to the improvement of different areas of the construction sector (Industrialised Construction Council, Strategic Construction Council).

Radiography of Innovation in Construction at the Company Level (CChC, 2024)

Regarding the current situation of innovation in the sector, it is interesting to review some of the findings and conclusions of *Radiography of Innovation in Construction* recently published by the Chilean Chamber of Construction due to its scope and timeliness.

The spectrum of companies included in the study corresponds to the sum of companies engaged in the construction of buildings, civil engineering works, specialised construction activities, and real estate activities. The study also presents a compilation and analysis of the main sources of supply to support the innovation efforts of construction companies.

In conducting the study, the definition of the following terms was considered, which is based on the standard of the National Institute of Statistics (ENI, National Innovation Survey form (2020), INE, 2021):

- Innovation: a new or improved good or service (or combination of both) that is significantly different from the company's previous goods or services, and which has already been introduced to the market or made available to potential users.
- Research and development (R&D): R&D comprises creative work carried out systematically to increase the volume of knowledge, including knowledge of people, culture, and society, and the use of this knowledge to create new applications.
- Product innovation: a good or service that is new or significantly different from the goods and/or services that the company has previously introduced to the market.
- Process innovation: a new or improved internal process for one or more of the company's functions, which differs significantly from previous internal processes, and which has already been implemented or put in place within the company.

The performance of R&D activities by a company is one of the main indicators used by companies to measure their innovation effort (CChC, 2024). From studies carried out by institutions related to innovation, it was found that 41.1% of companies consider the application of R&D in companies to be low or scarce, 20.6% consider it is extremely low or slow, and only 2.7% consider it is correct (CDT, 2022). On the other hand, the percentage of companies that performed R&D in construction during the period 2019–2020 reached 2%, which compares negatively with the average of other industrial sectors, which reached 6.7% (Diario Financiero, 2024). To reach the level of innovative companies, it addresses the need for more collaboration with the ecosystem, startups, and within the sector's value chain. This is because there are structural factors in this industry that make it more difficult to innovate, such as the predominance of small companies with limited resources, an extensive value chain, and regulatory aspects for the area that 'limit product innovation' (Diario Financiero, 2024).

Concerning company size (in terms of total sales) (Figure 4.1), the study shows that of the total number of companies surveyed by size, those that innovate the most are the largest companies (23.1%), and that innovation in new products is still low, at an average of only 8% for the three groups of companies.

On the other hand, according to the economic sector (Figure 4.2), the companies that innovated the most in new products and processes were in information and

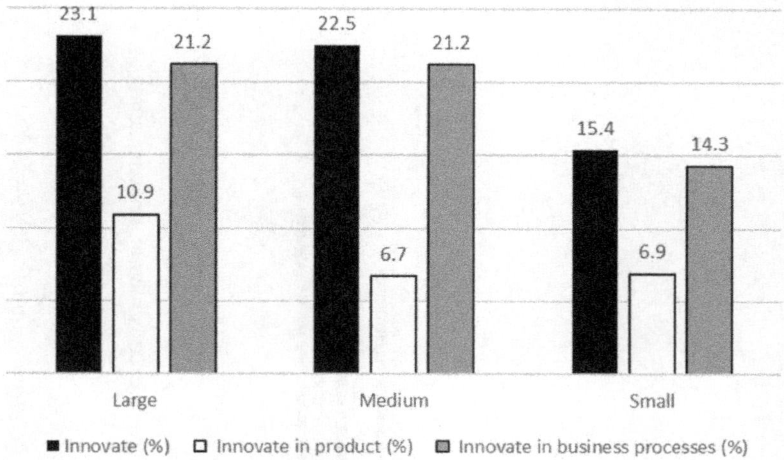

Note: The bars are segmented by company size: small, medium, and large.
Source: Based on data from INE (2021).

Figure 4.1 *Percentage of innovative companies relative to the total number of companies, by size, 2019–2020*

communication, with values close to 31%, and the least in real estate activities, with only 8% on average. Most companies also innovate the most in business processes.

Regarding the type of innovation, this same study indicates that 4.8% of construction companies carried out product innovation, which is lower than in other sectors, and 15.8% carried out process innovation, a percentage that is on par with other industries (INE, 2021). This indicates that the sector is looking for new ways to carry out its business activities and to modernise itself to respond to the current needs of the market, keeping on a par with other industries in this aspect (Diario Financiero, 2024).

On the other hand, the percentage of construction companies that requested public financing for innovation reached only 1.4% in the 2019–2020 period. A possible explanation for this figure would be that the type of innovation carried out by construction companies is generally in the field, as part of the work, and incremental, which would not require special funding.

Regarding the impact of innovation on construction companies in different business areas, the study provides the following data, presented in Table 4.1.

Another aspect addressed by the study refers to the supply of innovation consulting capabilities. In this regard, in 2022, 16.9% of innovation consulting projects were contracted by companies in the construction industry. Only 5% of the startups that participated in incubation or acceleration processes were construction companies, and 2% were companies from other markets that targeted the construction sector as

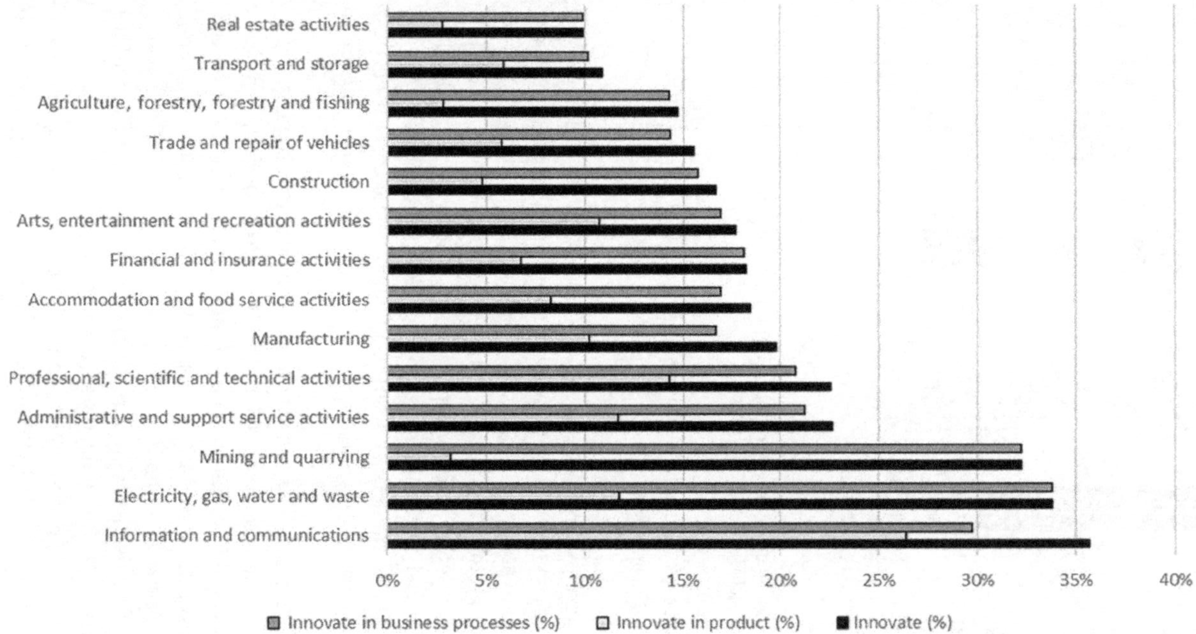

Note: Each bar represents the type of innovation in various economic sectors such as manufacturing, construction, services, among others.
Source: Based on data from INE (2021).

Figure 4.2 Percentage of innovative companies relative to the total in their economic sector, 2019–2020

Table 4.1 Perception of significant effects of innovation on companies according to respondents

Type of effect	Percentage perception surveyed, 2019–2020
Expanding the range of goods and services	71.5
Reduce environmental impact or improve occupational health and safety	52.8
Reduce response time to customer or supplier needs	82.2
Maintain or increase market share	54.2
Improve the quality of goods and services	81.7
Increase and improve capabilities to produce goods and services	65.3
Reduce costs per unit produced	62.6
Improve communication and/or information-sharing within your company and/or with other organisations	63.6
Introduce products in new market segments and/or geographically new markets	17.0

Note: The table describes the perception of the significant effects of innovation on companies' performance based on a study of construction companies.
Source: CChC, 2024; INE, 2021.

their primary market. Within the supply, the study considers incubators or developers, multidisciplinary technology centres, construction technology centres, and university master's degree programmes.

Finally, the study provides general conclusions and recommendations, including the following:

- Companies are beginning to adopt new ways of doing their work while maintaining their traditional culture.
- One of the main ways of measuring innovation has been the development of R&D activities. This indicator is most probably not the most appropriate for characterising innovation in the construction industry due to the characteristics of the sector, which has many small and medium-sized companies with limited resources, is partially fragmented in its different stages of the value chain, and has a significant number of subcontracted personnel working on a project basis, which makes the development of R&D activities difficult.
- Process innovation would be more likely to represent the innovation effort of the construction sector. Indicators for this need to be developed.

- Construction companies make very limited use of the supply of innovation support services, as well as the funds available for this purpose.
- Construction companies should measure the so-called hidden innovation (Loosemore, 2015) that occurs when solving problems that occur on a day-to-day basis in fieldwork.

Examples of Innovation of Some Construction Companies in Chile

In recent years, various innovation activities have been developed in some construction companies that are pioneers in this activity. Table 4.2 shows a variety of these activities.

Table 4.2 Examples of innovation in construction companies

Company	Innovation focus	Product – Process
Echeverria and Izquierdo (Reporte Sostenible, 2023)	Sustainability	A pioneer in measuring the carbon footprint of the company's business units.
Rio Claro and FLESAN (2022)	Sustainability	Revaluation of inert construction wastes.
AXIS (Consejo de Construcción Industrializada, 2023)	Industrialisation	CINTAXIS – a galvanised steel building with three apartments of 55 m^2 each, which can be assembled in 10 days thanks to the industrialised construction system.
RyV Foundations (RyV, n.d.)	Industrialisation	Ecopiles, bolts for an industrialised foundation.
Melon. Ready-mixed concrete supplier (Construye 2025, 2022)	Digital transformation	Optimix: dispatch optimisation system. Infomix: a sensor for online strength estimation and concrete demolding time.
RedMaestras (RedMaestras, 2024)	Human capital	Offering programmes to specialise women in construction trades.
Baumax, E2E, CTEC, and Echeverria Izquierdo (Consejo de Construcción Industrializada, 2023)	Research and development	Prolab: twin buildings to measure structural aspects, and operational indicators (cost, time, and quality).

Note: This table describes examples of innovations carried out by several construction companies.

An interesting case of innovation management in a construction company corresponds to the company Salfa Montajes, which obtained first place in the category of Industrial Assembly in the Most Innovative Companies Chile 2023 ranking. They received this award because they have carried out various initiatives and programmes to promote creativity, disruptive thinking, and innovative ideas in their processes (Diario El Mercurio, 2023). In an interview, the company's general manager stated that this recognition is concrete evidence of a transformation that today allows his company to generate greater value for its clients, the company, and the construction industry in general. As part of the company's strategic vision, he stated that it is only possible to be sustainable as a company if it has innovation management as a transversal competence, which allows agile adaptation of the company to the current challenges in the construction sector.

The examples presented in this section occur in large companies with a well-developed culture of innovation and are not widespread.

EVALUATING THE READINESS OF CONSTRUCTION COMPANIES FOR INNOVATION

In the mid-2010s, based on studies on innovation in construction companies, it was determined that, despite the efforts of some companies to innovate, these efforts did not normally lead to lasting results, nor was there a systematisation or formalisation of the innovation process. Based on general observations in 2010–2015, it was possible to confirm a growing concern for innovation but, at the same time, a lack of initiatives to carry it out properly and shortcomings in the organisational conditions to manage it. This is why many of the innovation efforts in construction companies continue to be isolated activities without a lasting impact on the companies.

According to Samson (2010), being innovative has clear rewards for a company: achieving quality, staying in business, and achieving growth, differentiation, and profitability. Thus, it is possible to state that innovation is increasingly important for the development of companies and industries in all areas, and clearly in construction. This requires that companies develop the conditions so that innovation is integrated as a permanent focus throughout the company.

It is known that innovating in construction is difficult, and its risk is almost unacceptable (Blázquez, 2005). Studies indicate that in the industry, there is a lot of distrust towards ideas and proposals for change and innovation (Serpell, 2002). Most companies do not consider investments in R+D+I to be attractive because they have not understood it as a key factor of competitiveness (Correa et al., 2007). Furthermore, Ocaranza (2001) concludes that in Chile, it is necessary to have an innovation system for construction, an efficient and structured system. Additionally, the analysis of the Encuesta Nacional de Innovación (ENI) (National Survey on Innovation) revealed that 2% of companies in the construction sector reported having conducted R&D activities in 2019–2020, a slightly higher percentage than in the previous survey (1.8%) (Diario Financiero, 2024).

Thus, in this scenario, a deficit was evident in the performance of innovation management in companies, as well as the need to have an evaluation system for this function in construction, which can identify shortcomings and gaps in performance, and allows generating and proposing improvement plans. In response to this challenge, an Innovation Management Evaluation System was developed that can identify shortcomings in companies and generate and propose improvement plans (Álvarez, 2013). In general, this system is aimed at helping companies overcome the resistance of the organisational culture to change. Its main components are shown in Figure 4.3.

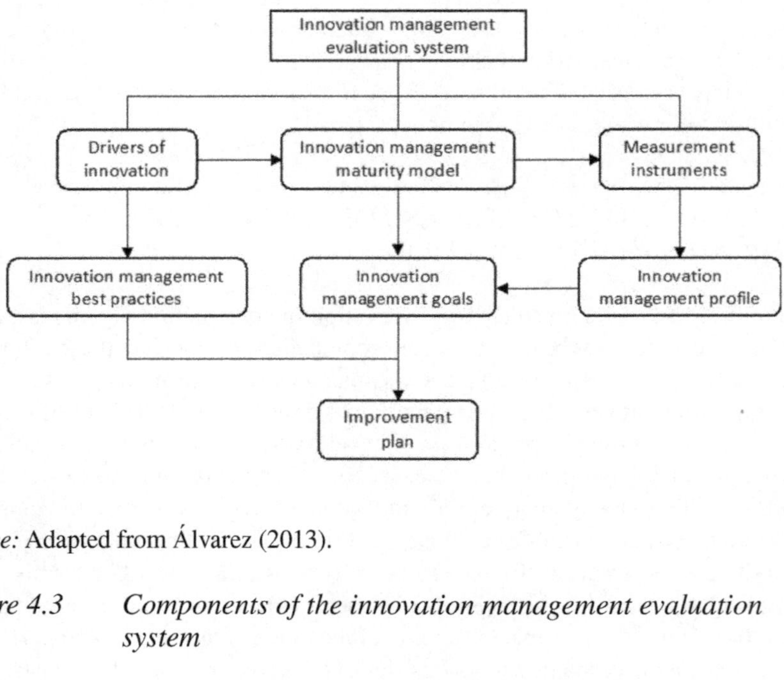

Source: Adapted from Álvarez (2013).

Figure 4.3 Components of the innovation management evaluation system

Innovation drivers correspond to the factors considered most relevant for an organisation to carry out effective innovation management. They were obtained from an exhaustive study of the literature on the subject, and there are a total of six factors (Álvarez, 2013):

- Knowledge management: This involves internal and external organisational knowledge, emphasising the importance of organisational learning and communication for tracking external best practices and adapting to changes.
- Research and development: R&D is analysed in terms of its form and purpose within organisations, highlighting the exploration of innovative ideas and using special innovation funds to build a conducive infrastructure for innovation.

- Culture and human capital: This focuses on attitudes towards change, commitment from top management, professional training, and teamwork enhancement, utilising strategic HR practices to influence employee behaviour and enhance innovation.
- Organisational structure: This section discusses whether an organisation's structure allows decision-making autonomy. A decentralised, informal structure supports innovation by promoting flexibility and generating innovative ideas.
- Partnering: This covers the development and impact of business alliances with external companies, universities, and research centres, aiming to synergise knowledge, share resources, and manage risks effectively.
- Technology: This section addresses the use and application of technology in construction processes, focusing on the type of technology used, its safety or innovation status, and its usage frequency.

The innovation management maturity model consists of five levels of maturity (example of technology factor in Table 4.3) for each of the drivers of innovation, levels that are clearly defined for each of the drivers, which allows generating a profile of the degree of readiness of the innovation. The measurement of the company's maturity of each driver of innovation is carried out through a questionnaire and the additional application of a focus group in case it is necessary to go deeper with respect to the driving factors of innovation. Finally, the system can propose a set of best practices to improve the company's profile and according to its maturity objectives.

The system was applied to three companies, whose characteristics are indicated in Table 4.4. The system is available to any company that is interested in evaluating its innovation management system.

The application procedure includes an internal validation step to verify the reliability of the questionnaire results. The internal validation consists of showing the results of the questionnaire to a panel of professionals and executives of the company (other than those who answered the questionnaire) and determining if they reflect the reality of the company. The results indicate the maturity levels for each innovation driver and for the total, along with a description of the situation of the factor. This panel then establishes whether the results are true to reality. If not, they are presented with five possible scenarios, where they should indicate which situation better reflects the reality of their company.

The results obtained by the companies are shown in Table 4.5. The table shows the total maturity level of each company and the maturity level of each innovation driver.

The maturity level of the three companies was extremely low. The professionals and managers who participated in the internal validation of the results considered that the results of the evaluation reflected the real situation of the companies regarding innovation management capability. They highlighted the value of this system as a tool to evaluate the current state of a construction company's innovation management capability and, starting from there, to propose best practices for its improvement to ensure future success.

Table 4.3 *Description of the five maturity levels of the technology innovation driver*

Factor	**Level 1** **Basic**	**Level 2** **Accepted**	**Level 3** **Conscious**	**Level 4** **Systematic**	**Level 5** **Synergised**
Use of technology in construction methods.	Use of basic technology in a narrow range.	Medium-level use of proven technology. Attempts to search for new technology as an informal and isolated process.	Use of more advanced proven technologies. Formal technology search process.	Use of advanced proven technology. Within the organisation, there is a constant search for new technology.	Use of advanced and innovative technology. Technology roadmaps are in place.

Note: The table describes each maturity level for measuring the technology factor or driver for innovation.

Table 4.4 *Characteristics of the construction companies that the innovation management evaluation system evaluated*

Company	Company size	Start of business	Trades	No of participants	Focus groups
1	Large	1960	Engineering and construction	16	2
2	Medium	1938	Real estate and construction	14	2
3	Medium	1989	Construction only	20	2

Note: This table shows the characteristics of the construction companies that participated in evaluating their innovation management systems.

Table 4.5 *Maturity levels as obtained from each company and for each innovation driver*

Innovation drivers	Company 1	Company 2	Company 3
Culture and human capital	3	3	3
Organisational structure	2	2	2
Knowledge management	2	2	2
Research and innovation	2	2	2
Technology	3	2	2
Partnering	2	2	2
Maturity level of the company	2	2	2

Note: This table shows the maturity level of each innovation driver and the general maturity level of the company.

For example, for the 'culture and human capital' factor, these practices include (Álvarez, 2013):

- The attitude of professionals and executives towards change: utilise recruitment tools to select employees with innovative traits, encouraging their commitment to organisational innovation. Performance evaluations that offer challenges, feedback, and rewards motivate employee involvement in problem-solving and decision-making.
- Workers' perception of management's commitment to innovation is heavily influenced by the organisation's actions. It is crucial for the organisation to clearly demonstrate its commitment to innovation, ensuring that innovative activities align with the organisation's ideologies.

- Training for professionals and executives: Offer various training programmes to develop the knowledge and innovation skills necessary for their roles, including formal training activities and problem-solving training.
- Promotion of teamwork by the company: To motivate employees towards innovation, the company should foster a supportive culture with rewards for innovation and use multi-skilled teamwork strategies. Teamwork should be encouraged and rewarded as it fosters innovation.

CONCLUSIONS

The Chilean construction industry is a sector that has long been characterised by its reluctance towards technological innovation, especially when compared with other industrial sectors. However, the pressure of the markets and technological development has pushed companies to advance and make increasing efforts to incorporate innovation within the focus of the business, which has been seen in different activities and dissemination instances. The construction industry in Chile is advancing slowly but surely towards innovation as one of the possible responses to the significant challenges it faces today, such as low productivity, quality problems, contractual conflicts, and integration problems, among others. The above does not imply that the work is completed. There is still much to do to achieve good development in the domestic construction industry. Additionally, construction companies must incorporate innovation management as an essential strategic factor.

In turn, the government and private organisations have created instruments and organisations to support companies' efforts, facilitating their innovation attempts. Despite this, significant barriers to innovation are still related mainly to organisational culture. To help detect and measure these barriers, this chapter presents an innovation management evaluation system that bases its methodology on measuring a company's readiness for innovation based on certain factors considered essential to promote it.

The information presented in this chapter shares the main aspects of the Chilean experience in innovation in construction, hoping that said knowledge can contribute to other realities in different countries. Also, the innovation management evaluation system can be used without major changes in other countries and by other construction companies.

REFERENCES

Álvarez, R. (2013). Evaluando la gestión de la innovación en empresas constructoras (Evaluating innovation management in construction companies). Tesis para optar al grado de Master of Science of Engineering, School of Engineering, Pontificia Universidad Católica de Chile.

ANID – Agencia Nacional de Investigación y Desarrollo (2024). Concursos ANID. ¿Para qué necesito financiamiento? (What do I need financing for?). https://anid.cl/concursos/

Blázquez, A. (2005). Innovation in construction: theory, status quo, prospects, and other considerations. *Informe de la Construcción, 57*(499–500), 111–132. doi: 10.3989/ic.2005.v57.i499-500.487

CDT (2022). *Consulta I+D en la Construcción* (Consultation R&D in Construction). Disponible en Consulta – I+D (cdt.cl).

Consejo de Construcción Industrializada (2023). Ganadores de concurso explican proyecto piloto de viviendas que se montan en solo 10 días (Competition winners explain pilot housing project that can be assembled in just 10 days). https://construccionindustrializada.cl/2023/07/17/ganadores-de-concurso-explican-proyecto-piloto-de-viviendas-que-se-montan-en-solo-10-dias/

Construye 2025 (2021). El sorprendente giro de la construcción chilena: duplicó su gasto en I+D durante 2020 (The surprising turn of Chilean construction: it doubled its spending on R&D during 2020). https://construye2025.cl/2021/09/29/el-sorprendente-giro-de-la-construccion-chilena-duplico-su-gasto-en-id-durante-2020/

Construye 2025 (2022). Digitalización en la construcción: La fórmula innovadora de Melón para mejorar su oferta de productos y servicios (Digitalisation in construction: Melon's innovative formula for improving its product and service offering). https://construye2025.cl/2022/07/29/digitalizacion-en-la-construccion-la-formula-innovadora-de-melon-para-mejorar-su-oferta-de-productos-y-servicios/

Construye 2025 and CORFO (2016). Programa Construye 2025. Hoja de Ruta 2022–2025. https://construye2025.cl/wp-content/uploads/2022/07/Hoja-de-Ruta-Construye2025-2022-2025.pdf

CORFO – Corporación de Fomento de la Producción (2024). Programas de Innovación 2024 (Innovation programs, 2024). https://www.corfo.cl/sites/cpp/convocatorias_programas_innovacion

Correa, C., Yepes, V. and Pellicer, E. (2007). Determinant issues and proposals for the management of innovation in construction companies. *Revista Ingeniería de Construcción, 22*(1), 5–14. doi:10.4067/S0718–50732007000100001

Diario El Mercurio (2023). Edición especial sobre el ranking Most Innovative Companies 2023 (Special edition on the Most Innovative Companies ranking 2023). Santiago de Chile.

Diario Financiero (2024). Primera radiografía de innovación en la construcción afirma que hay avances. Santiago de Chile. https://www.df.cl/df-lab/transformacion-digital/primera-radiografia-de-innovacion-en-la-construccion-afirma-que-hay

Flesan (2022). Residuos con Sentido: La alianza entre Flesan y Río Claro para el reciclaje de hormigón. (Waste with meaning: the partnership between Flesan and Rio Claro for concrete recycling). https://flesan.cl/residuos-con-sentido-la-alianza-entre-flesan-y-rio-claro-para-el-reciclaje-de-hormigon/

INAPI – Instituto Nacional de Propiedad Industrial – Chile (2023). Índice Global de Innovación 2023: Chile se posiciona como el segundo país en Latinoamérica del ranking mundial. https://www.inapi.cl/sala-de-prensa/detalle-noticia/indice-global-de-innovacion-2023-chile-se-posiciona-como-el-segundo-pais-en-latinoamerica-del-ranking-mundial

Innovación Chilena (2019). Industria de la construcción cuenta con su primer HUB de innovación (Industria de la construcción cuenta con su primer HUB de innovación). https://innovacionchilena.cl/industria-de-la-construccion-cuenta-con-su-primer-hub-de-innovacion/

INE – Instituto Nacional de Estadísticas – Chile (2021). Encuesta Nacional de Innovación (ENI) (National Innovation Survey). https://www.ine.cl/estadisticas/economia/ciencia-y-tecnologia/innovacion

Invest Chile (2023). Chile es la Economía más innovadora de Latinoamérica (Chile is the most innovative economy in Latin America). https://blog.investchile.gob.cl/bloges/chile-lidera-innovacion-america-latina

Loosemore, M. (2015). Construction innovation: fifth generation perspective. *Journal of Management in Engineering, 31*(6), 04015012.

Ministerio de Ciencia, Tecnología, Conocimiento e Innovación – Gobierno de Chile (2022). Inversión total de investigación y desarrollo en Chile se mantiene en un 0,34% del PIB y completa diez años sin mayores variaciones (Total research and development investment in Chile remains at 0.34% of GDP and completes ten years without major variations). https://www.minciencia.gob.cl/noticias/inversion-total-de-investigacion-y-desarrollo-en-chile-se-mantiene-en-un-034-del-pib-y-completa-diez-anos-sin-mayores-variaciones/

Ministerio de Economía – Gobierno de Chile, División de Innovación (2010). *Política Nacional de Innovación. Memoria 2006–2010.* https://www.economia.gob.cl/1540/articles-182927_recurso_1.pdf

Ministerio de Hacienda – Gobierno de Chile (2008). Ley 20241 Establece un incentivo tributario a la inversión privada en investigación y desarrollo (Law 20241 establishes a tax incentive for private investment in research and development). Modificación 2018. https://bcn.cl/2n8ff

Ocaranza, R. (2001). Análisis y evaluación de las acciones de innovación y transferencia tecnológica en la construcción (Analysis and evaluation of the actions of innovation and technological transfer in construction). Tesis para optar al grado de Master of Science of Engineering, School of Engineering, Pontificia Universidad Católica de Chile.

RedMaestras (2024). Página de inicio. https://redmaestra.cl/

Reporte Sostenible (2023). Echeverría Izquierdo: empresa constructora pionera en medir la huella de carbono en Chile (Echeverría Izquierdo: pioneering construction company in measuring the carbon footprint in Chile). https://reportesostenible.cl/blog/echeverria-izquierdo-empresa-constructora-pionera-en-medir-la-huella-de-carbono-en-chile/

RyV (n.d.). Ecopilotes. https://ryvfundaciones.cl/soluciones/ecopilote/

Samson, D. (2010). Innovation for business success: achieving a systematic innovation capability. (2nd) Australia: University of Melbourne. Recuperado en la página WEB: http://www.innovation.gov.au/Industry/IndustryInnovationCouncils/Documents/InnovationforbusinesssuccessTechstrat.pdf

Serpell, A. (1997). Innovación y Desarrollo Tecnológico: una Estrategia de Competitividad (Innovation and development: a competitiveness strategy). *Revista Ingeniería de Construcción, 2.*

Serpell, A. (2002). *Administración de Operaciones de Construcción* (Construction Operations Management). 2nd ed. Chile: Alfaomega.

Serpell, A. and Rubio, H. (2023). Evaluating project management (PM) readiness in construction companies: cases from Chile. *Procedia Computer Science, 219*, 1642–1649.

PART II

IMPLEMENTING INNOVATIONS IN ORGANISATIONS AND PROJECTS

5. Enabling innovation through systems thinking: insights from operational simulation in the construction industry

Mohammed Abdelmegid, Mani Poshdar, Vicente Gonzalez, Michael O'Sullivan and Cameron Walker

INTRODUCTION

The construction industry has transitioned from conventional approaches to embracing modern, sustainable practices, featuring two intertwined forms of innovation: technical and organisational. Technical innovation involves the adoption of advanced materials and technologies, with a strong focus on sustainability and efficiency (Abou-Zeid & Cheng, 2004). On the other hand, organisational innovation revolves around the transformation of management practices and operational strategies aimed at cultivating a responsive and agile organisational culture (Amarakoon et al., 2018).

This dual approach underscores the industry's commitment to modernisation. However, the dynamic interplay between these two domains of innovation is complex, necessitating a delicate balance between internal organisational capabilities and external insights to formulate effective innovation management strategies (Ayerbe et al., 2020). Furthermore, several challenges contribute to this complexity. These challenges encompass fragmented communication among diverse stakeholders, resistance to departure from traditional practices, substantial costs and risks associated with implementing new technologies, and regulatory constraints. Additionally, environmental sustainability concerns and the requirement for efficient resource management introduce further layers of complexity, posing significant hurdles to achieving seamless innovation in the construction sector.

Systems thinking, a comprehensive approach that examines all aspects of a system and their interconnections, plays a pivotal role in comprehending and fostering innovation within such a complex environment (Checkland & Poulter, 2020). This paradigm challenges conventional reductionist methods by acknowledging the complex and interconnected nature of construction projects. The contrast between 'hard' and 'soft' systems thinking provides a versatile toolbox for navigating the multifaceted challenges within the industry (Mingers, 2011). 'Hard' systems thinking revolves around problem-solving through the creation of systems that possess well-defined objectives and viable alternatives, using mathematical modelling and analysis (Checkland, 2011; Jackson, 2003; Yeo, 1993). In contrast, 'soft' systems thinking encompasses a collection of non-mathematical techniques that embrace an internal perspective on complex systems, emphasising the diversity and subjectivity of stakeholders (Mingers, 2011). This holistic viewpoint is crucial for tackling the ever-evolving and ambiguous characteristics of construction projects, steering the industry towards innovative solutions and strategies.

Applying systems thinking to construction management entails recognising and addressing the complexities and distinctive characteristics of construction projects. This approach goes beyond conventional production planning and control, encompassing a wide array of elements such as stakeholder engagement, environmental concerns, and technological progress. The growing adoption of systems thinking within the construction sector underscores its efficacy in tackling the industry's performance complexities and promoting a comprehensive approach to project management (Ayodele, 2021; Cobbinah et al., 2020).

Systems thinking serves as a catalyst for fostering innovation within the construction industry. By encouraging stakeholders to perceive challenges as interconnected elements within a broader system, it opens doors to inventive problem-solving approaches. This shift from conventional methods to embracing systems thinking nurtures creativity and empowers the industry to explore fresh and innovative solutions, ultimately enhancing its overall capacity for innovation (Jackson, 2003).

This chapter seeks to address the multifaceted barriers hindering innovation in construction. It does so by exploring the application of systems thinking and operational simulation. The objectives include demonstrating how implementing systems thinking through operational simulation can support the evaluation of innovative strategies to overcome these challenges, ultimately promoting a more integrated and innovative industry.

The scope of this chapter is specifically focused on examining the role of systems thinking within the construction industry's context. While recognising the broad potential of systems thinking, this exploration acknowledges research limitations, including available data and the practical applicability of theoretical concepts. Thus, while we delve deep into the use of operational simulation to implement systems thinking as an approach to enabling innovation, it is important to recognise that this is just one aspect of a vast and multifaceted field.

This chapter is organised to offer a thorough insight into how systems thinking, implemented through operational simulation, plays a pivotal role in driving innovation within the construction industry. It begins with an overview of the innovation landscape in construction and subsequently delves into the realm of systems thinking and its profound influence on fostering innovation. Then, it transitions to practical applications and case studies where theory meets practice. This structured approach ensures clarity and accessibility of concepts, offering valuable insights and strategies for industry professionals.

The chapter offers practical insights drawn from case studies and theoretical models, demonstrating the tangible benefits of this approach. It aims to provide a deeper understanding of the complexities involved in promoting innovation in construction management, offers innovative strategies to address these challenges, and highlights the pivotal role of systems thinking in driving industry-wide change. Furthermore, the chapter aspires to inspire professionals to adopt a more holistic and integrated approach to innovation in their practices.

BACKGROUND: INNOVATION IN CONSTRUCTION

The construction industry has undergone a significant transformation, shifting from traditional methods to modern, sustainable practices (Bajracharya, 2014; Hashim et al., 2022; Redmond & Alshawi, 2017). This transformative phase is characterised by two distinct but complementary forms of innovation: *technical innovation*, reflected in the adoption of advanced materials and methods, and *organisational innovation*, reflected in the evolution of management and operational strategies. This dual approach illustrates the industry's comprehensive strategy for embracing modernisation (Kumaraswamy et al., 2004; Redmond & Alshawi, 2017).

As explained, technical innovation primarily focuses on integrating advanced materials and technologies that promote sustainability and efficiency. It includes leveraging environmentally friendly materials and embracing technologies such as building information modelling (BIM) to optimise the construction process (Abou-Zeid & Cheng, 2004). In contrast, organisational innovation centres on the internal restructuring of construction firms, emphasising the modernisation of management practices and operational procedures. Its primary objective is to cultivate an agile and responsive organisational culture that effectively embraces and implements technological innovations while remaining adaptable to evolving industry standards (Amarakoon et al., 2018). However, the existence and nexus between these two realms of innovation merely exist without considerations and preparations. The success of innovation is intricately woven into the fabric of an organisation's structure and culture, extending far beyond a superficial endorsement of novel ideas. In this complex ecosystem, the way a company is organised, encompassing elements such as its hierarchy, communication pathways, and distribution of resources, plays a pivotal role in either fostering or hindering innovative efforts. For instance, a flexible structure that encourages open communication can lead to more collaborative and creative solutions. Similarly, an organisational culture rooted in shared values, beliefs, and behaviours significantly influences how innovation is perceived and pursued. A culture that embraces risk-taking and values creativity acts as a driving force behind innovative practices. Conversely, a culture that resists change can become a formidable barrier against new ideas and approaches.

This intricate interplay between structure and culture shapes the innovation landscape in a firm, setting the stage for an optimal mix of technological and non-technological innovations thoughtfully integrated into the company's structure and culture. This integration becomes the cornerstone of effective innovation management within construction firms, fundamentally shaping how these firms adapt, thrive, and navigate the ever-evolving landscape of the industry (Alves et al., 2018; Anderson et al., 2014; Azar & Ciabuschi, 2017; Damanpour, 1988; Damanpour et al., 1989).

Innovation management within the construction industry requires a careful balance between the strengths and resources within a firm (internal capabilities) and the insights and influences from outside the firm (external insights). The process begins by identifying areas where the firm can improve or evolve. This internal drive for change, when combined with external validation (such as customer feedback, market trends, or regulatory changes), plays a crucial role in shaping the direction of innovation. The overall innovation management process should find the right balance, aligning internal improvement goals with external opportunities and demands (Ayerbe et al., 2020; Birkinshaw et al., 2005; Birkinshaw & Mol, 2006).

Innovation at the firm level is shaped by several critical factors. Investments in research and development (R&D) and their impact on competitive performance are central. The synergy between innovation and R&D extends beyond idea generation, encompassing the strategic placement of innovations within the competitive landscape of the sector (Crépon et al., 1998; Hall & Sena, 2017; Miguel Benavente, 2006). Intellectual property protection and interdependent productivity emerge as pivotal elements, safeguarding innovations while amplifying their overall effectiveness and long-term impact.

At the industry level, the diffusion of innovation in construction extends beyond individual firms, embracing a broad socio-economic canvas where industry-specific and societal factors intricately intertwine. Here, the diffusion underscores the construction sector's adaptability and responsiveness. It reflects the industry's capacity to proactively embrace external societal shifts and evolving demands alongside internal innovation dynamics (Chen, 1983; Magnani, 2009; Semmler, 1994).

Effectively measuring innovation in construction entails a multifaceted process. This comprehensive approach assesses innovation effectiveness within a firm from both short- and long-term perspectives, while considering both tangible and intangible outcomes, such as immediate efficiency gains and cost reductions (tangible), and enhanced market position and adaptability to evolving industry trends (intangible) (Lichtenthaler, 2016; Tavassoli & Karlsson, 2015).

Beyond the boundaries of individual firms, measuring innovation effectiveness extends to assessing how innovations diffuse and shape industry trends. This dimension involves evaluating factors such as adoption rates, market acceptance, regulatory influences, and shifts in consumer preferences. Through this assessment, valuable insights are gained into the sustainability and broader impact of innovations, affecting not only individual firms within the industry but also the industry as a whole (Geldes et al., 2017).

Overall, the narrative of innovation in construction is a complex network of evolving practices, strategies, and structures. It is a story of an industry continually adapting to environmental challenges and societal needs, where the interplay between technical and organisational innovations forms the cornerstone of its resilience and competitiveness (Cainelli et al., 2006). This journey through the various facets of innovation in construction highlights the industry's dynamic approach. It emphasises the critical role of continuous adaptation and strategic foresight in maintaining its relevance and driving its future growth.

INTRODUCTION TO SYSTEMS THINKING

This section explains the definition of systems thinking in general practice and provides a brief historical overview of its development. It focuses on the theory of the concept without going into the details of the different tools that have been proposed to facilitate its implementation. The technical details of specific tools used in systems thinking are out of the scope of this chapter and can be explored in other, more specialised resources such as Brailsford et al. (2014).

Overview of Systems Thinking and its Emergence

Systems thinking is a transdisciplinary approach to problem-solving that challenges the reductionist approach of the traditional scientific method (Jackson, 2003). The reductionist approach, inspired by philosophers such as Rene Descartes, advocates a method of analysis based on dissecting complex issues into smaller, manageable parts to allow tackling them individually. Checkland (2012) emphasised that while this analytical method has proven useful in certain contexts, it fails to be universally applicable, especially when dealing with the complexity inherent in natural and human systems.

Systems thinking emerged in the mid-20th century when efforts were being made to develop civilian applications for operational research (OR), a term coined during the Second World War as a scientific approach to optimise war operations. Jackson (2003) indicated that OR was in crisis during this period as it was mainly concerned with 'engineering' a system to achieve pre-set objectives and failed to deal with the inherently complex environment of natural or human systems, leading to potential detrimental performance of the system by falling into sub-optimisation of its individual parts.

On the other hand, systems thinking offers a more holistic approach to tackling problems by recognising that complex systems consist of interconnected components, which are organised to perform a specific function for a particular purpose (Loosemore & Cheung, 2015). At its core, systems thinking incorporates four fundamental concepts: emergent properties; layered structure; processes of communication; and control (Checkland, 2011). The emergent properties in the first concept underline the holistic nature of systems thinking, where these properties manifest at the system level and cannot be reduced to the sum of individual components of the system. Thus, complex systems cannot be understood by analysing their individual components in isolation (i.e., the reductionist approach), confirming the notion that the whole is more than the simple aggregation of its components (Loosemore & Cheung, 2015).

Despite the fact that systems thinking lacks a universally accepted definition, common themes have been identified when describing the term. In their comprehensive review of literature on systems thinking definitions, Monat and Gannon (2015) concluded that systems thinking can be characterised as a perspective, a language, and a set of tools, collectively providing a framework for understanding complex environments. Therefore, systems thinking transcends disciplinary boundaries to support a comprehensive and integrated understanding of complex phenomena.

In the context of project management, systems thinking has been proposed to alter the view that projects are isolated entities with closed boundaries (Kapsali, 2011). Therefore, project managers are encouraged to shift from conventional linear thinking to a more holistic approach through systems thinking by focusing on outcomes instead of activities and on connections and structures between project entities (Meadows, 2008).

Hard vs Soft Systems Thinking

The distinction between 'hard' and 'soft' systems thinking stands as a pivotal theme in the literature on operational research and management science. 'Hard' systems thinking is reflective of the traditional post-Second World War view of OR and management science, which focused on solving problems by creating systems with clear objectives and viable alternatives that can be mathematically modelled and solved, whether manually or using advanced computing methods (Checkland, 2011; Jackson, 2003; Yeo, 1993). On the other hand, 'soft' systems thinking represents a collection of non-mathematical methods, which adopt an internal perspective on complex systems, placing value on the viewpoints and subjectivity of stakeholders (Mingers, 2011).

Soft systems thinking was initially met with scepticism within the OR society, especially in the US, even though soft systems thinking methods were proven successful in addressing complex and strategic problems in different fields such as healthcare, environmental management, and organisational risk management (Mingers, 2011). Checkland (2012) asserts that while hard systems thinking can be very successful in technological contexts, it falls short in addressing ill-defined and ambiguous situations, which often entail wicked problems. The pivotal difference between hard and soft lies in the attribution of systemicity, in which hard systems thinking assumes that systemicity can be observed in the world, while soft systems thinking assumes that only the process of inquiry into the world can be made systemic, not the world itself. From a philosophical standpoint, the term 'system' is considered ontologically in hard systems thinking and epistemologically in soft systems thinking.

In essence, the dichotomy between hard and soft systems thinking is not intended to create a rigid separation but rather to emphasise their complementary nature. Both paradigms should be considered to enrich the toolkit of systems thinkers, offering flexibility to draw on the strengths of each when navigating the complex, multifaceted challenges in the modern world.

SYSTEMS THINKING FOR CONSTRUCTION PROJECT MANAGEMENT

This section presents the use of systems thinking in the field of construction management. It explains why systems thinking is the way forward to deal with complexity in construction projects and reviews previous applications of systems thinking to support construction management.

Systems Thinking to Understand Complexity in Construction Projects

The field of construction management faces major challenges due to the growing complexity of construction projects. The complexity arises from several interconnected factors, such as substantial investments, advances in technology, and the involvement of multidisciplinary stakeholders. This complexity is compounded by the fragmented nature of construction operations and the unique nature of construction projects. This sophisticated environment requires the implementation of specialised management approaches that might need an extended time interval to assess their impact, consequently extending the implementation stage of the projects themselves.

Traditional production management methods adopted from OR, such as critical path method and Program Evaluation Review Technique (PERT), have seen some success in supporting the planning and control of construction operations. Recent advancements in operational simulation have extended the use of OR tools to develop sophisticated models for construction projects and test managerial interventions in a virtual environment. However, as the whole industry environment is becoming more complex and dynamic, project managers are required to address a diverse range of problems that extend beyond the traditional focus on production planning and control. Cobbinah et al. (2020) asserts that a paradigm shift is needed in the industry to foster a more holistic approach to construction project management.

The Institution of Civil Engineers (ICE) report on the systems approach to infrastructure delivery highlighted the need for a shift towards a more holistic perspective. It indicated that future construction projects will resemble software upgrade projects (Crudgington & Kitching, 2020). The report emphasised that a systems thinking approach that integrates social, environmental, economic, political, and technological aspects is necessary when dealing with problems in construction. This recommendation aligns with the prevalent idea in systems thinking literature that linear methodologies for project management are not capable of dealing with the increased complexity of projects. On the other hand, systems thinking is proposed as a more holistic approach that is not limited to the technical aspects of projects, embracing internal and external complexity as inherent aspects of projects (Kapsali, 2011). It has been well recognised that systems thinking provides a valuable approach to deal with ill-defined, dynamic, and multifaceted project environments (Checkland, 2011).

The application of systems thinking in construction projects has become increasingly popular in the last decade, with a realisation that the same level of thinking that created the problems related to industry performance cannot be used to solve them (Ayodele, 2021; Cobbinah et al., 2020). This aligns with the notion that systems thinking has proven very effective in industries driven by projects, such as aerospace, shipbuilding, and oil and gas (Crudgington & Kitching, 2020).

Applications of Systems Thinking in Construction Project Management

Recent studies have proposed the use of systems thinking to support construction management in different applications. A comprehensive list of previous research can be found in Ayodele (2021), which highlights the use of systems thinking in construction management fields such as information management, contract administration, waste management, prefabrication, skill shortage, health and safety, carbon emissions, and risk management. However, the investigations in these studies are mostly limited to using one systems thinking tool, system dynamics, to model the behaviour of construction systems within these fields, with limited research on how systems thinking can be systematically implemented at a broader and more strategic level. One study conducted an analysis of the adoption barriers for systems thinking in managing contractual risks in public–private partnership (PPP) projects (Loosemore & Cheung, 2015). They concluded that a transition to systems thinking requires challenging deeply ingrained ontologies, an appreciation of path dependencies, confrontational practices, and different ways of thinking.

The recent report by ICE on the systems approach to infrastructure delivery presents a more practical view of how systems thinking can be implemented to support the future delivery of infrastructure and construction projects (Crudgington & Kitching, 2020). This report proposed a systems approach to infrastructure delivery (SAID) to improve infrastructure project outcomes. The SAID approach applies the principles of systems thinking across the project lifecycle for different stakeholders. This approach represents a more strategic application of systems thinking compared with previous research, which has focused primarily on tools (e.g., system dynamics) instead of adopting a holistic systems thinking approach. The ICE report argues that a paradigm shift is required to challenge the current mindset and confront traditional practices in the industry in order to promote more holistic and outcome-focused delivery of infrastructure projects.

ENABLING INNOVATION THROUGH SYSTEMS THINKING

It has been well established that systems thinking can be a catalyst for holistic and creative thinking in addressing complex problems. Ayodele (2021) stressed that the systematic examination of problems encourages innovative thinking by seeking new definitions, alternatives, and evaluation measures. The departure from traditional problem-solving by prompting new problem definitions and alternative solutions has been linked to systems thinking innovative capabilities, as emphasised by Monga (2001).

Jackson (2003) asserted that simplistic solutions often fail in complex problems as they lack holism and creative depth. He further supported this argument with the fact that systems thinking can actively foster creativity by encouraging stakeholders to view problems as interconnected components within a larger system. Thus, an innovative approach to problem-solving is enabled. Moreover, implementing systems thinking can allow information to unfold, leading to unforeseen innovative behaviours. For instance, a systems thinking approach to interactive planning, as proposed by Ackoff (1999), aims to gain stakeholders' commitment to an 'idealised design' for a system. This approach encourages the active involvement of stakeholders in creatively addressing complex challenges and replacing them with sustainable solutions, as highlighted by Jackson (2003).

Similarly, Monat and Gannon (2015) highlighted that conventional problem-solving can lead to myopic and short-term solutions. They indicated three systems thinking questions reported in Weinberg (2001) as an enabler for deeper and creative solutions: 'Why do I see what I see?', 'Why do things stay the same?', and 'Why do things change?' Kapsali (2011) affirmed the same notion that systems thinking provides a flexible approach to enable innovation in complex and uncertain project environments.

From a construction management perspective, Loosemore and Cheung (2015) interviewed 16 senior construction professionals and concluded that systems thinking promotes creative thinking, emphasising the need for a paradigm shift in education and training to encourage more systems thinking implementation in the industry. On the other hand, the ICE report on systems thinking offers practical insight into the significance of systems thinking in improving innovation in construction projects (Crudgington & Kitching, 2020). They propose a strategic approach to dealing with the risk of disruptive technologies in a systematic manner to ensure the successful implementation of innovative solutions in construction projects.

LINKING SOFT AND HARD SYSTEMS THINKING THROUGH CONCEPTUAL MODELLING FOR COMPUTER SIMULATION

Simulation modelling is a powerful analytical approach to support decision-making in complex operational systems. In systems thinking literature, simulation modelling is perceived as a hard approach aligned with the technological application of systems thinking concepts (Jackson, 2003; Kapsali, 2011; Monat & Gannon, 2015). Simulation studies require a computer-free abstraction of the operational system to represent its key characteristics, such as the system entities, structure, and behaviour. The process of developing such an abstraction can be called 'conceptual modelling'. Within the intersection of systems thinking and OR research (i.e., soft vs hard systems approaches), conceptual modelling emerges as an imperative bridge (Gregory et al., 2020; Kapsali, 2011; Yeo, 1993). Checkland (2012) affirms that systems are essentially abstract concepts of real-life entities. Therefore, a structured approach is necessary to support conceptualising problems in a relatively acceptable way to stakeholders concerned about addressing these problems. He further explains that soft approaches such as the soft system methodology (SSM) act as a systemic process to design and implement solutions before moving into hard computer-based analytical tools.

In this chapter, we confirm the view that conceptual modelling is a key aspect to link soft and hard approaches for systems thinking. We further demonstrate that following a conceptual modelling framework for construction simulation studies enables innovation in addressing complex construction operations by stimulating creative problem-solving, which aligns with the findings of many studies in the OR literature (Ahmed, 2016; Chick, 2006; McHaney et al., 2018; Onggo, 2009; Pidd, 2014; Veeke, 2003). Figure 5.1 illustrates a summary of a novel conceptual modelling framework for construction simulation that is derived from an amalgamation of previous frameworks from other domains and adapted to suit the peculiar nature of construction projects. It should be noted that this framework utilises discrete event simulation, an approach that models and analyses dynamic system behaviour over time by representing individual events, such as arrivals, departures, or changes in system state. In this illustration, we adopt the distinction between the real-world and systems thinking domains as in the SSM (Checkland & Poulter, 2020). The process starts from the real-world domain by collecting information about the problem under investigation. This information is then developed into a conceptual model in the systems thinking domain. The conceptual model serves two purposes. First, it feeds back to the real-world domain by facilitating stakeholders' communication and engagement and providing them with a better understanding and insights for creative

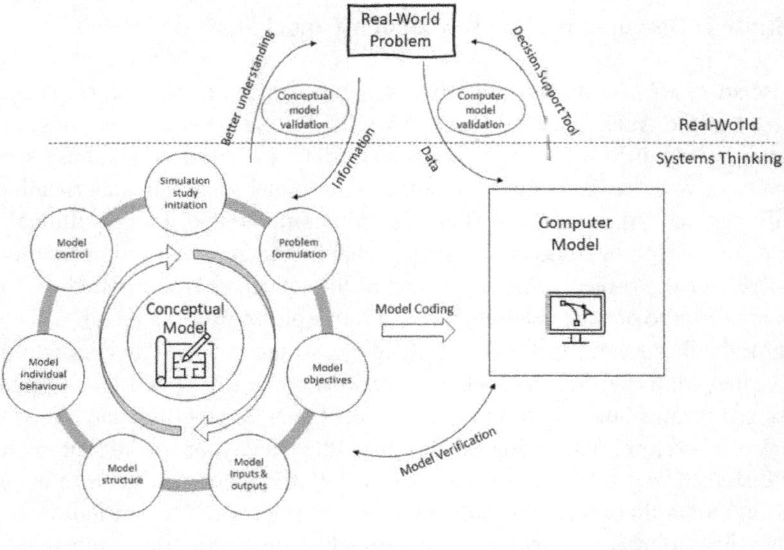

Figure 5.1 *A conceptual modelling framework for construction simulation*

solutions. Second, the conceptual model forms the basis of the design and coding of the computer model. The computer model requires reliable data from the real-world domain as inputs for the experimentation process. The output of the computer model is fed back to the real-world system as a decision support tool that can be used for operations design, risk management, and scenario testing. It should be noted that validation processes should be established between the real-world and systems thinking domains to ensure model credibility. In addition, verification within the systems thinking domain between the conceptual and computer models will ensure that the latter accurately reflects the agreed conceptual model, thus minimising the 'black-box effect' often described as one of the factors for lack of trust in modern simulation technologies in construction (AbouRizk et al., 2011; Kamat & Martinez, 2001).

CASE STUDIES

In this section, we present case studies of the implementation of the conceptual modelling framework in three construction projects to demonstrate how the application of systems thinking led to innovative solutions. For each case study, we first present background information about the project and the problem under investigation. Then, we explain the process of implementing the conceptual modelling framework to address the identified problem and discuss how innovation was enabled through systems thinking in these projects. The detailed technical design of the simulation models for the first two case studies is outside the scope of this chapter and can be found in Abdelmegid (2020).

Case Study 1: Design of the Site Layout of a Tunnel Shaft

This case study was conducted in 2016 at a launch shaft of a 292m storm water tunnel as part of the Auckland City Rail Link (CRL) project, one of the largest infrastructure projects in New Zealand. The launch shaft was located on a busy street in Auckland's city centre. It was used to launch two tunnel segments underneath existing buildings and infrastructure. These factors, compounded with traditional risks related to tunnelling operations and supply chain, created a very complex environment to deliver the project. The site layout at the launch shaft consisted of a small storage area for the precast concrete pipes, a muck pit for excavated soil, and a truck loading/unloading zone. Machines operating on the site included an excavator and a gantry crane. All these site components were working in a confined space with strict regulations to control the environmental impact of heavy tunnelling operations in the city centre. Therefore, optimising the design of the site layout was the main aim for this case study. We established the case study with the main contractor to develop a simulation model that represents site operations at the launch shaft to allow experimenting with different site layout designs to define the optimum arrangements and machinery configurations. Figure 5.2 shows a simple cross-sectional view of the site layout built on the commercial simulation software AnyLogic.

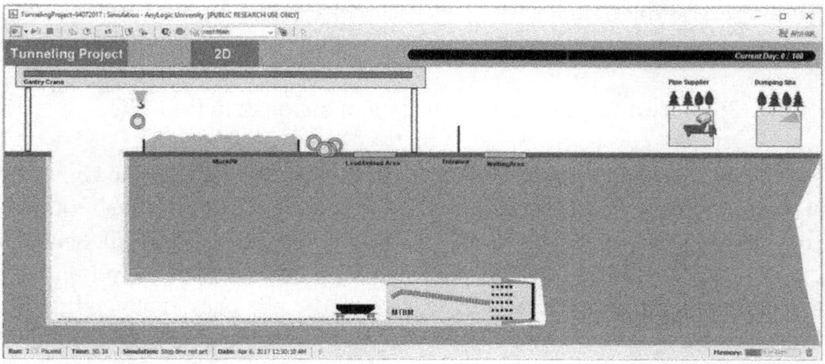

Figure 5.2 *Cross-sectional view of the launch site of Case Study 1*

The conceptual modelling framework in Figure 5.1 facilitated the application of systems thinking to support decision-making in this case study. The framework promoted organisational innovation by enabling stakeholders to be engaged in finding creative solutions to their managerial problems.

The implementation of the framework included four meetings to obtain stakeholders' input, including two site visits and two validation workshops. The key stakeholders involved in the process were the operations manager and the site manager. After our first visit to the site, we re-examined the layout by going through a phase of conceptual modelling. This step was crucial in clearly defining the problem that needed to be solved. It turned out that the construction team lacked experience in managing site logistics and operating tunnelling machinery as major tunnelling projects were relatively new to local contractors in New Zealand. A new site manager was recruited to utilise his rich experience in tunnelling from several European projects. He observed that the gantry crane, a substantial overhead crane crucial for lifting and moving heavy materials, had the potential to create a bottleneck in site operations. This was because it handled all loading tasks for both the pipes and excavated soil. A rich picture of the operations was developed through a non-computer simulation model to simplify the process for non-simulation experts as well as process maps using business process modelling notation (BPMN). These computer-free representations were used to facilitate a validation workshop, which helped all stakeholders gain a mutual understanding and confirm the validity of the conceptual model. In this workshop, a proposal was made to utilise an additional mobile crane to be located inside the shaft to allow continuous tunnelling operations regardless of the availability of the gantry crane. This unusual idea was well received by all stakeholders; thus, it was included as one of the scenarios to be tested in the computer simulation model. The next step was coding the computer model and experimenting with different scenarios to establish the optimum design for the site layout. A final report was provided to the construction company with the findings. A final workshop was held to present the findings to a wider audience and raise awareness about how systems thinking can be applied to support managing complex construction projects.

Case Study 2: Scheduling Piling Operations in a Construction Site

This case study was conducted in 2018 on a 19,500 sqm building for the University of Auckland with a cost of US$81 million. The foundation of that building required 183 piles for side walls and internal foundations. Piling operations were subcontracted to a specialised piling company, which was only required to deliver the piles and hand over to the foundation subcontractor to construct the pile caps. Thus, piling operations were on the critical path of the project and were under special consideration by the main contractor to ensure undisrupted performance. The piling contractor originally proposed an early finish of 72 days. However, the actual duration was 59 days, causing disruption to the activities downstream as materials and labour were not in place for such an early finish. The earlier-than-planned finish was attributed to inaccurate soil profiles that indicated that around 30% of the piles would penetrate through a rock layer requiring special configurations and longer drilling time. However, the rock layer was narrower than expected, leading to faster drilling time. Therefore, the objective of the study was to aid the main contractor in assessing the schedule reliability of piling operations and to overcome risks related to inaccurate soil profiles. Figure 5.3 shows the site layout of the piling project and how traditional site operations were planned using magnets to represent machinery.

The simulation study was initiated through a workshop with stakeholders from the main contractor, including a commercial manager, construction manager, site engineers, and site supervisors, in addition to the operations manager and site superintendent from the subcontractor's company. The first author of this chapter then engaged in several site meetings and full-day onsite shadowing to observe performance and collect information. A conceptual model was developed and validated with key stakeholders. The commercial manager of the main contractor flagged the problem of soil profile and enquired about creative ways to mitigate this risk. The operations manager from the subcontractor suggested changing piling sequencing from the traditional staggered piling drilling to a random sequencing that explores soil profile across the site (technical innovation). Although the concept was valid, it required testing through several virtual scenarios to ensure that other considerations, such as environmental and safety aspects, are not adversely impacted by the excessive movement of machinery around the site. An advanced computer model was developed accordingly to represent the complex piling operations and all internal and external factors. The findings of the model indicated that a creative solution to the piling sequencing problem is to divide the site into zones and operate two rigs across different zones to confirm soil profile across the whole site. This solution minimises machinery movement compared with the suggestion of random sequencing and supports learning from the experience of the piling team. This solution was perceived positively by stakeholders and provided insights for future projects.

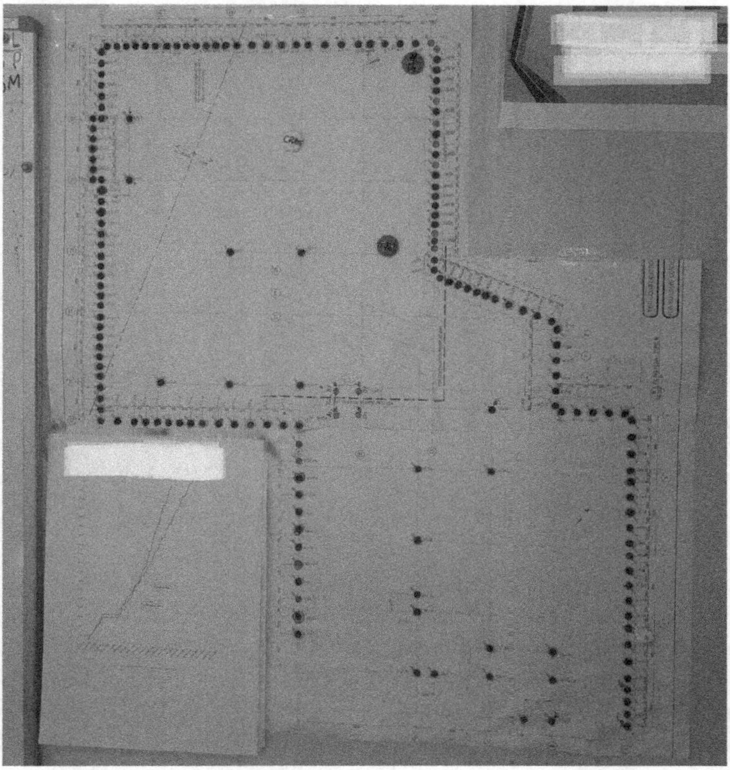

Figure 5.3 *Site layout of Case Study 2 with machines as magnets*

Case Study 3: Design of Supply Chain of Doors and Windows of a Building

This case study focuses on the construction of 17 reinforced masonry residential blocks in Iran, primarily highlighting three key activities: offsite fabrication of windows and doors, their transportation to the site, and subsequent installation. Figure 5.4 illustrates the project's process flow.

The offsite work involved fabricating windows and doors to specific architectural specifications. These products were transported to the construction site using two types of trucks: small and medium-sized. Onsite activities centred on precisely fitting these products into designated wall openings.

During the execution phase, several challenges emerged, including inaccuracies in product dimensions, site openings, misalignment of bottom plates, and uneven floor levelling that could lead to frame misalignment and distortion. Delays in fabrication, construction, or transportation disrupted the installation timeline, creating a misalignment between offsite fabrication and onsite readiness.

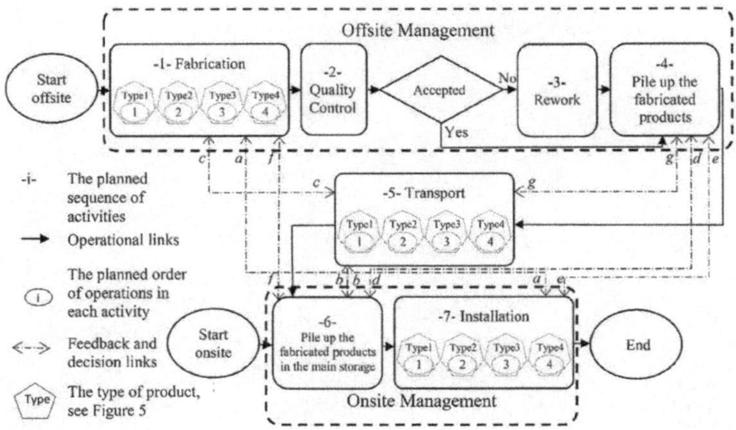

Figure 5.4 *Project process flow diagram for Case Study 3*

To effectively address these challenges and maintain project timelines, project managers had to adapt activity sequences and adjust task priorities. Recognising the need for an innovative solution, a conceptual simulation model was introduced as part of a systems thinking approach. This approach fostered organisational innovations through an SSM. The model considered several factors, such as available inventory space, waiting times, and delays affecting work efficiency, all while prioritising simplicity.

The system model also incorporated parameters such as realistic estimates of inventory size, completion time, and productivity, and exploring potential activity rearrangements to address project execution issues. It provided a comprehensive perspective, considering stakeholders' viewpoints and actions, aligning with the participatory nature of SSM. It also accounted for subjective challenges that might arise during project execution, promoting adaptability to changing circumstances.

The modelling process for this project consisted of three essential stages. First, the system structure was modelled by determining the activities, entities, and attributes that characterise the system. For instance, let us take a closer look at the 'fabrication' process, which involved several associated elements, such as 'fabrication drawings', 'raw fabrication materials', 'fabrication crew', 'fabrication machinery', and 'product types 1 to 4'. To set the 'fabrication' activity in motion, certain conditions, such as the availability of 'drawings', 'raw materials', 'fabrication crew', and 'fabrication machinery', needed to be met. In this context, entities could take on various forms, whether mobile, like the fabricated products, or stationary, as exemplified by the primary offsite storage area. Each entity was characterised by specific attributes that provided vital information, including submission dates, active timeframes, and working shift schedules. To illustrate the subjectivity introduced by modelling the structural components, consider a scenario where the 'fabrication' activity faced delays due to missing drawings. The model proved invaluable in identifying that such a delay would have a cascading effect on subsequent activities, including 'transportation'. As a result, the transportation crews and trucks would have been predicted to be idle, awaiting product availability.

Second, the individual behaviour of the system components was modelled to understand how entities interact within the system. Let's consider the example of 'fabricated products Type 1 to 4'. These products underwent a distinct lifecycle. They commenced as 'fabricated', transitioned to a 'wait for inspection' stage, proceeded to the 'inspection activity', and could either receive an 'approval' or 'rejection'. In the event of rejection, they underwent a 'rework process'. Alternatively, they proceeded to the 'storage area offsite' and awaited 'packing and piling up'. This sequence continued through various stages, encompassing 'transport', 'unloading', 'piling up in the main storage area onsite', 'moving to the installation spot', 'waiting for installation activity', culminating in 'installed'.

Finally, the model addressed system-level behaviour, focusing on managerial interventions and control decisions. For instance, let us consider a scenario where a delay occurred in 'offsite warehousing', subsequently affecting 'transportation' and 'installation'. Here, managerial intervention became imperative to recalibrate schedules and allocate resources, thereby preserving project timelines. This pivotal step enabled the capture of how macro-management strategies influenced the project's holistic behaviour. Throughout these modelling phases, our overarching objective was to provide a comprehensive perspective of the entire system. The designed model was inspected by two experienced project experts, whose extensive experience ensured the model's accuracy.

The conceptual model enabled the project team to embrace systems thinking and ushered in organisational innovations. For instance, the model enabled the project team to identify scenarios where fabrication activities could proceed ahead of schedule, deviating from the original plan. The model suggested bypassing offsite and onsite storage steps in such cases, streamlining the transportation process. This systems thinking approach promoted effective communication and information exchange between onsite installation managers and their transportation counterparts, allowing them to synchronise fabrication with onsite construction progress. Moreover, macro-management communication was established between onsite warehousing and fabrication. This interaction informed offsite fabrication managers about available warehousing space, prompting adjustments to the fabrication sequence. This proactive approach minimised overstocking in offsite storage and reduced waiting times in transportation and installation activities.

CONCLUSION AND RECOMMENDATIONS

This chapter has illustrated the key role of systems thinking in catalysing innovation and enhancing project delivery within the construction industry. The case studies presented demonstrate how integrating systems thinking into simulation modelling practices can promote collaboration, transparency, and creative problem-solving.

The following are key lessons that emerged from the case studies:

- The use of a conceptual modelling framework for construction simulation as a bridge between soft and hard systems thinking aided communication between stakeholders, which ensured mutual understanding of the problem context. This structured approach enabled stakeholders to contribute insights and propose creative solutions even if they were not familiar with the technicalities of coding a computer simulation model.
- The participation of stakeholders from different backgrounds and disciplines in simulation modelling studies facilitated interdisciplinary cooperation and explored alternative perspectives. Such a collective environment promoted transparency and enabled stakeholders to challenge assumptions.
- The experimentations through the computer simulation models provided a platform for stakeholders to test unconventional solutions and innovations, which minimised the bias of modellers who usually lack experience in the technical engineering knowledge of the system being modelled.
- The insights from the simulation models provided the stakeholders with objective and evidence-based solutions that consider the complexity and interdependencies inherent within construction systems, which eventually helps align innovation with the broader construction project outcomes.

The following are our recommendations from this study to promote the adoption of systems thinking to boost innovation across the industry:

- Adoption of systems thinking: Acknowledging that complex systems comprise interconnected components is crucial, with a focus on emergent properties, communication processes, and control structures. It is recommended that companies embrace a holistic approach to comprehend and address the interconnected elements in construction projects.
- Integration of technical and organisational innovation: A balance is needed between the adoption of technical innovation and the transformation of management practices and operational strategies (organisational innovation). Deep integration of these innovations into the company's structure and culture is essential.
- Encouragement of interdisciplinary cooperation: Participation from stakeholders of various disciplines in the application of systems thinking methodologies should be facilitated. This approach broadens perspectives, enhances transparency, and enables the challenging of assumptions.

- Consolidation of soft and hard systems thinking: Employing both soft and hard systems thinking approaches as needed is advisable. Their complementary nature is valuable in addressing complex challenges. Such integration can enable experimentation with unconventional solutions, which encourages creativity and innovative problem-solving approaches.
- Measurement and evaluation of innovation effectiveness: Developing comprehensive metrics to assess both the tangible and intangible outcomes of innovation efforts is important, as well as understanding their impact on the industry at large.
- Promotion of transparency and collaboration: Fostering an environment of transparency and collaboration in decision-making processes is key to building trust and collective ownership of projects.
- Incorporating systems thinking into engineering curricula: Systems thinking should be integrated throughout core engineering courses instead of considering it a separate discipline. For example, multidisciplinary capstone projects, modelling tools, and socio-technical systems can promote higher-order thinking skills and shape tomorrow's engineers to navigate complex real-world problems.

The integration of systems thinking into the practices of construction management is proving to be very important in addressing complex challenges in construction projects. By adopting a holistic approach, construction companies can capitalise on innovation in responding to the industry's dynamic environment. We believe that these recommendations will enable the construction industry to ingrain innovation into the DNA of its practices and increase the value of project outcomes.

REFERENCES

Abdelmegid, M. A. (2020). A conceptual modelling framework for construction simulation (Doctoral dissertation, ResearchSpace@ Auckland).

Abou-Zeid, E., & Cheng, Q. (2004). The effectiveness of innovation: A knowledge management approach. https://doi.org/10.1142/S1363919604001052

AbouRizk, S., Halpin, D., Mohamed, Y., & Hermann, U. (2011). Research in modeling and simulation for improving construction engineering operations. *Journal of Construction Engineering and Management*, *137*(10), 843–852. https://doi.org/10.1061/(ASCE)CO.1943-7862.0000288

Ackoff, R. L. (1999). *Ackoff's best: His classic writings on management*. John Wiley & Sons.

Ahmed, F. (2016). *Using structured analysis and design technique (SADT) for simulation conceptual modelling*. Loughborough University.

Alves, M., Galina, S., & Dobelin, S. (2018). Literature on organizational innovation: Past and future. https://doi.org/10.1108/INMR-01-2018-001

Amarakoon, U., Weerawardena, J., & Verreynne, M. (2018). Learning capabilities, human resource management innovation and competitive advantage. https://doi.org/10.1080/09585192.2016.1209228

Anderson, N., Potočnik, K., & Zhou, J. (2014). Innovation and creativity in organizations. https://doi.org/10.1177/0149206314527128

Ayerbe, C., Dubouloz, S., Mignon, S., & Robert, M. (2020). Management innovation and open innovation: For and towards dialogue. *Journal of Innovation Economics & Management*. https://doi.org/10.3917/JIE.032.0013

Ayodele, O. A. (2021). *Addressing labour turnover in the New Zealand construction sector: A system dynamics model.* The University of Auckland.

Azar, G., & Ciabuschi, F. (2017). Organizational innovation, technological innovation, and export performance: The effects of innovation radicalness and extensiveness. *International Business Review*. https://doi.org/10.1016/J.IBUSREV.2016.09.002

Bajracharya, A. (2014). A qualitative systems thinking approach in understanding the implementation of innovation on construction projects. *International Journal of System Dynamics Applications (IJSDA)*, *3*(4), 87–108. https://doi.org/10.4018/978-1-4666-9619-8.ch064

Birkinshaw, J., Hamel, G., & Mol, M. (2005). Management innovation. *IEEE Engineering Management Review*. https://doi.org/10.2139/SSRN.1306981

Birkinshaw, J., & Mol, M. J. (2006). How management innovation happens. *MIT Sloan Management Review*, *47*(4), 81–88.

Brailsford, S., Churilov, L., & Dangerfield, B. (2014). *Discrete-event simulation and system dynamics for management decision making*. John Wiley & Sons.

Cainelli, G., Evangelista, R., & Savona, M. (2006). Innovation and economic performance in services: A firm-level analysis. *Cambridge Journal of Economics*, *30*, 435–458. https://doi.org/10.1093/cje/bei067

Checkland, P. (2011). Systems thinking and soft systems methodology. In R. D. Galliers & W. L. Currie (eds), *The Oxford handbook of management information systems: Critical perspectives and new directions*. Oxford University Press. https://doi.org/10.1093/oxfordhb/9780199580583.003.0006

Checkland, P. (2012). Four conditions for serious systems thinking and action. *Systems Research and Behavioral Science*, *29*(5), 465–469. https://doi.org/10.1002/sres.2158

Checkland, P., & Poulter, J. (2020). Soft systems methodology. In S. Holwell & M. Reynolds (eds), *Systems approaches to making change: A practical guide* (pp. 201–253). Springer.

Chen, E. K. Y. (1983). The diffusion of technology. In E. K. Y. Chen (ed.), *Multinational corporations, technology and employment* (pp. 69–93). Palgrave Macmillan UK. https://doi.org/10.1007/978-1-349-06106-8_4

Chick, S. E. (2006). Six ways to improve a simulation analysis. *Journal of Simulation*, *1*(1), 21–28. https://doi.org/10.1057/palgrave.jos.4250006

Cobbinah, J. F., Aigbavboa, C. O., Thwala, W. D., & Banson, K. E. (2020). A Systems Thinking Approach to construction project management. The Construction Industry in the Fourth Industrial Revolution: Proceedings of 11th Construction Industry Development Board (CIDB) Postgraduate Research Conference 11.

Crépon, B., Duguet, E., & Mairesse, J. (1998). Research, innovation, and productivity: An econometric analysis at the firm level. National Bureau of Economic Research.

Crudgington, A., & Kitching, R. (2020). A systems approach to infrastructure delivery. A review of how systems thinking can be used to improve the delivery of complex infrastructure projects. ICE, UK.

Damanpour, F. (1988). Innovation type, radicalness, and the adoption process. https://doi.org/10.1177/009365088015005003

Damanpour, F., Szabat, K. A., & Evan, W. (1989). The relationship between types of innovation and organizational performance. https://doi.org/10.1111/J.1467-6486.1989.TB00746.X

Geldes, C., Felzensztein, C., & Palacios-Fenech, J. (2017). Technological and non-technological innovations, performance and propensity to innovate across industries: The case of an emerging economy. *Industrial Marketing Management*, *61*, 55–66. https://doi.org/10.1016/j.indmarman.2016.10.010

Gregory, A. J., Atkins, J. P., Midgley, G., & Hodgson, A. M. (2020). Stakeholder identification and engagement in problem structuring interventions. *European Journal of Operational Research*, *283*(1), 321–340. https://doi.org/https://doi.org/10.1016/j.ejor.2019.10.044

Hall, B. H., & Sena, V. (2017). Appropriability mechanisms, innovation, and productivity: Evidence from the UK. *Economics of Innovation and New Technology*, *26*(1–2), 42–62. https://doi.org/10.1080/10438599.2016.1202513

Hashim, R., Cooper, S., Salleh, N., & Mohd Nawi, M. N. (2022). The influence of regulatory pressure in shaping construction firms' decision to adopt green innovation. *Journal of Advanced Research in Applied Sciences and Engineering Technology*, *28*, 301–310. https://doi.org/10.37934/araset.28.2.301310

Jackson, M. C. (2003). *Systems thinking: Creative holism for managers*. John Wiley & Sons.

Kamat, V. R., & Martinez, J. C. (2001). Visualizing simulated construction operations in 3D. *Journal of Computing in Civil Engineering*, *15*(4), 329–337. https://doi.org/10.1061/(ASCE)0887-3801(2001)15:4(329)

Kapsali, M. (2011). Systems thinking in innovation project management: A match that works. *International Journal of Project Management*, *29*(4), 396–407. https://doi.org/10.1016/j.ijproman.2011.01.003

Kumaraswamy, M., Love, P. E., Dulaimi, M., & Rahman, M. (2004). Integrating procurement and operational innovations for construction industry development. *Engineering, Construction and Architectural Management*, *11*(5), 323–334. https://doi.org/10.1108/09699980410558511

Lichtenthaler, U. (2016). Toward an innovation-based perspective on company performance. *Management Decision*, *54*, 66–87. https://doi.org/10.1108/MD-05-2015-0161

Loosemore, M., & Cheung, E. (2015). Implementing systems thinking to manage risk in public private partnership projects. *International Journal of Project Management*, *33*(6), 1325–1334. https://doi.org/10.1016/j.ijproman.2015.02.005

Magnani, E. (2009). How does technological innovation and diffusion affect inter-industry workers' mobility? *Structural Change and Economic Dynamics*, *20*(1), 16–37. https://doi.org/10.1016/j.strueco.2008.12.003

McHaney, R., Reychav, I., Zhu, L., Lindstrom, M. M., & Arbel, Y. (2018). Iterative conceptual modeling: A case study in cardiac patient survival simulation. *Operations Research for Health Care*, *19*, 57–65. https://doi.org/10.1016/j.orhc.2018.03.002

Meadows, D. H. (2008). *Thinking in systems: A primer*. Chelsea Green Publishing.

Miguel Benavente, J. (2006). The role of research and innovation in promoting productivity in Chile. *Economics of Innovation and New Technology*, *15*(4–5), 301–315. https://doi.org/10.1080/10438590500512794

Mingers, J. (2011). Soft OR comes of age – but not everywhere! *Omega*, *39*(6), 729–741. https://doi.org/10.1016/j.omega.2011.01.005

Monat, J. P., & Gannon, T. F. (2015). What is systems thinking? A review of selected literature plus recommendations. *American Journal of Systems Science*, *4*(1), 11–26. http://article.sapub.org/10.5923.j.ajss.20150401.02.html

Monga, P. (2001). *A system dynamics model of the development of new technologies for ship systems*. Virginia Tech.

Onggo, B. S. S. (2009). Towards a unified conceptual model representation: A case study in healthcare. *Journal of Simulation*, *3*(1), 40–49. https://doi.org/10.1057/jos.2008.14

Pidd, M. (2014). *Tools for thinking: Modelling in management science*, 3rd Edition. Wiley.

Redmond, A. M., & Alshawi, M. (2017). Applying system science and system thinking techniques to BIM management. 2017 10th International Conference on Developments in eSystems Engineering (DeSE).

Semmler, W. (1994). Information, innovation and diffusion of technology. *Journal of Evolutionary Economics*, *4*(1), 45–58. https://doi.org/10.1007/BF01200837

Tavassoli, S., & Karlsson, C. (2015). Innovation strategies and firm performance: Simple or complex strategies? *Economics of innovation and New Technology*, *25*, 1–20. https://doi.org/10.1080/10438599.2015.1108109

Veeke, H. P. M. (2003). *Simulation integrated design for logistics.* Delft University of Technology.

Weinberg, G. M. (2001). *An introduction to general systems thinking* (Silver Anniversary ed.). Dorset House Publishing Co.

Yeo, K. T. (1993). Systems thinking and project management: Time to reunite. *International Journal of Project Management*, *11*(2), 111–117. https://doi.org/10.1016/0263-7863(93)90019-J

6. Business model innovation and its impact on the diffusion of innovation in construction business organizations

Shoeb Ahmed Memon, Nipuni Sumanarathna, Bismark Duodu and Steve Rowlinson

INTRODUCTION

Business model innovation (BMI) is gaining momentum and presents an important channel to commercialize innovation triggered by technology, climate change, and a competitive business environment. Although much of the interest has been driven by technology, historically, business models have been deeply rooted in economic behavior (Massa & Tucci, 2013).

Massa and Tucci (2013) describe BMI as the interaction of business models (BMs) and innovation, two complementary fields, where BMs allow the commercialization of innovative technologies and are the source of innovation. In either case, they contend that BMI leads to a change in the existing setting of the organization in which BMs allow the commercialization of innovative technologies and are the source of innovation for new BMs.

The new BMs emerge because of entrepreneurial activity. The core objective is to create and capture value by designing new organizations, structures, products, and services in cooperation with partner networks, specifically through the exchanges and activities across their supply chain network, known as boundary-spanning activities (Chesbrough, 2010). Amit and Zott (2015) highlight boundary-spanning activities as an important aspect of new model designs that enable organizations to conduct business. They identified four antecedents of model designs, namely: (1) goals: to create and capture value for the organization in responding to customer needs; (2) templates: as a source of inspiration for model designers to borrow, copy, or develop an understanding of business activities in a similar domain or successful examples; (3) stakeholders: represent collaboration among partners as a result of new model design to complete new products and services; and (4) constraints: which account for internal and external constraints for an organization to conduct a business. Here, internal constraints may refer to existing resources and capabilities of individuals within an organization to streamline processes per new model design. External constraints refer to a variety of general environmental constraints to which a business is exposed; these may include economic, legal, socio-political, regulatory, and cultural environments. The new BMs are often subject to internal and external pressures that influence the practice and even threaten its sustainability (Amit & Zott, 2015).

On the other hand, reconfiguration suggests an existing business model that is reconfigured to provide value. For example, during COVID-19, the food and beverage (F&B) industry invested in reconfiguring existing BMs to suit market needs. In this, Michelin Star restaurants moved from traditional dine-in services to home delivery, which significantly changed how they provided value to customers. Although the core business and value to consumers remained the same, the dining experience was significantly altered. One of the significant issues that organizations faced was internal inertia. When coping with COVID-19, several industries faced identity and internal inertia issues, where reconfiguration led to a significant change in the existing business. Chesbrough (2010) considers two barriers that may affect reconfiguration implementation. The first relates to the structural changes within organizations opting for BMI, which affects organizational structures, operational processes, and how they deliver value. The second is the cognitive barrier, which relates to people working in one type of BM for a long time and seeing no benefit in the change.

More recently, BMI has gained attention due to significant technological developments. Organizations are increasingly focusing on environmental, social, and governance (ESG) factors and a competitive business environment. Therefore, we see more organizations in various industries, including construction, opting for sustainable practices, materials, and technologies that may improve efficiency while addressing broader objectives. However, many of these initiatives face challenges in implementation within organizations and thus lead to extensive investments that do not result in improved performance. Chesbrough (2010) pointed out that BMI often does not occur in organizations as they face the challenges in identifying appropriate BMs for new technologies and solutions.

Geissdoerfer et al. (2018) identified four types of sustainable BMIs that focus on a long-term perspective: proactive stakeholder management, sustainable solutions, and sustainable development for organizations and stakeholders. A similar trend can be seen in the construction industry, where sustainable startups are emerging using various technologies to produce solutions. Three prominent examples from Hong Kong are where these startups aim to re-engineer BMs through technologies such as artificial intelligence (AI), building information modeling (BIM), geographical information systems (GIS), the Internet of Things (IoT), cloud computing, and blockchain:

1. Smart Site Safety System (4S) addresses the pressing issue of construction site safety through the supervision of digital works. Multiple startups can be found in this domain to provide real-time monitoring of construction sites to avoid incidents and accidents (e.g., AutoSafe, n.d.; isBIM, n.d.; viACT, n.d.).
2. Common data environment, touted as the single source of truth to address all issues from design to execution and facilities management (e.g., AutoCDE, n.d.).
3. A digitalized e-inspection system is used for real-time monitoring of the construction supply chain from iLab, Faculty of Architecture, University of Hong Kong (Remote e-inspection system, n.d.).

These startups offer new BMs for managing site safety and for how projects are planned, designed, and managed. Even though traditional organizations need to make significant changes in their business models, startups are leading a slow transformation in the industry. This is because the costs and implications of business model transformation or acquisition are significant for construction organizations to survive in a highly competitive environment.

The construction industry is also experiencing a significant shift toward industrialized construction and integrated project delivery. This change is driven by the client through public policy measures and a desire for construction organizations to diversify and provide more value to customers. As a result, the industry is moving from a project-based approach to a product-based approach, focusing on platform-based BMs and ecosystems. Li et al. (2024), in their extensive review of BMI in the construction industry, identified six archetypes for increased value to customers using integrated approaches while considering issues of fit between model components and the entire supply chain. The work provides exciting insights into how newly emerging archetypes, coupled with Industry 4.0 technologies, can transform businesses and the industry.

BUSINESS MODELS: DEFINITIONS AND CATEGORIZATIONS

A business model is defined as an essential tool for organizations to plan and elaborate on profit-making logic. Several researchers have attempted to define BMs at different levels. Earlier definitions have been developed focusing on the framework and elements in relation to a business to sustain and enable competitive advantage (Fielt, 2013). Subsequent definitions are more on value-creation and value-capturing (i.e., value logic) of an organization (Fielt, 2013). Researchers such as Shafer et al. (2005), Chesbrough et al. (2006), Johnson et al. (2008), and George and Bock (2011) have incorporated the notion of 'value' to define a BM. Chesbrough (2006) emphasizes a BM as follows:

> At its heart, a business model performs two important functions: value creation and value capture. First, it defines a series of activities that will yield a new product or service in such a way that there is net value created throughout the various activities. Second, it captures value from a portion of those activities for the firm developing the model. (p.108)

The concept of 'value' can be vague, especially since the definition of a BM is to be perceived in a tangible manner. For instance, an initial definition such as Timmers' (1998) is straightforward and easy to grasp:

> (a) an architecture for the product, service and information flows, including a description of the various business actors and their roles; and (b) a description of the potential benefits for the various business actors; and (c) a description of the sources of revenues. (p.4)

Similarly, Rappa's (2000) view of a BM is 'the method of doing business by which a company sustain[s] itself—that is, generate[s] revenue—indicates that a BM simply informs how a business is positioned to sustain and achieve continuous and new development'.

Nevertheless, the notion of value when defining a BM has indicated 'value for the customer' or mere 'value creation' without clarifying the context (Fielt, 2013). To define a BM as 'the value logic of an organization in terms of how it creates and captures customer value' (p.92), Fielt (2013) put forward the integral view of the value logic of an organization as an amalgamation of 'value for the customer/customer value, value-creation with business value/value added during the production process, and capturing created value for and by the customer'.

To look at categorizations, most BM definitions are developed at the firm or strategic level; the rest are at the operational and economic level (Morris et al., 2005). Initial definitions are activity/role-oriented, and the others are value/customer-oriented (Osterwalder, 2004; Osterwalder et al., 2005).

Wirtz et al. (2016), in their analysis of BMs and future research perspectives, discuss BM dynamics in relation to BMI. The authors perceive BMs as a phenomenon that is subjected to constant modifications according to external (and internal) conditions. In a dynamic perspective, Wirtz et al. (2016) define a BM as:

> a simplified and aggregated representation of the relevant activities of a company. It describes how marketable information, products and/or services are generated by means of a company's value-added component. In addition to the architecture of value creation, strategic as well as customer and market components are taken into consideration, to achieve the superordinate goal of generating, or rather, securing the competitive advantage. To fulfill this latter purpose, a current BM should always be critically regarded from a dynamic perspective, thus within the consciousness that there may be the need for business model evolution or BMI due to internal or external changes over time. (p.41)

This definition aligns with the emergence of 'digital' BMs in the 'new economy'; rather than recognizing them as new BMs, Krumeich et al. (2013) perceive these BMs as transitions of traditional BMs. Similarly, Al-Debei et al. (2008) define a BM as:

> The business model is an abstract representation of an organization, be it conceptual, textual, and/or graphical, of all core interrelated architectural, co-operational, and financial arrangements designed and developed by an organization, as well as all core products and/or services the organization offers based on these arrangements that are needed to achieve its strategic goals and objectives. (p.8)

Al-Debei et al. (2008) further highlighted that the BMs in the new world, which is enriched with technology, are intermediate. The BMs defined by Al-Debei et al. (2008) connect business strategy and processes in the context of a digital business. Hence, as the definition indicates, BMs in digital businesses mostly consist of similar elements, although nuances of such elements might vary based on the context.

ELEMENTS OF BUSINESS MODELS

Elements of BMs vary depending on their types or the way they are conceptualized. The elements of a BM are often illustrated when the basic component view is adopted in the conceptualization. The 'core strategy', which shapes the vision and mission of an organization, has been identified as a major element of a BM irrespective of the conceptualization or perspective. In addition, material and intangible resources are recognized as the most obvious components (Wirtz et al., 2016). When the network-oriented view and market perspective (i.e., value proposition) are concerned, networks, partnerships, customers, and rivals should be considered, respectively, as elements of a successful BM (Wirtz et al., 2016).

The value proposition is incorporated in developing the business model component framework where the equilibrium of five models (i.e., components)—value-creation, value-offering, value-capturing, cooperation, and financial models—are conceptualized (Krumeich et al., 2012, 2013). Krumeich et al. (2013) further emphasized the need for considering all the components and their equations in upgrading the BM architecture. Changing or isolating one component would change the whole dynamic of the BM. Model components may not have direct functional relations; however, some equations can exist in the form of interdependencies and influences, which should be taken into account. More details on the business model component framework can be found in Krumeich et al. (2012, p.6).

For value-creation, an efficient organizational structure that clearly denotes the roles and responsibilities of individuals in achieving goals is important. In addition to organizational structure, tangible and intangible resources, including human resources, are important. It is worth noting that having resources itself is not sufficient for organizations to create value, as individuals need to have a particular set of competencies to offer value. These competencies help carry out organizational activities and processes in achieving value proposition, which is the benefit offered to customers.

The value-offering model consists of a value proposition, product and service offerings, competitive advantage, and a competitive model. Product and service offerings are related to the value proposition since the quality of products and services determines the value of the organization. On the other hand, an organization's competitive model explains how competitive advantage is created and sustained. It distinguishes an organization from competitors by emphasizing the unique value proposition offered to its customers.

Value-capturing explains customer focus and market orientation. This element is vital in recognizing customer requirements based on market conditions. Maintaining productive customer relations through effective communication and distribution channels is a major component of a successful business model, which is why several organizations introduce customer loyalty programs to remain in touch with their customer base.

The financial model, in its literal meaning, explains the cost aspects of an organization. Strong positioning of funding mechanisms, revenue and pricing avenues, profit strategies, and distribution and positioning of funding, investments, revenue, and profits are important for value-creation (Krumeich et al., 2012, 2013).

The cooperation model is similar to the network-oriented view, where coordination with partners, cooperative relations with external parties, and the maturity of such relationships are necessary to sustain the BM (Wirtz et al., 2016). Although all BM components play a role in business operations, the cooperation model is of utmost importance, suggesting the significance of BM. Early studies such as Shafer et al. (2005) identified the elements of a BM in a similar way, highlighting the first three models of the BM component framework by Krumeich et al. (2012) as value-creation, value network (similar to value-offering), and value capture. Shafer et al.'s (2005) fourth element—strategic choices—somewhat aligns with Krumeich's cooperation and financial models.

BUSINESS MODELS AND BUSINESS MODEL INNOVATION IN CONSTRUCTION

As explained earlier in this chapter, a major stream of research in BMI has centered around finding a robust business model that may create and capture value. More recently, businesses have been keen to alternate BMs that may efficiently cater to business goals and create value for customers, resulting in a variety of new BMs.

According to Chesbrough (2010), organizations should build the capability to innovate their BMs. Indeed, BMI starts as a change in strategy dependent on dynamic capabilities and environmental turbulence, which requires organizations to prepare to take advantage of opportunities and provide responses to contingencies (Zhao et al., 2016).

Construction is noted for the delivery of projects that are often bespoke and one-off by temporary teams. Thus, value is created and captured through the organization of the parties involved in the project, with teams disbanded on completion. This one-off model, with its focus on individual projects, is deemed to be a major cause of fragmentation in the industry and in the business logic (Hall et al., 2022). Hall et al. (2022) noted that concerning the traditional business model in construction, fragmentation occurs across horizontal, vertical, and longitudinal dimensions. Horizontal fragmentation is explained as how work is divided among different types of contractors (i.e., main contractors, subcontractors), where each contractor operates independently. Thus, construction projects are often affected by issues such as logistic bottlenecks due to a lack of coordination. Vertical fragmentation is explained at the project level, as each project has a different set of stakeholders to achieve different goals and values. Due to this discontinuity, the agency over the project is also discontinued and the end-user's interest may not be prioritized. Longitudinal fragmentation occurs when project teams are chosen for future projects by competitive bidding. The uncertainty of working with the same team ignores the tacit knowledge embedded within team members. This has caused what is termed as a learning disability in the industry. Thus, the inability to continuously build on ideas through inter-organizational cooperation has resulted in slower innovation diffusion in construction (Dubois & Gadde, 2002; Hall et al., 2022).

Particularly due to vertical fragmentation, neither the main contractor nor the designer (e.g., architect) acts as the system integrator. When contractors are concerned, due to the lack of control over the design phase, they tend to pay more attention to the construction phase and suggest business strategies to reduce costs. Nevertheless, it is required to adopt innovative strategies/BMs to create new value for customers rather than paying attention only to cost minimization (Hall et al., 2022).

As a solution, developing new product architectures by collaborating with supply chain stakeholders is suggested. This new form of construction project delivery method includes informal or formal alliancing, and a successful case would also include coordination through BIM, early involvement of key stakeholders, the last planner system, and team co-location (Hall et al., 2018; 2022).

These practices result in more integration in orientation without causing a major change in the business logic of firms. Theoretically, innovative procurement strategies such as integrated project delivery (IPD) should solve the challenges of horizontal, vertical, and even longitudinal fragmentations (Zhao et al., 2021; Hall et al., 2014). However, IPD can fail to address the issue of longitudinal fragmentation given the disbandment of project teams after individual project completion (Hall et al., 2022). The above suggests the need for a shift from the traditional to new business models (BMs) to better create, deliver, and capture value.

It is fascinating to witness the movement of the construction industry, which has frequently been recognized as 'traditional' and 'slow-moving', toward adopting BMs to deliver construction projects (Das et al., 2023a). As rigorously explained by Hall et al. (2022), industrialized construction has been recognized as the key factor that opens up doors to new BMs (also called project delivery models) in the construction industry. Das et al. (2023a) further highlight that industrial construction is not a mere change from a discontinuous and temporary production approach to reorganizing, to continuous and industrialized production systems. It should also explain the components other than production. BMs that integrate supply chain management and its logistics, continuous improvement of manufacturing and assembly, customer focus, and understanding of the market have arisen. Huge investments in technologies such as 3D printing and onsite robotics have empowered industrialization, which aims at delivering Industry 4.0.

Depending on the production system and processes, two BMs are common in the construction industry. The first business model, *vertical integration*, involves construction firms conducting construction activities in an offsite factory where design, manufacturing, transport, and assembly are coordinated. Organizations that deliver modular houses are examples of this business model (Hall et al., 2022). The second business model, *digital systems integration*, does not own manufacturing facilities, unlike the first business model; instead, these digital integrators manufacture the construction components from supply chain partners and suppliers. Digital system integrators maintain long-term relationships within the supply chain and develop a system to deliver a project (Hall et al., 2022). Similarly, Aitchison et al. (2023) discuss the three most common BMs: product platforms, digital platforms, and hybrid architecture. Product platforms are discussed in relation to the BoKlok concept, a collaboration between IKEA and the construction company Skanska that specializes in delivering sustainable and quality houses at a low price. This model aligns with Hall et al.'s (2022) vertical integration, digital platforms (cf., Aitchison et al., 2023), and eco digital systems integration (cf., Hall et al., 2022). In addition, Aitchison et al. (2023) emphasize a hybrid architecture business model, denoting the practice of the last decade, which combines physical and digital components to deliver construction projects. This practice involves mapping the physical world to the digital world, visualized through analyzing big data collected from sensors attached to the physical components.

Another (third) business model is establishing a 'spinoff' factory, which is derived from the existing project-based firm to address the demands of the project in a factory setting (Hall et al., 2022, p.304).

Several studies have discussed BMI in the construction industry context. In their organizational-level study, Liu et al. (2017) explored the impact of a favorable business environment and entrepreneurial cognition (attitude to technological innovation) on BMI in the Chinese construction industry context. While both drivers were found to positively impact BMI, the study further provides a stance on how technological shifts in the context of modular prefabrication raise construction firms' revenue as they deviate from conventional construction methods. Similarly, Han et al. (2022) examined the mediating role of BMI between entrepreneurial orientation and corporate performance. BMI is conceptualized and explained by multiple items, such as the ability to prevent errors in the transaction process, level of applicability to handle large- to small-scale transactions, ability to summarize information, ability to attain faster transaction efficiency, ability to reintegrate the output services and products, use of innovative incentive measures to increase the enthusiasm of stakeholders, and a high number of quantities in terms of outputs and novel nature. Researchers such as Wang et al. (2023) studied the impact of organizational characteristics on BMI. For instance, dual attributes (flexibility and efficiency) of the organizational structure were found to have a positive impact on BMI.

Ruiter et al. (2021) have perceived transitioning traditional BMs into sustainable BMs—as BMI. In this qualitative study, the role of management control in this transition process (i.e., the BMI) in the context of the construction industry is elaborated upon by adopting the circular economy (CE) concept. The BMI has been designed based on CE principles, namely, the reuse, recycling, and recovery of materials. In line with this, the authors emphasized that the evaluation of the outcome—sustainable BMs—should be based on their contributions to the organization's mission rather than quantitative measures of predefined performance indicators. This study differs from other studies that focused on BMI in construction, where BMs are evident in all stages of the project lifecycle. As Table 6.1 illustrates, the elements of BMI are the organizational belief system (vision and mission, exploration of knowledge, drive for radical innovation), the interactive control system (internal communication and relationships with external partners), the boundary systems (meaningful code of conduct

Table 6.1 Levels of management control in facilitating BMI

—	**Simons (1994) Considering traditional BMs**	**Analytical focus points in this study concerning circular BMs**
Belief systems	Formal and systematic in nature	The belief systems: • rule over the other levers; the organization's vision and mission form the cornerstone of the management control system • drive radical innovations even beyond the frame of existing CE-oriented BMs.
Interactive control systems	Regulated. Mainly bottom-up	The interactive control systems: • are the mechanisms to explore the emergent strategies in the dialogue between organization members • may be useful for cooperation between partners in a long-term coalition.
Boundary systems	The organization's strategic boundaries are predefined	The boundary systems: • are meaningful as a code of conduct to direct day-to-day activities • help to find the opportunity space to connect the circular BMs with the current organization's strategic approach.
Diagnostic control systems	Measuring output, comparing this with preset goals, and taking corrective actions if appropriate	The diagnostic control systems: • reveal that measuring output to compare this with preset goals can be problematic • support the vision-driven evaluation and discussion to understand how new BMs can deliver a valuable contribution.

Note: Illustrates the levels of management control in traditional and circular business models.
Source: Adopted from Ruiter et al. (2021).

to carry out daily activities), and the diagnostic control systems (contemporary performance assessment).

Zhao et al. (2016) examined the effect of the business model on zero-carbon buildings (ZCB) and explored how BMI can facilitate delivering ZCBs. They highlighted three integrated components—value-offering, value-creation and delivery, and value capture—in the process. These components represent the optimization of the organization's vision, networking, internal capabilities and resources, and customer orientation toward improving building energy performance (Zhao et al., 2016, p.258).

Li et al. (2024) developed a BMI inspired by the bathpod manufacturing process. The BMI is developed by considering two aspects: (a) how the components of the business model fit the organization internally; and (b) interactions with the external environment. From the findings of a case study, six archetypes of the BMI—project-driven production, integrated project service delivery, design and make, one-shop, modular enthusiast, and bathpod components catalogues—are developed. The first and second archetypes are project-oriented, and the remaining types are product-oriented, which is more favorable in terms of efficiency. The following propositions are established in relation to the developed BMI:

- Project-driven production can evolve into design and make (achieving economies of scale) through capability and resource-building.
- For head contractors, an integrated project service delivery business model can create a pathway to expand market coverage and offer built-to-order solutions.
- The design and make business model provides an intricate understanding of customer requirements and creates the opportunity to evolve into mass customization.
- Companies that can offer design flexibility while ensuring increased production based on forecasts are likely to be more competitive in the market.
- The trend toward customer-centricity will continue toward capturing customer and client expectations and establishing market standards.
- Companies will leverage the strengths of their partners through integration and new service providers to deliver better solutions to customers (Li et al., 2024, pp.11–12).

Aitchison et al. (2023), in their building 4.0 CRC report, present the platform ecosystem model in delivering buildings where each element of the value chain (i.e., design, engineering, consultant, land, finance, assembly, etc.) interacts with the platform owner (see Figure 6.1). This approach is similar to digital system integration, allowing for a 'capital lite' approach. Unlike the traditional pipeline model, which operates project-to-project, the platform ecosystem aligns with growing industry capacity. A platform ecosystem is explained by three types of platforms: internal platforms, industry platforms, and supply chain platforms. As explained by Gawer (2014), the concept of a platform is viewed and understood as something more than internal company structures. For instance, pricing and specifications are considered the industry platforms and are managed through governance mechanisms and regulations.

Researchers have been keen on developing BMIs to improve businesses and investigating the impact of BMI on businesses. A few studies have focused on investigating the role of BMI in innovation diffusion, particularly digitalization. For instance, in their systematic review, Das et al. (2023a) point out the potential of BMIs as a medium for the strategic transformation of businesses through technology integration and strategy formulation. Sumanarathna et al.'s (2023) Hong Kong case study on implementing innovation through exploration and exploitation (i.e., radical and incremental innovation) highlights the case study organization's business model that integrates multiple organizational levels (i.e., integrated level) to facilitate both radical

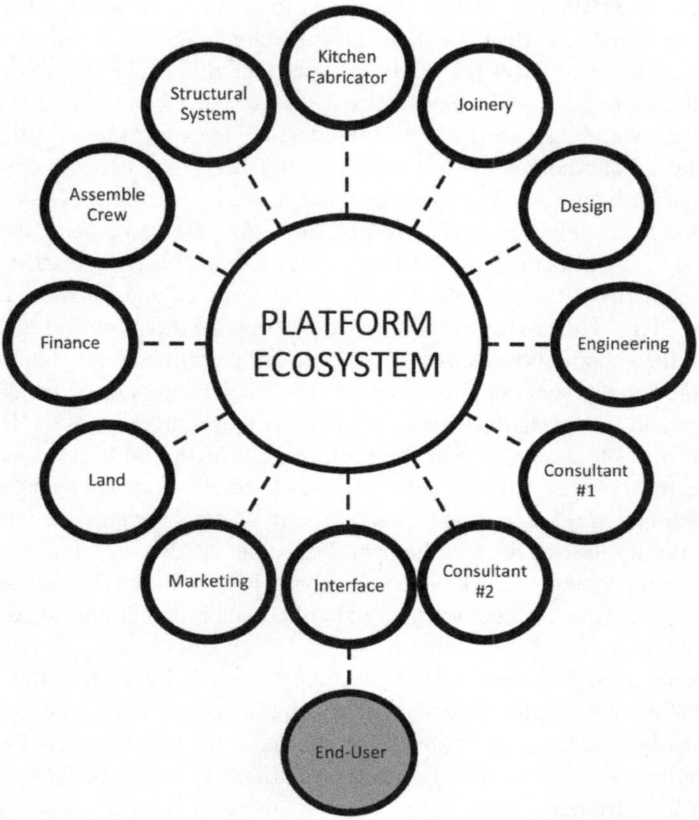

Source: Adopted from Aitchison et al. (2023, p.3).

Figure 6.1 *The platform business model*

and incremental innovation. The explained BMI of the case study organization is a testament that facilitates innovation diffusion and digitalization across the organization. On the contrary, digitalization has been recognized as a driver for disruptive innovation and BMIs (i.e., circular BMs) (Neligan et al., 2023).

DIGITALIZATION AND BMI IN CONSTRUCTION

There is an emerging consensus regarding the relevance of digital technologies as drivers of novel business models (BMs) characterized by decision-making that depends on data and is customer-centered (Das et al., 2023b). Industry 4.0 technologies are deemed to have significant effects on existing BMs in construction. These technologies use data and intelligent systems to create complex enterprises. Some studies have noted the effects of Industry 4.0 technologies in changes to the value chain, which shifts to networks with configurations that enable flexibility, cooperation, and coordination—a departure from the typical models that are linear and hierarchical (Nottbrock et al., 2023). These technologies include AI, IoT, 3D printing, robotics, cloud computing, etc. The existing BMs may become obsolete with advances in technology that present challenges, as increasing industry disruption shortens the lifecycles of BMs (Pan & Li, 2016). The fragmented and project-based nature of the construction ecosystem undergoes a radical transformation characterized by a shift from fragmentation to more standardized and integrated processes (Das et al. 2021). However, changes caused by digitalization are said to be in their infancy in the construction industry, with industry practitioners gradually realizing the tangible benefits that could be accrued from such technologies, which can connect design and supply chains with automation in orders, payment, as well as systems for monitoring site assembly and the performance of the built assets (Das et al., 2021). The technologies, through automation and operations driven by data, facilitate the creation of novel value propositions, expand customer segments, impact channels, and modify the activities of the firm (Das et al., 2023b).

Digital technologies (e.g., AI, big data analysis, IoT) facilitate the exchange of data in real time, improve communication, and aid collaboration among stakeholders in the value chain.

Digitalization is a customer-focused strategy that helps firms offer enhanced value. The ability to grow through digitalization offers opportunities to transform the BMs of firms in addition to the internal efficiency benefits (Paiola et al., 2022). Digitalization facilitates servitization, which is the 'transformational process of shifting from a product-centric business model and logic to a service-centric approach' (Kowalkowski et al., 2017, p.7). Servitization thus causes changes in firms' value propositions as well as how they create and capture value. These technologies, including automation and additive manufacturing, have also been known to impact the ability to scale up value chains and advance production techniques by enabling effective scaling up to respond to market demands. According to Bouwman et al. (2017), big data affects business processes in terms of monitoring processes and process quality through its ability to trigger business model experimentation, with firms trying out changes to the existing models with an allocated budget. IoT is deemed to make it possible for firms to innovate their BMs due to the opportunities and threats they present, which organizations are unable to address through their existing BMs (Paiola et al., 2022). These technologies allow organizations to offer more services or add new features to existing products and services that generate more revenue, reduce costs, enhance customer relationships, and improve offerings.

Digital transformation impacts firms' resources, capabilities, and activities. Digitalization is deemed to play a major role in firms' BMI, with technologies such as big data crucial to production monitoring as well as production quality (Bouwman et al., 2017). With digitization comes new entrants that offer unique and novel value propositions to clients that are often disruptive and driven by data (Christensen et al.,

2015). The use of digital technologies such as social media and big data affects firms' ability to experiment with their business models. A creative approach to digitalization results in new business models with novel functions where the digital components mainly drive value (Krotov, 2017). Bouwman et al. (2017) observed that big data utilization can create new customer segments, bring improvement in channels as well as the optimization of business activities. Digitalization facilitates value-creation through the orchestration of ecosystems (Parida et al., 2019). Digitalization enables the achievement of efficiency and effectiveness in offerings through its facilitation of scalable platforms that utilize modularity (Parida et al., 2019). With digitalization, there is continuous improvement in routine integration and centralization of service activities and processes, which enhances service to customers. With process costs reduced due to digital platforms, firms are able to aggregate large amounts of data and utilize AI to learn from such data—hence, the development of BMs that allow exponential scaling across different markets (Winter, 2023). Through product or service smoothing, digitalization increases value through transaction cost reduction (Parida et al., 2019). Winter (2023) reports on how the utilization of digital platforms and a collaborative approach can ensure reliability and profit on projects by assembling relevant stakeholders on the project. Thus, the coordination among project participants enhances productivity and drives costs down (Winter, 2023). These platforms also have the potential to enhance sustainability along the lifecycle of buildings through the optimization of the capacities of firms and, by extension, influence existing models.

Studies have identified patterns of BMI driven by Industry 4.0. Weking et al. (2020) identified integration, servitization, and expertization. While integration consists of utilizing new processes to integrate the supply chain, servitization is the combination of novel products and services. Expertization, on the other hand, is a mixed approach consisting of product and process models. Ibarra et al. (2018) also identify three Industry 4.0-based BMs: network ecosystems, service orientation, and customer orientation. In such instances, digital platforms are the catalysts for cooperation among firms and facilitate the co-creation of value (Das et al., 2023a). According to Veile et al. (2022), digital platforms enhance BMs through customer integration and the formation of new relationships between buyers and suppliers; partnerships characterized by trust and frequent communication; platform creation; development of ecosystems for recurring tasks; technological and digital knowledge; improved offers, value extension and digitalization, and data-driven value; and direct or indirect channels through platforms and the ecosystem.

While the impact of digital technologies is increasingly becoming apparent, particularly in their transformation of BMs and organizational processes, there is a challenge in identifying the steps to fully take advantage of their potential, with enterprise leaders deemed to lack understanding or the requisite preparation (KPMG, 2020). Employee competence and organizational capacity challenge the ability of firms to digitalize. In the construction sector, the possibility for the integration of digital technologies is low due to the lack of know-how, low research and development, and low buy-in from clients (Osunsanmi et al., 2020). Hoch and Brad (2021) observed the lack of expertise and experience in the usage and application of digital technologies (sensors, IoT, robotics) among employees as a challenge to their ability to innovate the existing business model from a traditional construction machine rental company into a digital ecosystem and multi-sided platform provider. According to Newman et al. (2020), implementation is hampered by residual practices despite the optimism about the potential of the technologies. These challenges seem to have restricted the attention on what these technologies can do, with little focus on their utilization to effect change in the BMs and strategies of firms (Das et al., 2021; Osunsanmi et al., 2020).

BUSINESS MODEL INNOVATION AND DIFFUSION OF INNOVATION IN CONSTRUCTION BUSINESS ORGANIZATIONS

Moving away from the traditional focus on a business model, the new conceptualization provides compelling evidence for construction organizations to innovate toward integrated approaches. In this, several new BMs are identified, such as digital, product-based, and sustainable BMs for construction (see Aitchison et al., 2023; Hall et al., 2022; Geissdoerfer et al., 2018). Construction organizations could reconfigure or redesign the entire BM to take advantage of the competitive business environment and offer better value to customers.

Das et al. (2023a) viewed BM as a driver of strategic transformation within construction organizations. Through an extensive literature review, they suggest BM can be used for strategy reformulation, shaping strategic partnerships, and acting as a bridge between technology integration and strategy in construction. However, their explanation of a customer-centric strategy is oversimplified and may not relate to the variety of projects and stakeholders within the construction business. For example, in the case of property developers, the end-users may be easily identifiable as tenants/individual property owners. Considering the complex nature of the construction industry, where construction contractors rely heavily on one-off and different projects with extensive supply chains, they may face the challenge of identifying value to end-users in every single case. Thus, it would be difficult for such organizations to adjust their BMs and instead focus on a generic template (see Amit & Zott, 2015) to help shape their business model. On the other hand, literature on BMI in shaping strategic partnerships and integrating technology and strategy has long been argued (Massa & Tucci, 2013). Nevertheless, their work sheds important light on the construction organization from a broader perspective.

Neligan et al. (2023) found that when considered part of BM, digital technologies help in resource optimization. They suggest digital technologies may be employed at the process, product, or whole lifecycle levels to improve efficiency. Since construction organizations have been recognized as slow in the adoption of technology, there will likely be resistance to BMI and a tendency to keep business as usual for apparent reasons of cost and familiarity with existing practices. Nevertheless, as a common practice, organizations opt for the least disruptive technologies to improve the process(es), to begin the transformation, since a radical change in processes may lead to unintended consequences. A similar practice can be witnessed in construction organizations, where organizations were reluctant to adopt BIM, common data environment (CDE), GIS, and many other Industry 4.0 technologies in projects and overall business processes, with a somewhat limited focus on fulfilling mandates. Hence, no significant change in the BM is seen.

As argued in this chapter, the literature on innovation diffusion is limited in its focus on the creation and development of innovation. Rogers' innovation diffusion perspective, where innovation is communicated through certain channels over time among social systems (Rogers, 2003, p.5), implies that diffusion takes time and is often dependent on the type of innovation under consideration. Construction organizations face a similar dilemma in BMI and innovation diffusion, where they encounter an increasingly competitive environment, lower profit margins, and higher quality standards; adoption and diffusion take a lower priority, especially when it is a radical innovation in BM. Thus, there is an increased reliance on ad hoc measures as quick fixes to respond to change with no lasting impact or significant change in business processes and transformation.

In this, Widén (2006) found the role of innovation brokers (such as professional institutions, industry-led organizations, and governmental agencies) was to drive innovation in construction. Innovation brokers are usually at higher levels and negotiate innovations through public policy and regulatory mechanisms, as presented in the case of the Hong Kong Housing Authority.

The Case of Hong Kong Housing Authority

The Housing Department (HD), an executive branch of the Hong Kong Housing Authority (HKHA), is a premier government organization with a vision to help low-income families with housing needs gain access to affordable housing. Established in 1972, HD constitutes four divisions, including the Development and Construction Division (DCD). DCD is responsible for planning, design, project management, and contract administration of public housing projects for almost a third of Hong Kong residents (Housing Department, n.d.).

Since the early 1980s, HD has been focused on promoting innovation within and across organizations through public policies (see Figure 6.2) to instigate change in public housing projects and the construction industry. Kuitert et al. (2019) argued that public policy is crucial for driving innovation in construction. Public sector clients, as innovation brokers in a complex product system, promote innovation through interaction and competition among the network of actors (Havenvid, 2015). This network of actors includes clients and service providers, generalized here as construction organizations, regulators, and professional institutions, as an innovation superstructure presented by Winch (1998).

Though interaction and competition in the innovation superstructure have a significantly different theoretical basis, interaction as a driver of innovation promotes policies that are communicated through systems integrators (Lindblad and Guerrero,

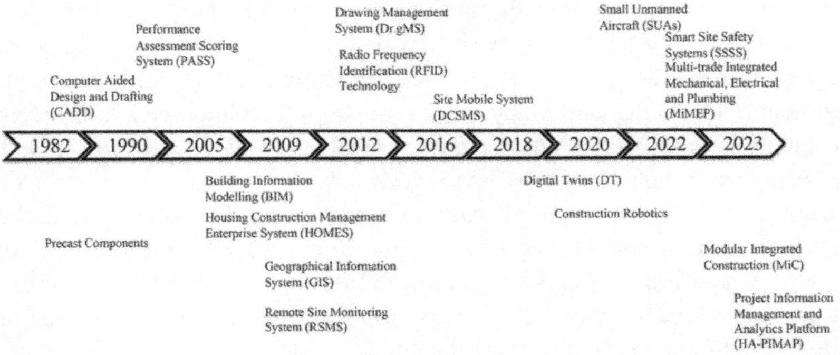

Source: Hong Kong Housing Authority (2019a, 2019b, 2023).

Figure 6.2 *Application of innovation and technology in planning, design, and construction management in public housing projects*

2020), which has the potential to drive BMI in construction organizations. Public sector clients as innovation brokers play an active role in establishing demands from construction organizations through policy measures, for example, requiring innovative technologies and systems in new projects. This, in essence, establishes the need for BMI in construction organizations. A significant amount of literature can be found on the client's role in promoting innovation, most pointing toward the power to influence a change in traditional BMs.

A similar approach can be witnessed in the case of the HD, where the HD, as an innovation broker, promotes innovative technologies and systems through a series of policy measures over time to drive change. For example, the HD promoted the use of 4S, Multi-trade Integrated Mechanical and Electrical and Plumbing (MiMEP), Modular Integrated Construction (MiC), Project Information Management, and an Analytics Platform (HA-PIMAP) for newly procured projects. Not only do these technologies and systems improve productivity, but they also demand that construction organizations recalibrate their business model canvas, as the key activities, resources required, channels, and, most importantly, cost structure are significantly altered when compared with traditional BMs.

Despite the significance of the client's role as an innovation broker and their influence on BMI in construction organizations, these organizations may still face challenges. It is believed that the problem is not of individual organizations but rather the nature of how the construction industry operates, which requires change at multiple levels. In the first instance, a change in mindset may lead organizations to accept reality and consider a broader perspective toward BMI and lasting change in construction organizations' processes.

CONCLUDING REMARKS

The global construction industry continues to grapple with the aftermath of the COVID-19 pandemic, facing significant challenges in its recovery. The industry's current BMs are deemed inadequate, necessitating the implementation of substantive reforms.

This pivotal transition necessitates resolute leadership from government entities in procurement and policy. Public sector clients across various jurisdictions contribute a substantial proportion, often exceeding 40%, to the industry's overall turnover. Consequently, there exists an imperative to recalibrate the prevailing fiscal paradigm, redirecting its emphasis from cost minimization to the maximization of overall benefit.

This is an institutional issue that cannot be addressed by tinkering with individual contractors' business model approaches. There is a need for governments to impose an approach on the industry that forces change within a structured yet rewarding environment for those who adopt a collaborative approach.

Thus, the paradigm shift needed will probably take a decade to implement, but the lessons learned from government leadership should be institutionalized in terms of the approach to industry governance. All in all, as Canon and Hillebrandt argued in the 1960s, there still is an urgent need for change to a broken business model. To fix the current broken business model, construction organizations could adopt a multi-pronged approach using technology-based innovations, where a dedicated innovation team must be formed. The innovation team could start by reviewing the existing model and broader environmental, social, and, most importantly, industry-specific parameters that may hinder effective functioning, outcomes, and innovation diffusion. After due consideration of existing circumstances, they should evaluate potential technologies, systems, and new designs that may drive better value to customers. Lastly, employing innovation champions to ensure the diffusion of innovation within an organization may turn the new business model into a reality.

REFERENCES

Aitchison, M., Knapp, C., Maxwell, D., Das, P., Godbole, S., & O'Leary, C. (2023). *Business Model Innovation: Unlocking the Next Chapter of Building through a Platform Ecosystem Model*. Building 4.0 CRC Project #40, Commissioned by Prefab Hub, the Advanced Manufacturing Growth Centre. Retrieved May 18, 2024, from https://building4pointzero.org/wp-content/uploads/2023/12/40-Business-Model-Innovation-Report-040823.pdf

Al-Debei, M.M., El-Haddadeh, R., & Avison, D. (2008). *Towards a Business Model for Cellular Network and Telecommunication Operators: A Theoretical Framework. Proceedings of the UK Acedamy of Information Systems (UKAIS), 13th Annual Conference*, Bournemouth, pp. 1–15, April.

Amit, R., & Zott, C. (2015). Crafting business architecture: The antecedents of business model design. *Strategic Entrepreneurship Journal*, 9(4), 331–350.

AutoCDE (n.d.). AutoCDE. Retrieved May 18, 2024, from https://www.llewellynandpartners.com/autocde-1.

AutoSafe (n.d.). About Us. Retrieved May 18, 2024, from https://autosafe.ai/about-us/

Bouwman, H., de Reuver, M., & Nikou, S. (2017). The impact of digitalization on business models: How IT artefacts, social media, and big data force firms to innovate their business model. 14th Asia-Pacific Regional Conference of the International Telecommunications Society (ITS): 'Mapping ICT into Transformation for the Next Information Society', Kyoto, Japan, 24–27 June, International Telecommunications Society (ITS), Calgary.

Chesbrough, H., Vanhaverbeke, W., & West, J. (Eds.). (2006). *Open innovation: Researching a new paradigm.* Oxford University Press.

Chesbrough, H. (2010). Business model innovation: Opportunities and barriers. *Long Range Planning*, 43(2–3), 354–363.

Christensen, C.M., Raynor, M.E., & McDonald, R. (2015). What is disruptive innovation? *Harvard Business Review*, Special Feature.

Das, P., Hijazi, A.A., Maxwell, D.W., & Moehler, R.C. (2023a). Can business models facilitate strategic transformation in construction firms? A systematic review and research agenda. *Sustainability*, 15(17), 13022.

Das, P., Godbole, S., Maxwell, D.W., & Moehler, R.C. (2023b). Exploring the role of Industry 4.0 in the building industry: A case study of emerging business models in the Australian construction ecosystem. In Creative Construction e-Conference 2022 (p. 42). Diamond Congress Ltd.

Das, P., Perera, S., Senaratne, S., & Osei-Kyei, R. (2021). Developing a construction business model transformation canvas. *Engineering, Construction and Architectural Management*, 28(5), 1423–1439.

Dubois, A., & Gadde, L.E. (2002). The construction industry as a loosely coupled system: implications for productivity and innovation. *Construction Management and Economics*, 20(7), 621–631.

Fielt, E. (2013). Conceptualizing business models: Definitions, frameworks, and classifications. *Journal of Business Models*, 1(1), 85–105.

Gawer, A. (2014). Bridging differing perspectives on technological platforms: Toward an integrative framework. *Research Policy*, 43(7), 1239–1249.

Geissdoerfer, M., Vladimirova, D., & Evans, S. (2018). Sustainable business model innovation: A review. *Journal of Cleaner Production*, 198, 401–416.

George, G., & Bock, A.J. (2011). The business model in practice and its implications for entrepreneurship research. *Entrepreneurship Theory and Practice*, 35(1), 83–111.

Hall, D., Algiers, A., Lehtinen, T., Levitt, R.E., Li, C., & Padachuri, P. (2014). The role of integrated project delivery elements in adoption of integral innovations. In Engineering Project Organization Conference 2014, Devil's Thumb Ranch, Colorado, 29–31 July (pp. 1–20). Engineering Project Organization Society (EPOS).

Hall, D.M., Algiers, A., & Levitt, R.E. (2018). Identifying the role of supply chain integration practices in the adoption of systemic innovations. *Journal of Management in Engineering*, 34(6), 04018030.

Hall, D.M., Lessing, J., & Whyte, J. (2022). New business models for industrialized construction. In: Bolpagni, M., Gavina, R., & Ribeiro, D. (eds) *Industry 4.0 for the Built Environment. Structural Integrity*, vol. 20. Springer. https://doi.org/10.1007/978-3-030-82430-3_13

Han, W., Zhou, Y., & Lu, R. (2022). Strategic orientation, business model innovation and corporate performance—Evidence from construction industry. *Frontiers in Psychology*, 13, 971654.

Havenvid, M.I. (2015). Competition versus interaction as a way to promote innovation in the construction industry. *IMP Journal*, 9(1), 46–63.

Hoch, N.B., & Brad, S. (2021). Managing business model innovation: An innovative approach towards designing a digital ecosystem and multi-sided platform. *Business Process Management Journal*, 27(2), 415–438.

Hong Kong Housing Authority (2019a). Quality for Housing: Partnering for Change, Off-site Manufacturing and Modular Integrated Construction (BC 7/2019). Retrieved May 17, 2024, from https://www.housingauthority.gov.hk/en/common/pdf/about-us/housing-authority/ha-paper-library/BC7-19en.pdf

Hong Kong Housing Authority (2019b). Application of Building Information Modelling in Public Housing Projects (BC 26/2019, and 35/2019).https://www.housingauthority.gov.hk/en/common/pdf/about-us/housing-authority/ha-paper-library/BC35-19EN.pdf

Hong Kong Housing Authority (2023). Application of Innovation and Technology in Public Housing Development and Construction (LC Paper No. CB (1)887/2023(04)). Retrieved May 17, 2024, from https://www.legco.gov.hk/yr2023/english/panels/hg/papers/hg20231003cb1-887-4-e.pdf

Housing Department (n.d.). About Us. Retrieved May 17, 2024, from https://www.housingauthority.gov.hk/en/about-us/housing-department/index.html

Ibarra, D., Ganzarain, J., & Igartua, J.I. (2018). Business model innovation through Industry 4.0: A review. *Procedia Manufacturing*, 22, 4–10. https://doi.org/10.1016/j.promfg.2018.03.002

isBIM (n.d.). About Us. Retrieved May 18, 2024, from https://www.isbim.com.hk/about-us

Johnson, M.W., Christensen, C.M., & Kagermann, H. (2008). Reinventing your business model. *Harvard Business Review*, 86(12), 50–59.

Kowalkowski, C., Gebauer, H., Kamp, B., & Parry, G. (2017). Servitization and deservitization: Overview, concepts, and definitions. *Industrial Marketing Management*, 60, 4–10.

KPMG (2020). *The 2020 Fourth Industrial Revolution Benchmark. Digital and Disruption.* KPMG Digital Delta in Collaboration with Faethm. Retrieved from https://assets.kpmg/content/dam/kpmg/au/pdf/2020/fourth-industrial-revolution-benchmark.pdf

Krotov, V. (2017). The Internet of Things and new business opportunities. *Business Horizons*, 60, 831–841.

Krumeich, J., Werth, D., & Loos, P. (2013). Interdependencies between business model components: A literature analysis. *Proceedings of the Nineteenth Americas Conference on Information Systems*, Chicago, Illinois, August 15–17.

Krumeich, J., Burkhart, T., Werth, D., & Loos, P. (2012). Towards a component-based description of business models: A state-of-the-art analysis. *AMCIS 2012 Proceedings*, 19. https://aisel.aisnet.org/amcis2012/proceedings/EBusiness/19

Kuitert, L., Volker, L., & Hermans, M.H. (2019). Taking on a wider view: Public value interests of construction clients in a changing construction industry. *Construction Management and Economics*, 37(5), 257–277.

Li, Y., Das, P., Kuzmanovska, I., Lara-Hamilton, E., Maxwell, D.W., & Moehler, R. (2024). Business model innovation in the construction industry: Emerging business model archetypes from bathpod modularization. *Journal of Management in Engineering*, 40(2), 04023066.

Lindblad, H., & Guerrero, J.R. (2020). Clients' role in promoting BIM implementation and innovation in construction. *Construction Management and Economics*, 38(5), 468–482.

Liu, G., Li, K., Zhao, D., & Mao, C. (2017). Business model innovation and its drivers in the Chinese construction industry during the shift to modular prefabrication. *Journal of Management in Engineering*, 33(3), 04016051.

Massa, L., & Tucci, C.L. (2013). Business model innovation. In: Dodgson, M., Gann, D.M., & Phillips, N. (eds) *Oxford Handbook of Innovation Management* (pp. 420–441). Oxford University Press.

Morris, M., Schindehutte, M., & Allen, J. (2005). The entrepreneur's business model: toward a unified perspective. *Journal of Business Research*, 58(6), 726–735.

Neligan, A., Baumgartner, R.J., Geissdoerfer, M., & Schöggl, J.P. (2023). Circular disruption: Digitalisation as a driver of circular economy business models. *Business Strategy and the Environment*, 32(3), 1175–1188.

Newman, C., Edwards, D., Martek, I., Lai, J., Thwala, W.D., & Rillie, I. (2020). Industry 4.0 deployment in the construction industry: A bibliometric literature review and UK-based case study. *Smart and Sustainable Built Environment*, 10(4), 557–580.

Nottbrock, C., Van Looy, A., & De Haes, S. (2023). Impact of digital Industry 4.0 innovations on interorganizational value chains: A systematic literature review. *Business Process Management Journal*, 29(1), 43–76.

Osterwalder, A. (2004). The business model ontology: A proposition in a design science approach (Doctoral dissertation, Université de Lausanne, Faculté des hautes études commerciales).

Osterwalder, A., Pigneur, Y., & Tucci, C.L. (2005). Clarifying business models: Origins, present, and future of the concept. *Communications of AIS*, 16(1), 1–40.

Osunsanmi, T.O., Aigbavboa, C.O., Oke, A.E., & Liphadzi, M. (2020). Appraisal of stakeholders' willingness to adopt construction 4.0 technologies for construction projects. *Built Environment Project and Asset Management*, 10(4), 547–565.

Paiola, M., Agostini, L., Grandinetti, R., & Nosella, A. (2022). The process of business model innovation driven by IoT: Exploring the case of incumbent SMEs. *Industrial Marketing Management*, 103, 30–46.

Pan, W., & Li, K.J. (2016). Clusters and exemplars of buildings towards zero carbon. *Building Environment*, 104, 92–101.

Parida, V., Sjödin, D., & Reim, W. (2019). Reviewing literature on digitalization, business model innovation, and sustainable industry: Past achievements and future promises. *Sustainability*, 11(2), 391.

Rappa, M. (2000). Business models on the web. Retrieved from https://digitalenterprise.org/models/

Remote e-inspection system (n.d.) Projects. Retrieved May 18, 2024, from https://www.viact.ai/contactsales

Rogers, E.M. (2003). Diffusion of innovations (5th ed.). Free Press.

Ruiter, H., De Feijter, F., & Wagensveld, K. (2021). Management control and business model innovation in the context of a circular economy in the Dutch construction industry. *Sustainability*, 14(1), 366.

Simons, R. (1994). *Levers of control: How managers use innovative control systems to drive strategic renewal.* Harvard Business Press.

Shafer, S.M., Smith, H.J., & Linder, J.C. (2005). The power of business models. *Business Horizons*, 48(3), 199–207.

Sumanarathna, N., Duodu, B., Memon, S.A., & Rowlinson, S. (2023). Exploring innovation deployment of construction contracting firms through the lens of exploratory–exploitative learning: A Hong Kong case study. *Construction Innovation*, 25(2), 641–664.

Timmers, P. (1998). Business models for electronic markets. *Electronic Markets*, 8(2), 3–8.

Veile, J.W., Schmidt, M.-C., & Voigt, K.-I. (2022). Toward a new era of cooperation: How industrial digital platforms transform business models in Industry 4.0. *Journal of Business Research*, 143, 387–405. https://doi.org/10.1016/j.jbusres.2021.11.062

viACT (n.d.). Solutions. Retrieved May 18, 2024, from https://www.viact.ai/contactsales

Wang, Z., Lin, S., Chen, Y., Lyulyov, O., & Pimonenko, T. (2023). Digitalization effect on business performance: Role of business model innovation. *Sustainability*, 15(11), 9020.

Weking, J., Stöcker, M., Kowalkiewicz, M., Böhm, M., & Krcmar, H. (2020). Leveraging industry 4.0: A business model pattern framework. *International Journal of Production Economics*, 225, 107588. https://doi.org/10.1016/j.ijpe.2019.107588.

Widén, K. (2006). Innovation diffusion in the construction sector (Doctoral dissertation, Division of Construction Management, Lund Institute of Technology).

Winch, G. (1998). Zephyrs of creative destruction: Understanding the management of innovation in construction. *Building Research & Information*, 26(5), 268–279.

Winter, J. (2023). Business model innovation in the German industry: Case studies from the railway, manufacturing and construction sectors. *Journal of Innovation Management*, 11(1), 1–17. https://doi.org/10.24840/2183-0606_011.001_0001
Wirtz, B.W., Pistoia, A., Ullrich, S., & Göttel, V. (2016). Business models: Origin, development and future research perspectives. *Long Range Planning*, 49(1), 36–54.
Zhao, D., Duva, M., Mollaoglu, S., Frank, K., Garcia, A., & Tait, J. (2021). Integrative collaboration in fragmented project organizations: Network perspective. *Journal of Construction Engineering and Management*, 147(10), 04021115.
Zhao, X., Pan, W., & Lu, W. (2016). Business model innovation for delivering zero carbon buildings. *Sustainable Cities and Society*, 27, 253–262.

7. A conceptual framework to understand the implementation of innovation on construction projects

Arun Bajracharya and Mohammed Fadhil Dulaimi

INTRODUCTION

Innovation is a driver of progressive change in any sector of the economy. It is also an important means of gaining a competitive advantage in the market. Nations, industries, and firms are constantly under increasing pressure to perform better, incrementally, or radically, with the implementation of innovative ideas. The construction sector has its own peculiarities in the context of innovation. At the industry level, construction firms and related institutions hardly interact with each other unless they are engaged in construction projects where a limited number of firms work together. Such loose relationships at the industry level are not conducive to collective learning and innovative activities. On the other hand, at the production level, that is, on construction projects, the involved firms become tightly interdependent within the rules of the adopted procurement and contractual frameworks. A small change in one component of a firm's work creates ripple effects on other firms' work in such tightly coupled project systems (Sterman, 1992). Moreover, as construction is a one-of-a-kind site-based production carried out by a temporary project organization (Ninan et al., 2022; Cox and Thompson, 1997), the involved firms work under various project-related pressures (Kumaraswamy and Dulaimi, 2001). With such industrial and production settings, the behavior of the firms in construction is bound to be oriented toward increasing their short-term benefits from the project instead of preferring long-term learning and innovation (Dubois and Gadde, 2002). Therefore, the construction industry has its specific kind of institutional base and production processes that have to be taken into consideration for understanding innovation in construction (Heaney et al., 2015; Bygballe and Ingemansson, 2014; Slaughter, 1998; Tatum, 1986).

The production platform in construction is an individual and unique project comprising different firms with separate roles and interests with different functional and/or contractual relationships (Barlow, 2000). The main relevant firms would be the client, consultants, main contractor, subcontractors, and suppliers, and, at times, universities and research and development (R&D) collaborators. The timing and duration of involvement of the different firms could differ according to their roles in the project. Construction innovation can be taken as a subset of construction project activities, and the project itself can be taken as a subset of business activities of the relevant firms.

Further, the specificity of construction innovation can be explained by discussing the following attributes: (a) area of innovation; (b) source of innovation; (c) demand for innovation; (d) types of innovation; (e) process and stages of innovation; (f) challenges for innovation; (g) opportunities for innovation; and (h) state of innovation in construction:

(a) **Area of innovation**: Generally, the area of innovation would be within the possible construction services, management, processes, and in different spectrums of part or all of the product of construction. Specifically, Tatum (1984) identified some potential domains of innovation in construction, such as: (i) changes in construction methods or sequences; (ii) expansion or new applications of a special technique; (iii) scale-up of previous methods; (iv) changes in design to accomplish the same function at a lower cost, in a manner similar to value engineering; (v) simplification of design requirements and construction operations; (vi) changes in planned construction methods to allow more favorable engineering techniques; and (vii) use of alternative materials or equipment. Bernstein and Lemer (1996) explained some other possible areas of innovation, such as using high-performance materials, automation and instrumentation, and green technologies. They viewed the constructed product development process as a whole lifecycle process from planning, design, construction, operation, and maintenance to renewal/demolition, and they suggested that innovation can take place at any stage of the development process.
Sawhney et al. (2020) pointed at the physical and digital streams as the two emerging innovation streams that have the potential to transform the design and construction process. They mentioned that the former stream is a bit slow in the industry, but the latter, with the support of emerging technologies such as robotics, industrialized construction, additive manufacturing, artificial intelligence (AI), augmented reality (AR), unmanned aerial systems, digital fabrication, building information modeling (BIM), and others can potentially move the design and construction process forward, provided that they operate on the right platform.

(b) **Source of innovation**: The primary source of innovation in construction is the knowledge and expertise of different firms involved in the construction project. It could be the contractor's innovative construction and management techniques, the consultant's bank of innovative designs and services, or the supplier's innovative equipment/materials. Some Japanese contractors have a separate R&D function at the firm level so that any developed innovative idea could be used in projects in hand (Yamazaki and Ueda, 2003). One specific aspect in construction is that most of the innovative ideas are not generated in parallel with the production process; rather, some well-tested new ideas would be adapted as innovations in order to avoid any risk of absolute newness. In general, the "available technology guides problem" type of innovation is more frequent than the "problem leads solution" type of innovation in construction (Nam and Tatum, 1992b).

(c) **Demand for innovation**: There is a host of factors that trigger demand for innovation in construction. Some of the main factors that create demand for an innovative idea in construction include the owner's requirement, technological development in constructed product, technological development in the construction process, regulatory requirements, challenges of specific projects and problems during construction, and the need for a firm to maintain a competitive position in the construction industry (Nam and Tatum, 1992b; Tatum, 1989; Tatum, 1987).

(d) **Types of innovation**: Innovation in construction could be categorized into several types, such as product, process, radical, incremental, technical, or administrative innovations. Following Henderson and Clark (1990), Slaughter (1998) explained distinct types of construction innovation such as incremental, modular, architectural, system, and radical innovation. The categorizations are based on the scale and impact of innovation – and as such, Slaughter (1998) argued that the implementation of each type of innovation requires a different level of management and supervision.
Another categorization suggested by Harty (2005) is based on the modes of innovation itself. He divided construction innovations into two modes: bounded and unbounded. In the bounded mode, the implications of innovation are restricted within a single coherent sphere of influence, whereas in the unbounded mode, the effects of innovation spill over beyond the single entity. Many firms' collaboration and hence their inter-organizational interaction is essential for the implementation of the unbounded mode of innovation.

(e) **Process and stages of innovation**: The whole normal process of innovation would start from the recognizing or discovering stage to the final implementation and feedback stage. Tatum (1987) proposed a sequential process model of innovation for the construction firm which includes: (i) recognizing forces and opportunities for innovation; (ii) creating a climate for innovation; (iii) developing the necessary capabilities; (iv) providing new construction technologies; (v) experimenting and refining; and (vi) implementing.
Bernstein and Lemer (1996) identified four generic stages of construction innovation: (i) research or discovery; (ii) development; (iii) production, application, or demonstration; and (iv) technology transfer, dissemination, or commercialization.

(f) **Challenges for innovation**: There are various propositions on challenges to fostering innovation in construction. The main propositions are:

i. As construction is a one-time, on-site, project-based activity, there is a lack of knowledge transfer and learning mechanisms from one project to another (Gann, 2000; Dubois and Gadde, 2002).

ii. The high variation of the works in construction (Tatum, 1986) and the uniqueness of the constructed product itself limit economies of scale.
iii. As the project organization consists of several independent firms, fragmentation becomes a major problem due to the underlying institutional and production processes of construction (Sawhney et al., 2020; Gann, 2000; Winch, 1998; Tatum, 1986).
iv. There are substantial liability and litigation threats in the case of failure (Nam et al., 1991; Tatum, 1986).
v. There are structural impediments, such as a large number of small firms operating in limited markets with low profit margins, market fluctuations, and a multitude of regulatory codes and standards (Gann, 2000; Bernstein and Lemer, 1996; Tatum, 1986).
vi. There are cultural impediments, such as lowest-cost procurement practice, adversarial culture, reluctance to invest in R&D, and risk-averse attitude (Gann, 2000; Powell, 1995; Nam et al., 1991; Ball, 1988).
vii. There is a loose patent system and very low direct R&D intensity in construction (Newton, 1999; Seaden, 1996).

(g) **Opportunities for innovation**: Despite the difficulties, there are some inherent potential advantages in fostering innovation in the construction sector, especially at the project level. Compared with the manufacturing sector, the construction industry has some unique and advantageous production processes for innovation. In the manufacturing sector, innovation is viewed as a departure from existing technologies and managerial practices, and it connotes that innovation may bring challenges of re-establishing the whole physical and organizational structure of production processes (Knoke, 2000). As construction is a project-based activity, each new construction offers a new setting to implement new ideas (Dulaimi et al., 1996; Tatum, 1986) without the need to dismantle existing production processes and without the need for extra capital investment.

(h) **State of innovation in construction**: Some authors argue that construction is not less innovative than other sectors. According to Slaughter and Shimizu (2000), in the minute works at the project level, many innovative ideas are developed and implemented. Such micro-innovations can be taken as the normal problem-solving processes in construction projects. Winch (2003) argued that the construction sector is not slower than (nor is it ahead of) other sectors (compared with the car manufacturing industry) in the rate of increasing innovation. He emphasized that the standard comparisons of the rates and levels of innovation in construction with other industries do not equally consider their respective full value chain/systems, and such imbalanced comparisons are inherently flawed. Therefore, he concluded that there is no firm evidence that the construction sector's performance is worse (or better) than that of any other sector (such as motor vehicles).

Contrary to these arguments, most of the authors argue that the construction industry is lagging behind in innovation when compared with other sectors of the economy (Wang et al., 2023; Xue et al., 2014; Lenard, 2001; Bernstein and Lemer, 1996; Nelson and Winter, 1977). The concrete impact of innovation in the construction industry, in terms of the relative increase in overall productivity, has not been realized. The productivity growth of construction is lower than that of the manufacturing sector (Bahr and Laszig, 2021; Warren, 1993). Therefore, it seems that so far, the challenges for innovation have been stronger than the available opportunities in the construction industry.

In such a specific implementation context, with the temporary participation of multiple organizations and their individual interests, knowledge of the processes that determine the behavior of relevant firms is highly important to explore the leverage for managing innovation at the project level. This chapter, therefore, presents a framework that will help identify the key processes that inform the probable behavior of and the interactions among relevant parties during the implementation of innovation on construction projects.

THE CONCEPTUAL FRAMEWORK

Background

When an innovative idea, especially an unbounded one as categorized by Harty (2005), is to be implemented on a construction project, it is most likely that the client, main consultant, and main contractor get involved with the relevant specialist consultants, subcontractors, and suppliers. The team of all the value-adding organizations involved in an innovation can be termed as the innovation value chain, and it would be a subset of the whole project value chain. Most often, innovation in a construction project is implemented as a part of the whole project, and the value of innovation depends on the relative impact that it can impart to the project in terms of various advantages.

In order to generalize the role of the organizations in the innovation value chain, they can be divided into two broad groups: origin organization (OO) and supporting organizations (SOs) (Dulaimi et al., 2003). OO is the organization in the innovation value chain that proposes innovative ideas in a construction project. SOs are all the relevant organizations in the value chain whose support is needed for the OO to implement the proposed innovation. The innovation value chain might consist of various combinations of OO and SOs. At times, an SO might also need the association of its own SOs (such as SO2-1 or SO3-2) to support the innovative idea proposed by the OO. One of the possible combinations of OO and SOs in an innovation value chain is shown in Figure 7.1.

Normally, the decision power pertaining to innovation (whether the innovation is top-down or bottom-up) lies at the management level of a firm (Kanter, 1988). In construction firms, the central management looks after the business interests of the firm,

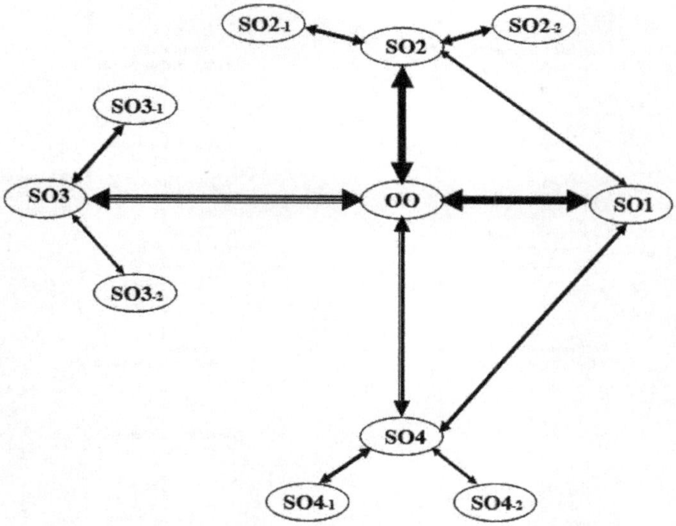

Figure 7.1 *A typical innovation value chain*

and the projects, which are the business delivery platforms of the firm, remain within the purview of the management (Gann and Salter, 2000). Therefore, in the conceptual framework, the decision hierarchy of each of the OO and SOs is divided into two levels: management level and project level. The resources and information regarding the decisions on innovation are presumed to be directed from the management level to the project level, and once the implementation process starts, all the information regarding the implementation feedback is taken care of at the project level.

The information and resource flow decisions for innovation are governed by cognitive and behavioral processes of different organizations. In the conceptual framework (Figure 7.2), the behavioral processes of individual OO and SOs are shown in box X and box Y respectively. Box Z shows the inter-organizational interaction between OO and SOs.

With this background, a detailed discussion of the whole conceptual framework is now presented below in three main sections: (a) Innovation process in OO (Box X); (b) Innovation process in SO (Box Y); and (c) Inter-organizational interaction process (Box Z). Reference to each construct in the framework is given by the single-letter identification in parentheses.

Innovation Process in Origin Organization (Box X)

Innovation is a purposive but inherently stochastic process (Nelson and Winter, 1977). Nightingale (1998, p. 689) stated that "innovation starts with a desired end result and attempts to find the unknown starting conditions that will achieve it". This

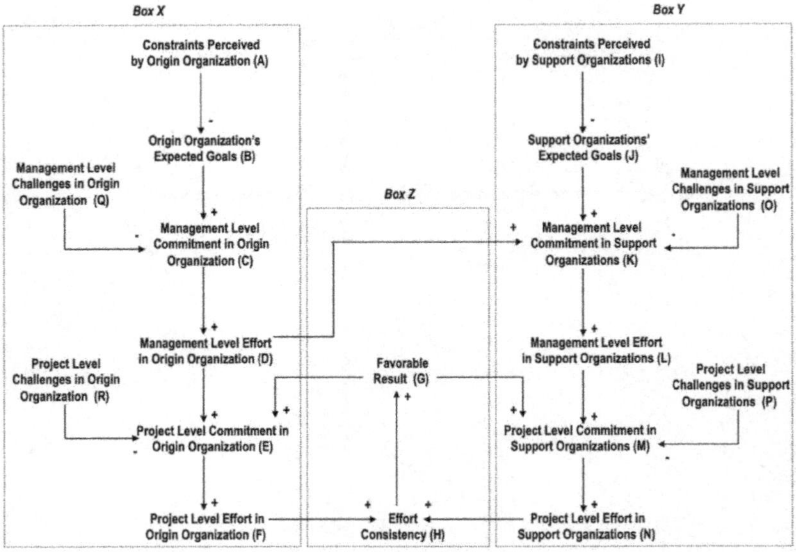

Figure 7.2 Conceptual framework

implies that every organization implementing innovation is driven by its individual motives amidst various constraints and consequent uncertainties. The motivation of an organization in the context of the implementation of innovation can be explained by the goal-seeking theory and instrumental motivation theory (Repenning, 2002). Although these theories are complementary to each other in explaining the motivation behind innovation, the goal-setting theory is more relevant at the initiation or pre-implementation stage of innovation, while instrumental motivation is relevant during and after the implementation stage.

In the explanation of the innovation process in OO, the goal-setting process is first explained in detail. At the initiation stage, the OO comes up with an innovative idea and proposes it in project organization. The "interest" of the OO in initiating the innovation can be explained by its aspiration to achieve certain goals out of the innovation. It means that the goals direct the behavior (Locke and Latham, 2002) of the OO while initiating and proposing the innovation. At the same time, the choice and level of goals are influenced by some relevant determining factors such as self-efficacy, expectancy, and equity (Locke et al., 1988).

Self-efficacy reflects the individual's ability, past performance, prior successes, and failures, and most importantly, the conclusions the individual has reached about total capability based on such information (Bandura, 1977; Bandura, 1997; Locke and Latham, 1990). Expectancy is defined as a momentary belief concerning the likelihood that a particular act (innovation) will be followed by a particular outcome (from the innovation), and this notion of calculative motivation is explained by the expectancy theory (Vroom, 1964; Katzell and Thompson, 1990). The concept of equity (equity theory) corroborates the expectancy theory in that the act (innovation) is expected to be rewarded equitably by the outcome (out of innovation), otherwise, the action would be adjusted (downward) to match the reward (Adams, 1965).

This combined explanation, using the essence of goal-setting theory, expectancy theory, and equity theory, suggests that the "operational/practical" goal behind any act of an organization is the aspiration influenced by cognitive values such as self-efficacy, expectancy, and the sense of equity perceived by the organization. For the initiation of an innovation, the cognitive mechanism is important in determining the further course of action because instrumental expectation makes a stronger contribution to behavior, particularly in a cultural climate that places a high value on individual need gratification (Wiener, 1982), and such a culture is more relevant in the case of construction projects.

In the conceptual framework (Figure 7.2), the operational goal-setting process of an OO is represented by a negative relational link between the constructs "Constraints Perceived by OO (A)" and "OO's Expected Goal (B)" in box X. The "Perceived Constraints (A)" indicates the level of perceived self-efficacy and consequent sense of expectancy and equity. Higher perceived constraints mean a lower sense of self-efficacy, which would cause low expectancy and thus a low level of "Expected Goal (B)". Therefore, the perceived constraints (A) (or the sense of less self-efficacy/expectancy/equity) would adversely affect the level of the expected/operational goal of OO (B). The operational goal-setting process thus includes the cognitive selection of the "purpose" behind innovation.

After setting the goal for innovation, Repenning (2002) showed that the "commitment" toward the goal plays an important role in determining the extent of effort that an organization would put into the innovation. He stated that goal commitment "refers to one's attachment to or determination to reach a goal" (Locke et al., 1988, p. 24), and the higher the level of expected goal, the more the commitment that would be needed. Locke and Latham (1990) stated that once the individual has a goal and once s/he chooses (gets committed) to act on it, the three direct mechanisms – effort, persistence, and direction – are brought into play more or less automatically. This also implies that goal commitment is a type of motivating force (Scholl, 1981) that affects the level of effort in innovation. Therefore, the expected goal, commitment, and effort are related concepts, and the relationships are shown by the consecutive positive links between "OO's Expected Goal (B)", "Management Level Commitment in OO (C)", and "Management Level Effort in OO (D)" in the conceptual framework (Figure 7.2).

The functional organization of a construction firm can be divided into management (business) level and project levels (Gann and Salter, 2000) – the former deals with the strategic interests of the firm, whereas the latter focuses on the projects at hand. As such, the effort at the management level of OO would mainly focus on making its project unit adequately resourceful, organized, and committed to implementing the innovation. This means the management-level effort would attempt to generate normative commitment in its project unit toward the expected goal behind the innovation. The normative commitment would then be the motivational force for the project-level unit in putting effort into the innovation. This process of the transfer of effort from management level to project level of OO is shown by the positive links between "Management Level Effort in OO (D)", "Project Level Commitment in OO (E)" and "Project Level Effort in OO (F)" in the conceptual framework.

One of the main characteristics of innovation is that it is an inherently challenging, risky, and uncertain undertaking (Seaden et al., 2003; Christensen, 2000; Quinn, 1985). As previously described, the "Perceived Constraints (A)" at the initiation stage of innovation are the cognitive or attitudinal constraints that reflect the initial sense of low self-efficacy of an organization and thus affect the "Expected Goal (B)" of innovation. Besides the cognitive constraints at the initiation stage, there would be other challenges at the pre-implementation and implementation stage of innovation. In the conceptual framework, the challenges faced by OO at the pre-implementation stage are represented by "Management Level Challenges in OO (Q)". These "Management Level Challenges in OO (Q)" would adversely affect the "Management Level Commitment in OO (C)", and therefore these two constructs are negatively related in the conceptual framework. Similarly, the constraints faced by OO at the implementation stage are represented by "Project Level Challenges in OO (R)", which would adversely affect the "Project Level Commitment in OO (E)". These two constructs at the project level are also negatively related in the conceptual framework.

Innovation Process in Support Organization (Box Y)

The innovation process in OO, as explained in the previous section, is a generic process that can be used to explain the behavioral process of any organization in the innovation value chain. The main features of the process, such as (a) expected goal-setting processes, (b) the management-level goal commitment, (c) management-level effort, (d) normative commitment in the project unit, (e) project-level effort, (f) operational constraints/challenges at management and project levels, and (g) the attitudinal constraints at the initiation stage, are generic in essence. It is so because the underlying concepts are based on some classical organizational theories which are tailor-made to explain the behavioral innovation process of a construction firm in the innovation value chain. Therefore, the innovation processes of other SOs (Box Y) are similarly portrayed as that of OO (Box X) in the conceptual framework. However, even under similar processes, the judgment and decisions of each of the organizations are bound to be different, and this difference represents the challenging structure of the implementation of innovation in construction.

Inter-organizational Interaction Process (Box Z)

The innovation processes as described in the previous sections are applicable to individual OOs and SOs. However, in order to explain the overall innovation value chain, the innovation process of an individual organization has to be extended to the inter-organizational level.

As shown in Figure 7.1, the innovation value chain can be disaggregated into different dyadic relationships between the OO and related SOs, and also between an SO and other related SOs. Each of the dyadic relationships is presumed to be contractually or functionally bound in the value chain, and therefore, the dyadic contractual or functional relationships would actually generate the inter-organizational interactions between two parties in the innovation value chain.

Inter-organizational Interaction at Pre-implementation Stage

When the OO proposes innovation in a construction project, part of its effort is directed toward identifying the relevant SOs and negotiating with them to gain their commitment toward the innovation. This pre-implementation effort of the OO is shown by the positive relational link between the constructs "Management Level Effort in OO (D)" and "Management Level Commitment in SOs (K)" in the conceptual framework (Figure 7.2). The strength of this relational link depends on the strength of the OO's management effort and its receptivity (extent of commitment toward the proposed innovation) from the SOs' management side. This relational link represents the inter-organizational interaction between OO and related SOs at the pre-implementation stage, and it can be interpreted with the concepts of inter-organizational theories like (a) transaction cost theory, (b) dependency theory, and (c) principal–agent theory.

(a) **Transaction cost theory**: Here, transaction cost theory as such is not directly applicable for explaining the mode of transactions between the OO and SOs. Winch (1998) stated that the general mode of transaction in construction is the market-based contractual transaction. Therefore, the inter-organizational relationships between the OO and SOs, which are contractually related, can be explained within the contractual transaction mode by using some relevant concepts of transaction cost theory. The relevant concepts would be the main characteristics of a transaction, such as uncertainty, asset specificity, frequency, and behavior in the transaction, like opportunism and bounded rationality.
The OO would be committed and put in more effort in initiating innovation if the perceived uncertainties (attitudinal constraints (A) and pre-implementation operational challenges (Q)) pertaining to the expected goal of the innovation are less. The high asset specificity against the value of innovation would discourage the OO from putting in effort to initiate innovation. The perceived uncertainties and the high asset specificity would limit the expected goal (B) (with a low level of self-efficacy), the management commitment (C), and then the management effort (D). In order to avoid the uncertainties and high asset specificity, and to attain its own goals out of innovation, the OO's effort would also be guided by its opportunistic behavior in using or gaining relevant SOs' support. The rationality of the OO in the judgment of uncertainties, asset specificity, and opportunistic space would be based on the limited knowledge and information possessed by the OO. Therefore, bounded rationality would also be a key factor

in determining the commitment and thus the strength of the OO's effort toward the innovation.

These same effects of the characteristics of a transaction would also be applicable to all the SOs in supporting the innovation proposed by the OO. The SOs would also prefer fewer uncertainties and less asset specificity. They would be opportunistic in supporting the OO, and their decisions would also be based on their bounded rationality. Therefore, the commitment and effort of the SOs would depend on their conditioned judgment of the OO's proposal. If the OO's effort in initiating the innovation is trustworthy in attaining the SOs' expected goal, they would work with commitment toward the innovation.

(b) **Dependency theory**: Using the concepts of dependency theory, it can be stated that the relative resource dependency and consequent position and power of the OO and SOs in the innovation value chain would determine the strength of the OO's effort in initiating innovation and the corresponding receptivity from the SOs. Therefore, based on the dependency and power structure in the innovation value chain, the OO may persuade, cooperate with, or control the relevant SOs to initiate the innovation. The SOs, on the other hand, may also cooperate with or bargain with the OO to work on the proposed innovation. Therefore, this suggests that the interaction between the OO and SOs could be cooperative or conflictive at the pre-implementation stage of innovation.

(c) **Principal–agent theory**: Principal–agent theory deals with the design of the contract between the principal and the agent, the distribution of contractual risks between the two parties, and the probable behavior of the two parties in a contractual relationship.

The OO and SOs in the innovation value chain may have the role of either principal or agent. Likewise, the contract could be designed for the innovation either separately or as part of another project package (for which the innovation is a subset) – it may depend on the relative value of innovation in the project. However, regardless of the contract type, the principal would choose and design the contract type according to its interests and perceived risks and uncertainties. By nature, the implementation of innovation is uncertain and risky. Therefore, the principal tends to minimize the cost of the information system for monitoring the agent and/or to shift the risks to the agent at the minimum possible costs. The agent, on the other hand, perceives the risks from its own point of view and it would seek good compensation if it has to bear the risk. Therefore, the agreed mutual risk and incentive distribution process forms a contractual relationship between the two parties for implementing innovation. The extent of the mutual agreement can be represented by the strength of effort of the OO in initiating the innovation and the receptivity of SOs in supporting the OO's proposal.

The three theories described above are different organization theories with different settings of arguments. The dependency theory is a socio-political theory, whereas the transaction cost and principal–agent theories are economic theories (Eisenhardt,

1989). In the dependency theory, the goal conflicts are resolved by the power mechanism of political science, such as bargaining, negotiation, and coalition. On the other hand, in the transaction cost and principal–agent theories, they are resolved by the price mechanism and its influencing factors, such as incentives for alignment of different interests. Despite the differences, the three theories have some common issues, such as self-interest, goal conflict, bounded rationality, and consequent behavior of the parties engaged in an effort (Pfeffer, 1981; Barney and Ouchi, 1989). The relational link between the constructs, "Management Level Effort in OO (D)" and "Management Level Commitment in SOs (K)", reflects the common concepts of the three theories.

Both of the constructs in this relational link are determined and affected by the behavioral processes of innovation in the OO and SOs (as described in the previous sections referring to Box X and Box Y in Figure 7.2). The behavioral processes were explained by the expected goals behind the innovation (B and J), the attitudinal constraints (perception of self-efficacy) (A and I) that determine the level of expected goals, and the pre-implementation constraints (Q and O). Therefore, these connected links explain how different parties would be motivated and how their motivation would shape the inter-organization interactions between the relevant parties at the pre-implementation stage of innovation in a construction project.

Inter-organizational Interaction at the Implementation Stage

As explained in the innovation processes in OO (Box X) and SOs (Box Y), the management of the respective organizations would put in effort (D and L) to make their project unit committed (F and Q) toward the innovation they are initiating. The commitment and the effort at the project (or implementation) level would therefore be guided by the normative pressure from the respective organization's management. At the same time, the management-level decisions would also be affected by inter-organizational interaction. This suggests that the effort at the project level would be directed by individual organizational motivation as well as by collective (common, agreed, or negotiated) project objectives (Bresnen, 1990). The issue of the interpretation of collective project objectives and the extent of internalization of the project objectives into individual organizational expected goals would depend on the individual organization's value judgment of the innovation, self-interest, and expected goal (Afuah and Bahram, 1995). In collective multi-organizational efforts, each party has every right for each party to pursue its own interest (Buckley and Casson, 1998). In such an implementation setting, the consistency of collective effort would depend on different vectors of project-level efforts put in by the OO and SOs in the innovation value chain. This process is represented in the conceptual framework (Figure 7.2) by the positive relational links from "Project Level Effort in OO (F)" and "Project Level Effort in SOs (N)" to "Effort Consistency (H)".

The level of effort consistency represents the quality of collaborative efforts of OO and SOs at the project level to achieve the unity of effort among them. This concept resembles the concept of "integration" as stated by Lawrence and Lorsch (1967, p.

47), Mouzelis (1967), and Lansley et al. (1974). Therefore, the level of "effort consistency" represents the level of "integration" in the innovation value chain.

Child (1984) stated that there is a high and positive correlation between integration and project performance. Focusing on innovation, Nam and Tatum (1992a) stated that a high degree of integration in the project organization has a close link to successful innovation. Therefore, the "Effort Consistency (H)" is positively linked to "Favorable Result (G)" in the conceptual framework.

As the expected goal of innovation could be different for different organizations in the innovation value chain, the perception of achieved favorable results would also be different for them. Nevertheless, the more the result is perceived to be favorable, the more the organization would be encouraged and committed to the implementation of the innovation. This process is shown in the conceptual framework by the relational links that positively relate "Favorable Result (G)", and "Project Level Commitment in OO (E)" and "Project Level Commitment in SOs (M)". This process is the reinforcement process of commitment at the project level, and it is consistent with the reinforcement (instrumental) theory of motivation as explained by Repenning (2002).

CONCLUSIONS

The framework presented in this chapter provides a conceptual basis to explore the processes of:

(a) how an organization would set the goal for innovation – how the attitudinal constraints (lack of self-efficacy and expectancy) would affect the goal-setting process of the organization;
(b) how the expected goal would generate the management-level commitment and consequent efforts in the organization;
(c) how the management-level effort would get translated into project-level commitment and efforts;
(d) how the management-level effort of the origin organization (OO) affects the management-level commitment of support organizations (SOs);
(e) how the operational constraints would affect the management- and project-level commitments;
(f) how the efforts of the OO and SOs would be combined to give rise to the concept of integration or effort consistency;
(g) how effort consistency would affect the result of innovation; and
(h) how the result of innovation would be instrumental in affecting the project-level commitment.

The main concept behind the framework is based on the cognitive and behavioral processes of the relevant organizations while implementing an innovation within a construction project. It is an aggregated generic framework that can be applied to capture a possible spectrum of different construction innovation cases. The concepts

Table 7.1 Research propositions

Related variables in the conceptual framework (Figure 7.2)	**Corresponding research propositions**
Variables A and B Variables I and J	*Proposition 1 (P-1-OO and P-1-SO)*: Perceived constraints in initiating innovation would influence negatively the level of the organization's expected goal of the innovation.
Variables B and C Variables J and K	*Proposition 2 (P-2-OO and P-2-SO)*: The level of expected goal of the innovation would influence positively the organization's management-level commitment to the innovation.
Variables C and D Variables K and L	*Proposition 3 (P-3-OO and P-3-SO)*: The organization's management-level commitment to the innovation would increase its management-level effort for the innovation.
Variables Q and C Variables O and K	*Proposition 4 (P-4-OO and P-4-SO)*: Challenges for the innovation at the management level would decrease the organization's management-level commitment to the innovation.
Variables D and K	*Proposition 5 (P-5)*: OO's management-level effort for the innovation would influence positively the SOs' management-level commitment to the innovation.
Variables D and E Variables L and M	*Proposition 6 (P-6-OO and P-6-SO)*: The organization's management-level effort for the innovation would positively influence its project-level commitment to the innovation.
Variables E and F Variables M and N	*Proposition 7 (P-7-OO and P-7-SO)*: The organization's commitment to the innovation at the project level would increase its effort for the innovation at the project level.
Variables R and E Variables P and M	*Proposition 8 (P-8-OO and P-8-SO)*: Challenges for the innovation at the project level would decrease the organization's project-level commitment to the innovation.
Variables F and H Variables N and H	*Proposition 9 (P-9-OO and P-9-SO)*: The organization's project-level effort for the innovation would enhance the overall effort consistency for the innovation.
Variables H and G	*Proposition 10 (P-10)*: The effort consistency for the innovation would positively influence the favorable results out of the innovation.
Variables G and E Variables G and M	*Proposition 11 (P-11-OO and P-11-SO)*: The favorable results out of the innovation would positively influence the organization's project-level commitment to the innovation.

Note: The table presents a set of 11 testable research propositions developed from the conceptual framework presented in Figure 7.2. The propositions have been actually tested with case study data in the subsequent research presented in Chapter 8.

of motivation alignment, integration (effort consistency), and their influence on the success of innovation (favorable result) are incorporated in the framework. Therefore, in essence, the framework represents a sequential model of the implementation of innovation (from initiation to the realization of results) on construction projects. In the next chapter, the framework is further tested as an overarching model with *ex post* data from an innovation case study by segregating it into testable research propositions for OO and SOs as shown in Table 7.1.

REFERENCES

Adams, J. S. (1965), Inequity in Social Exchange, in *Advances in Experimental Social Psychology*, Vol. 2, Berkowitz, L. (ed.), pp. 267–299, Academic Press, New York.

Afuah, A. N. and Bahram, N. (1995), The Hypercube of Innovation, *Research Policy*, 4, 51–76.

Bahr, M. and Laszig, L. (2021), Productivity Development in Construction Industry by International Comparison and the Example of Human Capital, *Civil Engineering and Urban Planning: An International Journal*, 8 (2), 1–20.

Ball, M. (1988), *Rebuilding Construction: Economic Change in the British Construction Industry*, Routledge, London.

Bandura, A. (1997), *Self Efficacy: The Exercise of Control*, W. H. Freeman, New York.

Bandura, A. (1977), Self Efficacy: Towards a Unifying Theory of Behavioral Change, *Psychological Review*, 84, 191–215.

Barlow, J. (2000), Innovation and Learning in Complex Offshore Construction Projects, *Research Policy*, 29, 973–989.

Barney, J. and Ouchi, W. (eds.) (1989), *Organizational Economics*, Jossey-Bass, San Francisco.

Bernstein, H. M. and Lemer, A. C. (1996), *Solving the Innovation Puzzle*, ASCE Press, New York.

Bresnen, M. (1990), *Organizing Construction: Project Organization and Matrix Management*, Routledge, London.

Buckley, P. J. and Casson, M. (1998), A Theory of Cooperation in International Business, in *Strategic Alliances*, Beamish, P. W. (ed.), pp. 91–110, Edward Elgar Publishing, Cheltenham, UK and Northampton, MA, USA.

Bygballe, L. E. and Ingemansson, M. (2014), The Logic of Innovation in Construction, *Industrial Marketing Management*, 43 (3), 512–524.

Child, J. (1984), *Organizations*, Harper & Row, London.

Christensen, C. M. (2000), *The Innovator's Dilemma*, Harper Business, New York.

Cox, A. and Thompson, I. (1997), "Fit for Purpose" Contractual Relations: Determining a Theoretical Framework for Construction Projects, *European Journal of Purchasing and Supply Management*, 3, 127–135.

Dubois, A. R. J. and Gadde, L. E. (2002), The Construction Industry as a Loosely Coupled System: Implications for Productivity and Innovation, *Construction Management and Economics*, 20 (7), 621–631.

Dulaimi, M. F., Ling, F. Y. Y., and Bajracharya, A. (2003), Organizational Motivation and Inter-Organizational Interaction in Construction Innovation, *Construction Management and Economics*, 21 (3), 307–318.

Dulaimi, M. F., Baxendale, A. T., and Langford, D. A. (1996), The Organization and Management of Construction: Shaping Theory and Practice, *The Construction Process as a Process of Innovation, Proceedings of the CIB W65,* Glasgow, UK, 572–577.

Eisenhardt, K. M. (1989), Agency Theory: An Assessment and Review, *Academy of Management Review*, 14 (1), 57–74.

Gann, D. M. (2000), *Building Innovation, Complex Constructs in a Changing World*, Thomas Telford, London.

Gann, D. M. and Salter, A. J. (2000), Innovation in Project Based, Service Enhanced Firms: The Construction of Complex Products and Systems, *Research Policy*, 29, 955–972.

Harty, C. (2005), Innovation in Construction: A Sociology of Technology Approach, *Building Research and Information*, 33 (6), 512–522.

Heaney, G., Murphy, M. E., and Perera, S. (2015), Innovation Management Model: A Tool for Sustained Implementation of Product Innovation into Construction Projects, *Construction Management and Economics*, 33 (3), 209–232.

Henderson, R. and Clark, K. (1990), Architectural Innovation: The Reconfiguration of Existing Product Technologies and the Failure of Established Firms, *Administrative Science Quarterly*, 35, 9–30.

Kanter, R. M. (1988), When a Thousand Flowers Bloom: Structural, Collective, and Social Conditions for Innovation in Organization, *Research in Organizational Behavior*, 10, 169–211.

Katzell, R. A. and Thompson, D. E. (1990), Work Motivation: Theory and Practice, *American Psychologist*, 45 (2), 144–153.

Knoke, D. (2000), *Changing Organizations*, Westview Press, Boulder, CO.

Kumaraswamy, M. and Dulaimi, M. F. (2001), Empowering Innovative Improvements through Creative Construction Procurement, *Engineering, Construction and Architectural Management*, 8 (5/6), 325–334.

Lansley, P. R., Sadler, P. and Webb, T. (1974), Organizational Structure, Management Style and Company Performance, *OMEGA, The International Journal of Management Science*, 2 (4), 467–485.

Lawrence, P. R. and Lorsch, J. W. (1967), *Organization and Environment*, Harvard University Press, Cambridge, MA.

Lenard, D. (2001), Promoting the Development of an Innovative Culture through the Strategic Adoption of Advanced Manufacturing Technology in Construction, *Proceedings of CIB World Building Congress: Performance in Product and Practice*, Vol. 1, Wellington, New Zealand.

Locke, E. A. and Latham, G. P. (2002), Building a Practically Useful Theory of Goal Setting and Task Motivation: A 35-Year Odyssey, *American Psychologist*, 57 (9), 705–717.

Locke, E. A. and Latham, G. P. (1990), *A Theory of Goal Setting and Task Performance*, Prentice Hall, Englewood Cliffs, NJ.

Locke, E. A., Latham, G. P., and Erez, M. (1988), The Determinants of Goal Commitment, *Academy of Management Review*, 13 (1), 23–39.

Mouzelis, N. P. (1967), *Organisation and Bureaucracy: An Analysis of Modern Theories*, Routledge and Kegan Paul, London.

Nam, C. B. and Tatum, C. B. (1992a), Non-Contractual Methods of Integration of Construction Projects, *Journal of Construction Engineering and Management*, 118 (2), 385–398.

Nam, C. B. and Tatum, C. B. (1992b), Strategies for Technology Push: Lessons from Construction Innovations, *Journal of Construction Engineering and Management*, 118 (3), 507–524.

Nam, C. H., Gasiorowski, J. G., and Tatum, C. B. (1991), Microlevel Study of Integration in High Strength Concrete Innovation, *Journal of Construction Engineering and Management*, 117 (2), 294–309.

Nelson, R. R. and Winter, S. G. (1977), In Search of Useful Theory of Innovation, *Research Policy*, 6, 36–76.

Newton, P. W. (1999), *Modeling Innovation in AEC: Understanding the Fourth Dimension of Competition*. http://www.ce.berkeley.edu/~tommelein/CEMworkshop/Newton.pdf

Nightingale, P. (1998), A Cognitive Model of Innovation, *Research Policy*, 7, 689–709.

Ninan, J., Sergeeva, N., and Winch, G. (2022), Narrative Shapes Innovation: A Study on Multiple Innovations in the UK Construction Industry, *Construction Management and Economics*, 40 (11–12), 884–902.
Pfeffer, J. (1981), *Power in Organizations*, Pitman, Marshfield, MA.
Powell, J. (1995), *Towards a New Construction Culture*, Chartered Institute of Building, Ascot.
Quinn, J. B. (1985), Managing Innovation: Controlled Chaos, *Harvard Business Review*, May–June, 73–84.
Repenning, N. P. (2002), A Simulation Based Approach to Understanding the Dynamics of Innovation Implementation, *Organization Science*, 13 (2), 109–127.
Sawhney, A., Riley, M., and Irizarry, J. (2020), Construction 4.0: Introduction and Overview, in *Construction 4.0: An Innovation Platform for Built Environment*, Sawhney, A., Riley, M., and Irizarry, J. (eds), pp. 1–20, Routledge, Abingdon.
Scholl, R. W. (1981), Differentiating Organizational Commitment from Expectancy as a Motivating Force, *Academy of Management Review*, 6 (4), 589–599.
Seaden, G. (1996), Economics of Innovation in the Construction Industry, *Journal of Infrastructure Systems*, September, 103–107.
Seaden, G., Guolla, M., Doutriaux, J., and Nash, J. (2003), Strategic Decisions and Innovation in Construction Firms, *Construction Management and Economics*, 21 (6), 603–612.
Slaughter, S. (1998), Models of Construction Innovation, *Journal of Construction Engineering and Management*, 124 (3), 226–231.
Slaughter, S. E. and Shimizu, H. (2000), "Clusters" of Innovations in Recent Long Span and Multi-Segmental Bridges, *Construction Management and Economics*, 18, 269–280.
Sterman, J. D. (1992), System Dynamics Modeling for Project Management, *A Working Paper*, Sloan School of Management, MIT, Cambridge, MA. http://web.mit.edu/jsterman/www/SDG/project.pdf
Tatum, C. B. (1989), Organizing to Increase Innovation in Construction Firms, *Journal of Construction Engineering and Management*, 115 (4), 602–617.
Tatum, C. B. (1987), Process of Innovation in Construction Firm, *Journal of Construction Engineering and Management*, 113 (4), 648–663.
Tatum, C. B. (1986), Potential Mechanisms for Construction Innovation, *Journal of Construction Engineering and Management*, 112 (2), 178–191.
Tatum, C. B. (1984), What Prompts Construction Innovation?, *Journal of Construction Engineering and Management*, 110 (3), 311–323.
Vroom, V. H. (1964), *Work and Motivation*, John Wiley & Sons, New York.
Wang, N., Xu, Z., and Liu, Z. (2023), Innovation in the Construction Sector: Bibliometric Analysis and Research Agenda, *Journal of Engineering and Technology Management*, 68, 101747.
Warren, M. (1993), *Economics for the Built Environment*, Butterworth-Heinemann, Oxford.
Wiener, Y. (1982), Commitment in Organizations: A Normative View, *Academy of Management Review*, 7(3), 418–428.
Winch, G. (2003), How Innovative is Construction? Comparing Aggregated Data on Construction Innovation and Other Sectors – A Case of Apples and Pears, *Construction Management and Economics*, 21 (6), 651–654.
Winch, G. (1998), Zephyrs of Creative Destruction: Understanding the Management of Innovation in Construction, *Building Research and Information*, 26 (4), 268–279.
Xue, X., Zhang, R., Yang, R. J., and Dai, J. (2014), Innovation in Construction: A Critical Review and Future Research, *International Journal of Innovation Science*, 6 (2), 111–125.
Yamazaki, Y. and Ueda, Y. (2003), Technology and Knowledge Fusions towards Construction Innovation, in *Proceedings, Knowledge Construction*, Ofori, G. and Ling, F. Y. Y. (eds), Vol. 1, pp. 40–64, Singapore.

8. Implementation of innovation on construction project: a case study

Arun Bajracharya and Mohammed Fadhil Dulaimi

INTRODUCTION

This chapter is a continuation of Chapter 7, "A Conceptual Framework to Understand the Implementation of Innovation on Construction Projects". In this chapter, the conceptual framework is tested by using the data from a construction innovation case on the first group housing project that was implemented in a developing country.

First, the general information on the innovation is presented in terms of the background, project team, technical features, and the implementation process. Then, the details of the innovation case are specifically analyzed with the constructs and propositions embedded in the conceptual framework. All the case observations and findings are then summarized and presented as the conclusions of this chapter.

Regarding the methodology, a proper case-study-based methodological approach was used for data collection and analysis processes. The primary data were collected from the key leaders and professionals involved in the project, whereas the secondary data were sourced from the relevant literature and technical reports. For the analysis of the case study data, suitable strategies and techniques were selected as suggested by Yin (2018). The author suggested three general strategies – relying on theoretical propositions, thinking about rival explanations, and developing a case description – and five specific techniques – pattern matching, explanation building, time series analysis, logic models, and cross-case synthesis – for the analysis of case study data.

As Yin (2018) suggested that none of the strategies or techniques could be used as a simple or mechanical procedure, the most suitable strategy combination was used for the data analysis in this research. The strategy of "relying on theoretical propositions" was adopted as the main strategy because the propositions (see Table 7.1 in the Conclusions section of the previous chapter) that were derived from the conceptual framework could be used as the theoretical propositions. The strategy of "thinking about rival explanations" was not used as a predefined strategy, but the concept of rival explanations was used to challenge the theoretical assumption during the analysis. The other strategy of "developing a case description" was not used as a main content of case analysis but was used as a supplementary part for the main strategy of "relying on theoretical propositions".

For analysis, the combined techniques of pattern matching and explanation building were used. The techniques had to be adapted and used according to the availability of the empirical data on the case.

The identity of the country, context, project, and the parties related to the case have been kept confidential in this research.

THE INNOVATION CASE: THE FIRST GROUP HOUSING PROJECT

The Background

General features of the housing sector in urban areas of a country can be explained on the basis of its sociocultural fabric, economy, regulatory regime, and physical development. The society in the country is multi-racial, multi-lingual, and stratified class-wise. It is traditionalist and relatively collectivist, tilted toward high masculinity, high power distance, and short-term orientation, and it prefers uncertainty avoidance. Descended from traditional feudal or semi-feudal values, land and housing are taken as major assets and symbols of economic well-being, prestige, and the power of a family. Almost all families in the country are paternalistic joint families characterized by the eldest male member as the head of the family.

Besides direct purchasing, the ownership of the land and housing in Nepal is inherited from generation to generation in a family. In the tradition of family succession, all the male offspring are entitled to the parental property including land and housing. As the family tree expands and as the joint family gets divided into several other related joint families, the family-owned land and housing also get divided and distributed as per the number of male members in the family.

Probably because of the traditional value system, the people in the country have a deep-rooted preference for building their own independent houses on their own land with their own resources. Procuring a house is therefore a long process with the owner's self-involvement in purchasing land, building the house by own resources in various phases, and then in providing infrastructure and utility services later. Such a housing system is termed the owner-builder system, in which a self-financing mechanism is the predominant form of housing finance. As such, about 92% of households in the country own their housing units, 5% are renters (pay rent), and 2% live in rent-free housing units.

As in other developing countries, there is a high attraction differential – in terms of infrastructure facilities, job opportunities, and securities – between the urban areas, especially the capital city, and other parts of the country. Such differentials have created high immigrant pressure in the capital. About 40% or more of the population in the city are migrants from outside, and this has highly increased the demand for and value of land and housing in the capital city. As such, there is a high deficit in the supply of housing in the city.

On the regulatory front, there were separate land and property laws that guided the regulatory policies pertaining to land and housing. However, the government policy on zoning and the distribution of land was not strong or clear. Furthermore, the state administration of land and housing standards was relatively relaxed, and there were no public sector housing programs or comprehensive/consolidated land development schemes that facilitated access to housing or land to any class of people, rich or poor.

The sociocultural, economic, and regulatory aspects have significantly influenced the physical development pattern of land and housing in the capital city. The "subdivide and inherit" practice within families, the preference for individual separate housing, the owner-builder system, the high demand, value, and consequent shortage of land and housing, and the weak regulatory system are some of the main reasons behind the typical fragmented land distribution and haphazard housing development patterns in the city. The tendency for a haphazard housing pattern is thus remarkably critical in the high-value urban center. Consequently, the government has been unable to provide adequate infrastructure facilities to the haphazardly developed houses. Infrastructure facilities such as access roads, drinking water, electricity, sewerage and wastewater drainage, arrangements for solid waste disposal, and open spaces have increasingly become deficient for dwelling units. At the same time, the quality of housing itself is not satisfactory – only half of the housing in urban areas is of permanent type, whereas the remaining houses are semi-permanent, temporary, traditional, and squatter types.

The Innovation

The fragmented, haphazard, and low-quality land and housing development pattern ingrained in the sociocultural, economic, and regulatory regimes of the country's urban areas has been a glaring example of the tragedy of the commons. Therefore, there was an urgent need to reform the basic concept of housing. However, changing the age-old housing concept in the complex setting of urban areas is a highly challenging task, and, in order to address it, a private business in the country came forward with a new idea of an organized group housing project in the capital city. The group housing project was a very important innovation in the whole concept of housing in the country. This significant "concept innovation in housing" has been chosen for a detailed case study in this research.

The Group Housing Project

The organized group housing project, which is the first of its kind in the country, was named "KR". It is basically an apartment-based housing project located in a suburb of the capital city. The project occupies about 1.76 acres (7,125 square meters) of land on which a total of 145 apartment units were built. It is an integrated and self-contained residential complex consisting of a range of apartment types, from studio apartments to one, two, and three-bedroom apartments, with all the required amenities such as abundant water supply, sufficient electricity and power backup, an efficient drainage system, private parking, 24-hour security, an in-house club, a crèche, convenient shopping, and enough open space and greenery. The total cost of the project was 300 million in the local currency.

The Project Team

The KR project was initiated by "CG" in the country in a joint venture (JV) with "AG". CG is one of the big businesses in the country, a conglomerate with more than 40 companies under its umbrella. AG is one of the major real estate and construction companies in a neighboring country, with experience in several projects in Thailand, Russia, Vietnam, Bangladesh, and Myanmar, in addition to its home country. AG and CG had 51% and 49% of shares in the JV, respectively. The established JV company was registered as ACDPL in the country.

The architectural and engineering consulting tasks of planning, designing, detailing, and quantity surveying of all the buildings and ancillaries of the apartment system were undertaken by AG itself. All the designs, drawings, and bill of quantities were supplied from AG's headquarters in its country. Supervision of the construction works was jointly done by ACDPL.

The ACDPL appointed a local contractor company as the main contractor of the project. The local contractor company was a joint venture of three local contractors – MBC (an A-class contractor), SB (a B-class contractor), and RMB (a C-class contractor) – led by a director of SB. The JV main contractor was selected by a local tendering process and appointed on the basis of a cost-plus-fee contract.

Implementation of the Project

Regulatory backdrop

Before 1998, the law related to land and housing in the country was traditional and was primarily based on an individual or family ownership system. There was no national housing act that could accommodate the concept of group housing. There was a law for the vertical division of land and houses, but there was no law that described the horizontal division of land and houses. In 1994, a group of three local businesses came up with the idea of group housing in the country. The group was influential in national politics, and they were successful in raising the issue of a group housing act in parliament. However, the process of formulating and enacting the new housing law took a long time, and it receded in the political turmoil of the country. Therefore, the group had to abandon their conceptualized group housing project. They had even procured the land for the project, but they had to sell off the land instead. The absence of the law and regulatory mechanisms meant their innovative move was a failure.

Even though the first attempt failed, it was a good push and eventually helped in formulating the first laws and by-laws pertaining to group housing in the country. In 1997, the country's first set of Building By-Laws and Apartment Act was promulgated.

Formation of the ACDPL

After the law was enacted, CG moved forward with the concept of the group housing project in Nepal. The president of the conglomerate set up a separate task group under his supervision, including the experts in group housing from the neighboring country. The task group started its work with a feasibility study of the project. A local market study and economic analysis of the project were conducted. Group housing practices in other countries were also observed by visiting some of the successful projects in India, Thailand, Malaysia, and Singapore. At the same time, the task group sought an experienced business partner to undertake the project. It contacted several real estate companies in a neighboring country, and eventually, it was successful in forming a partnership with AG. The partnership was formed for both the financial and technical aspects of the project, and it launched the ACDPL JV.

Initial marketing

The initial concept and building designs were prepared at the design cell of AG's office. Then the project was formally launched and publicized in the local media as a new kind of housing system named KR. As part of the publicity, wooden showpiece mock-up apartment blocks were displayed for marketing purposes. The initial marketing effort was successful in attracting customers, and they were willing to pay the booking advance, which was 30% of the total cost of the proposed apartment unit. ACDPL was successful in tapping a substantial amount of cash inflow from the booking advance.

Searching and appointing a main contractor for construction

After formally launching and publicizing the project, detailed designs, drawings, and bill of quantities (BOQ) were prepared by AG. As per the design and BOQ, a tender was called to select a main contractor who would undertake the whole project package. The tender form was distributed to 11 local contractors, including a Chinese one. The tender was not published to the public; rather, it was kept closed and selective. The tender amount was about 90.8 million in the local currency. The tender was eventually awarded to a local contractor joint venture team based on the negotiated lowest bid. It took around six months, from the time the first lot of customers made advance bookings, to appoint the local contractor joint venture team as the main contractor of the project.

Receiving the construction permit from the local authority

After appointing the main contractor, the process of receiving the construction permit from the local municipality was initiated. The ACDPL involved the main contractor and its project personnel in dealing with the municipality. As it was a new kind of project, the municipality took about a year to study and grant the construction permission.

Construction of the project

The construction of the project started after the main contractor was assigned the job. As the municipality took a long time to grant the construction permit, the construction actually began before the permission was granted. The main contractor completed Blocks 1 to 8 within 18 months, Blocks 9 to 24 within 26 months, and Blocks

29 to 32 within 29 months. The original schedule for the construction of the project was 24 months; however, it was delayed by about 20 months. The remaining Blocks 25 to 28 and 33 to 36 were departmentally constructed by ACDPL specifically for the employees at CG.

Project handover

Once the apartment units were completed with all the finishing work, they were handed over to the customers who booked them. The occupation process of the booked and completed apartment units and the construction of the remaining blocks were conducted side by side. All the apartment units were fully occupied six months after the completion of the whole apartment system.

Observation of the Innovation Case in Terms of the Conceptual Framework

The main parties involved in the innovation of the group housing project were CG, AG, and the main contractor with the JV of MBC, SB, and RMB. Although CG and AG worked together in the ACDPL JV, their individual roles were quite significant in the project. Therefore, their roles are separately analyzed in this research. Similarly, the roles of the three local contractors – who formed the JV main contractor – are also discussed separately.

CG, which initiated the project, was the origin organization (OO) of the innovation. The other parties – AG, MBC, SB, and RMB – were the main supporting organizations (SOs). The innovation value chain in the KR project is shown in Figure 8.1. There were two key parties in the value chain – ACDPL (the client JV) and the main contractor JV. These two key parties were contractually related in the project. In the contractual relationship, ACDPL was the principal, and the main contractor JV was the agent. There was only one layer of major principal–agent relationships in the value chain. ACDPL sought the function of construction from the market and appointed the main contractor. Another important function of architectural and engineering consultation, however, was internally arranged within the AG.

The client JV was formed with the alliance of AG and CG. They worked together in the JV under a set of terms and conditions of the alliance. On the other hand, the main contractor JV was formed with the alliance of SB, MBC, and RMB. These three local contractors created their JV with a formal set of terms and conditions and agreed to work together through their informal agreement.

A detailed observation of the case in terms of the constructs of the conceptual framework is now presented in Box 8.1. In this case study, the stage with all the activities undertaken prior to the actual construction process is taken as the initiation or pre-implementation stage, whereas the actual construction stage is taken as the implementation stage.

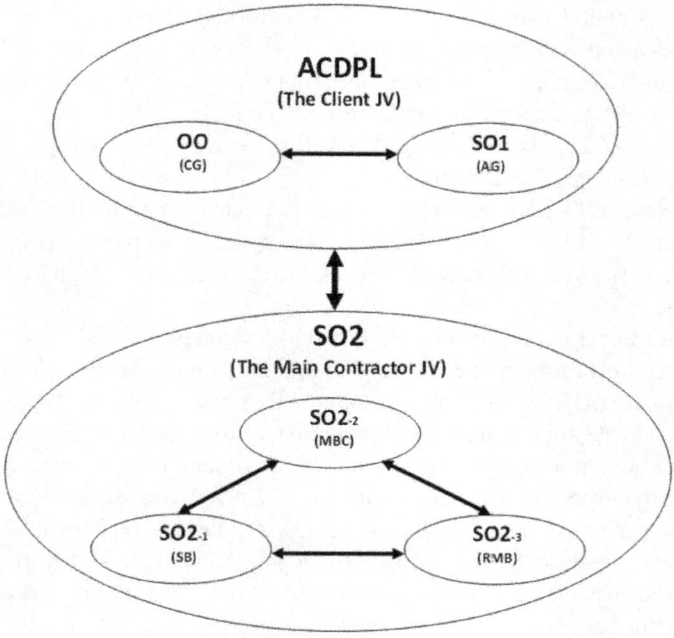

Figure 8.1 *The innovation value chain in the KR project*

BOX 8.1 CASE OBSERVATION OF THE KR GROUP HOUSING PROJECT IN TERMS OF THE CONSTRUCTS OF THE CONCEPTUAL FRAMEWORK

Key Party: CG (OO)

Perceived Constraints (CNST): (1) Constraints pertaining to the sociocultural, economic, regulatory, and physical aspects of urban housing in the context of the country; (2) Lack of core capability in the housing business; (3) Unstable political situation in the country.

Expected Goals (EG): (1) Completing a pioneering group housing project within the stipulated time and cost; (2) Establishing a new corporate portfolio in the housing and real estate business; (3) Securing profit out of the project; (4) Introducing the new concept of organized group housing in the local society.

Management Level Challenges (MLCh): (1) Forming and preparing the project organization; (2) Receiving the construction permission from the local municipality; (3) Dealing with all the project-level challenges that needed management-level attention.

Management Level Commitment (MLC): (1) High and persistent throughout the project.

Management Level Effort (MLE): (1) Resource investment in internal organization and in the project, including the land; (2) Establishment of ACDPL JV with AG; (3) Coordination and monitoring of all the corporate functions for the project; (4) Searching for and deploying the main contractor; (5) Tackling all the management-level challenges as well as the project-level challenges that needed management-level attention.

Project Level Challenges (PLCh): (1) Political turmoil in the country; (2) Errors in quantity estimate; (3) Price escalation and adjustment in the contract amount; (4) The main contractor's grievances; (5) Turmoil in the ACDPL's project team.

Project Level Commitment (PLC): (1) Intermittent.

Project Level Effort (PLE): (1) Formation of implementation team with AG under the ACDPL JV; (2) Coordination and monitoring of all the project activities; (3) Dealing with the local municipality for receiving the permits, and with other government utility offices for the apartment utility services.

Effort Consistency (EC): (1) With AG, the effort consistency was strong. However, during the construction of the project the two groups had several project-level crises to deal with; (2) With the main contractor JV, effort consistency was satisfactory for the first six months of the construction period. Later as the contractor could not perform well, CG had to react accordingly through the ACDPL and eventually the contract modality had to be changed. The unity of effort between the two parties was severed thereafter.

Favorable Result (FR): (1) The project was successfully completed; (2) At the corporate level, CG was highly successful; (3) The group could also harness a substantial profit from the project; (4) However, there was around a 20-month delay in the construction of the project; (5) The interaction with the main contractor team in construction was highly turbulent; (6) Despite the disturbances in construction, the group was also successful on the social front. The project proved to be a seminal project which triggered a booming housing sector in the country.

Key Party: AG (SO)

Perceived Constraints (CNST): (1) Local political, regulatory, and social risks in a foreign country; (2) Constraints pertaining to the sociocultural, economic, regulatory, and physical aspects of urban housing in the local context.

Expected Goals (EG): (1) Expanding its new business in a foreign country's housing and real estate market; (2) Securing profit out of the project.

Management Level Challenges (MLCh): (1) Forming and preparing the project organization; (2) Receiving the construction permission from the local municipality; (3) Dealing with all the project-level challenges that needed management-level attention.

Management Level Commitment (MLC): (1) Persistent throughout the project.

Management Level Effort (MLE): (1) Resource investment in internal organization and in the project; (2) Undertook the architectural and engineering consulting tasks; (3) Establishment of ACDPL JV with CG; (4) Coordination and monitoring of all the corporate functions for the project; (5) Searching for and deploying the main contractor; (6) Tackling all the management-level challenges as well as the project-level challenges that needed management-level attention.

Project Level Challenges (PLCh): (1) Political turmoil in the country; (2) Errors in quantity estimate; (3) Price escalation and adjustment in the contract amount; (4) The main contractor's grievances; (5) Turmoil in the ACDPL's project team.

Project Level Commitment (PLC): (1) Intermittent.

Project Level Effort (PLE): (1) Formation of implementation team with CG under the ACDPL JV; (2) Coordination and monitoring of all the project activities; (3) Dealing with the local municipality for receiving the permits, and with other government utility offices for the apartment utility services.

Effort Consistency (EC): (1) With CG, the effort consistency was strong. However, during the construction of the project the two groups had several project-level crises to deal with; (2) With the main contractor JV, effort consistency was satisfactory for the first six months of the construction period. Later as the contractor could not perform well, CG had to react accordingly through the ACDPL and eventually the contract modality had to be changed. The unity of effort between the two parties was severed thereafter.

Favorable Result (FR): (1) The project was successfully completed; (2) At the corporate level, AG was highly successful; (3) The group could also harness a substantial profit from the project; (4) However, there was around a 20-month delay in the construction of the project; (5) The interaction with the main contractor team in construction was highly turbulent.

Key Party: SB (SO2-1)

Perceived Constraints (CNST): (1) Lack of technical and managerial capabilities in carrying out the relatively large-sized local project individually.

Expected Goals (EG): (1) Securing a good business relationship with the client; (2) Improving its track record and securing good market reputation in the construction industry; (3) Securing reasonable profit out of the project.

Management Level Challenges (MLCh): (1) Dealing with the client; (2) Dealing with MBC and RMB in the JV; (3) Dealing with all the project-level challenges that needed management-level attention.

Management Level Commitment (MLC): (1) Persistent throughout despite the client changing the amount and modality of the contract.

Management Level Effort (MLE): (1) Establishment of the JV with MBC and RMB and taking leadership in the JV; (2) Financial investment for the project; (3) Coordination and monitoring of organizational functions for the project; (4) Tackling all the management-level challenges as well as the project-level challenges that needed management-level attention.

Project Level Challenges (PLCh): (1) The critical cash-flow problem; (2) Political turmoil in the country and consequent delay in procuring the labor and materials; (3) Errors in the quantity estimate; (4) Delay in receiving payment on running bills; (5) Price escalation; (6) Change in contract amount and contract modality.

Project Level Commitment (PLC): (1) Persistent throughout.

Project Level Effort (PLE): (1) Formation of implementation team with MBC and RMC under the JV; (2) Coordination and monitoring the actual implementation process of the project; (3) Tackling all the project-level challenges.

Effort Consistency (EC): (1) With the client, effort consistency remained smooth for the first six months of the construction period. Later, as the series of problems cropped up, the client showed uncooperative behavior, and they eventually changed the contract modality. The unity of effort between the two parties was severed thereafter. However, SB had to continue to work on the project, managing the labor and material supply, because the team leader from SB had to recover his investment in the project; (2) The unity of effort with the JV partners was severed after six months.

Favorable Result (FR): (1) SB could not satisfactorily achieve its expected goals. The project was not a successful undertaking for SB; (2) It could not complete the project as it was assigned in the initial contract; (3) Even under the "with material" contract sub-package, there was about a five-month delay; (4) SB could not establish a good business relationship with the client; (5) SB had to bear a considerable financial loss in the project; (6) However, SB was successful in improving its track record as it was involved in the project.

Key Party: MBC (SO2-2)

Perceived Constraints (CNST): (1) Lack of technical and managerial capabilities in carrying out the relatively large-sized local project individually.

Expected Goals (EG): (1) Securing a good business relationship with the client; (2) Improving its track record and securing good market reputation in the construction industry; (3) Securing reasonable profit out of the project.

Management Level Challenges (MLCh): (1) Dealing with the client; (2) Dealing with SB and RMB in the JV; (3) Dealing with all the project-level challenges that needed management-level attention.

Management Level Commitment (MLC): (1) Weak and dwindling; (2) MBC withdrew from the project after six months as the team leader of SB could not provide the return on investment to the partners in the JV.

Management Level Effort (MLE): (1) Establishment of the JV with SB and RMB and taking leadership in the JV; (2) Labor and equipment investment for the project; (3) Coordination and monitoring of organizational functions for the project; (4) Tackling the management-level challenges as well as the project-level challenges that needed management-level attention.

Project Level Challenges (PLCh): (1) Delay in receiving payment from SB.

Project Level Commitment (PLC): (1) Weak and dwindling; (2) MBC withdrew from the project after six months as the team leader of SB could not provide the return on investment to the partners in the JV.

Project Level Effort (PLE): (1) Formation of implementation team with SB and RMB under the JV; (2) Coordination and monitoring the actual implementation process of the project.

Effort Consistency (EC): (1) The unity of effort with the JV partners was severed after six months.

Favorable Result (FR): (1) MBC could not satisfactorily achieve its expected goals. The project was not a successful undertaking for MBC; (2) It could not continue its role in the project as it was initially agreed in the JV; (3) MBC could not establish a good business relationship with the client; (4) MBC had to bear a financial loss in the project; (5) However, MBC can claim that it participated in the project, which helped improve its track record.

Key Party: RMB (SO2-3)

Perceived Constraints (CNST): (1) Lack of technical and managerial capabilities in carrying out the relatively large-sized local project individually.

Expected Goals (EG): (1) Securing a good business relationship with the client; (2) Improving its track record and securing good market reputation in the construction industry; (3) Securing reasonable profit out of the project.

Management Level Challenges (MLCh): (1) Dealing with the client; (2) Dealing with SB and MBC in the JV; (3) Dealing with all the project-level challenges that needed management-level attention.

Management Level Commitment (MLC): (1) Weak and dwindling; (2) RMB withdrew from the project after six months as the team leader of SB could not provide the return on investment to the partners in the JV.

Management Level Effort (MLE): (1) Establishment of the JV with SB and MBC and taking leadership in the JV; (2) Technical workforce and labor investment for the project; (3) Coordination and monitoring of organizational functions for the project; (4) Tackling the management-level challenges as well as the project-level challenges that needed management-level attention.

Project Level Challenges (PLCh): (1) Delay in receiving payment from SB.

Project Level Commitment (PLC): (1) Weak and dwindling; (2) RMB withdrew from the project after six months as the team leader of SB could not provide the return on investment to the partners in the JV.

Project Level Effort (PLE): (1) Formation of implementation team with SB and MBC under the JV; (2) Coordination and monitoring the actual implementation process of the project.

Effort Consistency (EC): (1) The unity of effort with the JV partners was severed after six months.

Favorable Result (FR): (1) RMB could not satisfactorily achieve its expected goals. The project was not a successful undertaking for RMB; (2) It could not continue its role in the project as it was initially agreed in the JV; (3) RMB could not establish a good business relationship with the client; (4) RMB had to bear financial loss in the project; (5) However, RMB can claim that it participated in the project. It helped improve its track record.

DISCUSSION

The case study data are now further discussed for testing the propositions in detail.

Proposition 1: Perceived constraints in initiating innovation would influence negatively the level of the organization's expected goal of the innovation

The observations in the case study suggest that in order to inquire into this proposition, some aspects pertaining to the innovation, the involved parties, their expected goals and perceived constraints have to be considered.

It was observed that the relevant influence domain of innovation affects the domain of the expected goals and perceived constraints of the parties in the innovation value chain. The influence domain of innovation depends on the type of innovation, its scale, and impact. Group housing is a concept innovation that best resembles radical innovation as categorized by Slaughter (1998) and Coccia (2016).

The relevant influence domains of innovation for the clients were not only the construction project and their respective organizations, but the domains also covered the broader societal environment. Therefore, the clients also needed to deal with broad social, political, and economic constraints. As such, they extended their expected goals at the social level as well. In the case of all the other parties, however, the relevant influence domains of innovation were the construction project and their respective organizations. Therefore, the domain of their expected goals and perceived constraints remained within their relevant influence domains of innovation.

Irrespective of the domains of perceived constraints and expected goals, it was found that the perceived constraints do affect the expected goals at the initiation stage of innovation. The expected goals of all the parties (the OO and SOs) in the case were actually conditioned by the perceived constraints at the initiation stage of the respective innovation. All the parties participated in the respective innovations with certain acceptable levels of their expected goals with a sense of self-efficacy and expectancy amidst the perceived constraints. If the perceived constraints were more critical to the extent that they would challenge self-efficacy and expectancy, the parties would not have expected to get their goals fulfilled, and they would not have participated in the innovation. For example, if the clients had perceived frustrating societal constraints, they would not have launched their respective projects. Similar conditions apply to all the other parties as well. The exception could be that if the party is powerful or important, and it wants to levy a higher profit margin against the high risks associated with the constraints, then higher perceived constraints could induce demand for a higher profit goal. This situation applies to CG and AG in the project.

On the other hand, if the perceived constraints were less critical, the parties would have higher self-efficacy and expectancy. In such a situation, there is a possibility that the parties would upgrade their expected goals. For example, if CG had its own core capability in the housing business, it would not have contacted AG. Rather, it would have launched the project alone without the need to share the profit with the other party. A similar condition also applies to all the other parties in the three cases.

The above analysis indicates that the perceived constraints generally negatively influence the level of expected goals of different parties. However, in the specific situation in which if any powerful or important party wants to levy the risks pertaining to the perceived constraints in its profit, then more or less critical constraints might respectively induce higher or lower expected profit goal.

Proposition 2: The level of expected goal of the innovation would influence positively the organization's management-level commitment to the innovation

In order to discuss this proposition, the types of expected goals first need to be distinguished. The types of operative goals for an organization could be diverse. For instance, Katz and Kahn (1966) categorized them as organizational goals, individual goals, and society goals, Perrow (1961) as societal goals, output goals, system goals, product goals, and derived goals, Gross (1969) as output goals, adaptation goals, management goals, motivation goals, and positional goals, and Aguilera et al. (2023) as goal systems such as financial/non-financial goals, multiple substitute goals, multiple positive/negative complementary goals. However, in the context of construction innovation, the types of expected goals could be project-level and organizational-level goals, superordinate and subordinate goals, common and private goals, and short-term and long-term goals.

The parties involved in construction innovation are separate organizations, and their main interest would be the interest of their organizations. They come forward to work for innovation in order to achieve the project-level goals, which would eventually fulfill their organizational expected goals. The project-level goals could be superordinate or subordinate goals. In construction projects, the superordinate goal would be the client's project-level goal. Other parties' individual goals would be their respective subordinate goals. The superordinate and subordinate goals can also be termed respectively as common and private goals in the context of construction innovation. Therefore, the parties involved in construction innovation would work for the project-level superordinate goals in order to fulfill their project-level subordinate goals so that they could achieve their respective organization-level subordinate goals. In essence, the bottom line of each of the parties would be their individual organizational interests, and they work to fulfill these interests, whether they are short-term or long-term interests. Therefore, the commitment of the parties to innovation in real terms would be the commitment to their individual organizational interests. The critical question would be the extent to which a party would relate its commitment toward its organizational subordinate goals to the project-level superordinate goal.

In the case, the project-level superordinate goal was to complete a pioneering group housing project in the country, which was the developers' (CG's and AG's) project-level subordinate goal that could be instrumental in fulfilling their organizational subordinate goal of long-term corporate benefits and short-term profit. CG also had the socially expected goal of introducing the organized group housing concept in the country, which was again expected to be instrumental in strengthening its corporate image in the country. Therefore, the developers' commitment in real terms was directed toward their corporate interests.

The three parties (SB, MBC, and RMB) in the main contractor JV worked with their subordinate project-level goals (for instance, meeting the project schedule) to fulfill their short-term and long-term subordinate organizational goals.

The analysis suggests that the expected goal/s do positively influence management-level commitment toward innovation. However, it was found that the level of commitment would in real terms be influenced by the individual subordinate organizational goals. At the same time the nature of commitment would also be affected by the importance of the innovation for an organization and the consequent prioritized set of organizational expected goal/s.

Proposition 3: The organization's management-level commitment to the innovation would increase its management-level effort for the innovation

In this case, the broad types of management-level commitment of different parties were found to be (a) consistent throughout, (b) persistent throughout, and (c) weak and dwindling. If there are no major challenges during the course of initiation or implementation of innovation, the resulting smooth commitment is termed consistent commitment. If there are substantial challenges and the involved party persevered throughout, then such strong commitment is termed persistent commitment. If an involved party could not withstand the challenges during the course of initiation or implementation of innovation, then the commitment of such a party is termed weak and dwindling commitment.

It was observed that the parties – such as CG, AG, and SB – that worked with persistent management-level commitment put in their effort relatively strongly throughout the course of initiation and implementation. In order to deal with the challenges, these parties had to put in strong efforts. On the other hand, the parties – such as

MBC and RMB – that worked with weak and dwindling management-level commitment stopped their efforts upon facing the challenges and eventually left the project.

The above observation suggests that generally an organization's management-level commitment to the innovation positively affects its management-level effort for the innovation. Specifically, it was found that the parties which are consistent and persistent in their commitment would continue their effort in the innovation. The parties which have to be persistent to deal with the challenges would be more robust in putting their effort in comparison with the parties that worked without facing significant challenges.

Proposition 4: Challenges for the innovation at the management level would decrease the organization's management-level commitment to the innovation

After the case study, it seems that an assumption made in the conceptual framework regarding the jurisdiction of the management level of an organization in an innovation value chain needs to be relaxed. In the framework, it was assumed that the management-level and project-level challenges would be separately dealt with by the management and project teams of an organization. This might be true for a large organization in which one of its complex projects is dealing with an innovation that is not very significant for the management of the organization. However, for a small organization (such as MBC, SB, and RMB) dealing with a significant innovation, it was found that the domains of the management and project levels would not be as distinct as in a large organization. The critical project-level challenges could also be dealt with by the management level of such an organization, and thus the project-level challenges could also affect the management-level commitment. Even in large organizations, if the scale and impact of innovation are critical for the organization, the project-level challenges could be dealt with by the management level. The link between the project-level challenges and the management-level commitment is added in the modified conceptual framework as shown in Figure 8.2. The links are shown in dotted lines because the relationship was found to be conditional to some extent. Therefore, this Proposition 4 needs to be rephrased as Proposition 4a.

Proposition 4a: The challenges for innovation at the management and project levels would decrease the organization's management-level commitment to the innovation

CG and AG had to tackle several critical management-level challenges. Despite the host of challenges, the clients/developers were persistently committed to the project. Their strong priority on their organizational expected goals could be the reason behind their persistent commitment.

SB faced highly critical management-level challenges, such as dealing with an unfriendly client, its JV partners, and the critical project-level challenges. The local contractor, however, persistently worked on the project because its financial

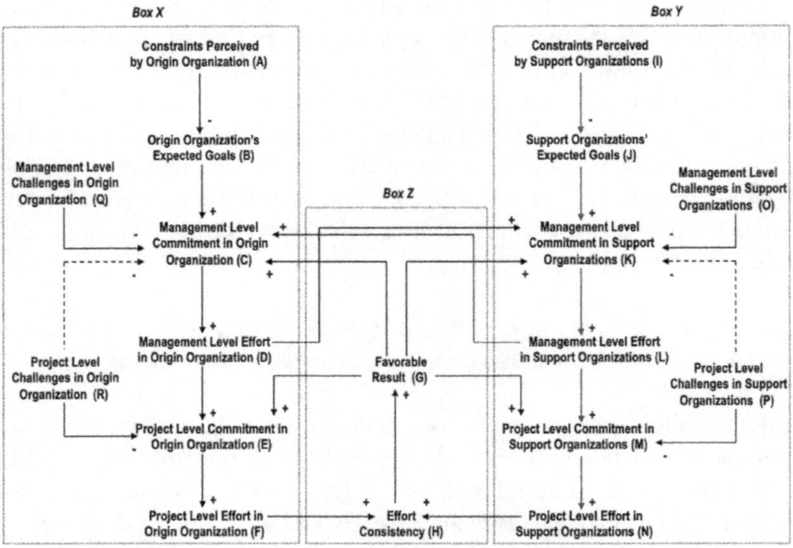

Figure 8.2 The modified conceptual framework

investment was confined to the project, and could not be retrieved unless the project was satisfactorily completed. The situation created critical pressure on the local contractor's goal of securing a reasonable profit out of the project.

The other two local contractors, however, left the project when they faced similar preliminary management-level challenges as SB. They showed weak and dwindling commitment because they did not have financial investment in the project. Additionally, their performance was not satisfactory, so they did not expect to profit out of the project.

The above analysis suggests that management-level challenges do not always adversely affect the management-level commitment. It would not be incorrect to logically infer that project-level challenges do not always adversely affect the management-level commitment. It was found that the more priority that is given to the expected goals and the more that is at stake in the project, the more the parties would work harder and persistently to deal with the challenges that crop up along the way. The management- or project-level challenges would adversely affect the management-level commitment if a party does not give priority to the expected goal and if the party has little stake in the venture. Moreover, if a party perceives lack of self-efficacy in dealing with the management-level challenges, there is a high possibility that the commitment of the party would turn weak and dwindling amid the challenges.

Proposition 5: OO's management-level effort for the innovation would influence positively the SOs' management-level commitment to the innovation

This proposition attempts to explain the inter-organizational interaction between the OO and SOs at the pre-implementation stage of innovation. It states that a part of the OO's management-level effort would be directed to identifying the relevant SOs and negotiating with them to gain their commitment toward the innovation. The strength of interaction depends on the strength of the OO's management effort and its receptivity (extent of commitment toward the proposed innovation) from the SOs' management side. In the conceptual framework, it was assumed that the OO would propose the innovation to the SOs and the SOs would respond to the OO with the level of commitment they could put into the innovation. However, from the case study, it was found that while the SOs respond to the OO's proposal, they would do so with their management-level effort, and their effort would also affect the commitment of the OO. The effort and commitment of the two parties would represent their two-way interactive relationship. Such interaction would not take place only between the OO and SOs, but it would also take place between all the functionally or contractually related SOs as well. Moreover, the interaction would continue to take place from the initiation stage to the completion or termination stage of the innovation venture. Therefore, Proposition 5 needs to be rephrased as Proposition 5a.

Proposition 5a: An organization's (OO or SO) management-level effort for the innovation would influence positively the other related organization's (OO or SO) management-level commitment to the innovation

The corresponding change in the relational link between the management-level effort of one organization and the management-level commitment of the other related organization(s) has been presented in the modified conceptual framework as shown in Figure 8.2.

The inter-organizational interaction starts from the initial negotiation stage and it continues in the subsequent implementation processes – such as organization, coordination, monitoring, and other required negotiation processes – of the innovation. Every effort of a party directed to affect the commitment of the other relevant party induces the interaction, as depicted by this hypothesis. The interaction terminates after the completion of the implementation of innovation, or, if one or both of the interacting parties do not show any more commitment toward the effort of the other party during the course of implementation. The latter scenario would be a case of premature withdrawal from the innovation.

From the case, it was found that in every effort to generate the commitment of the other party, power and position differentials of the involved organizations play a major role in their interaction. The choice of procurement framework and the mode of contract itself are highly influenced by the powerful party in the relationship. The choice and modality of whether to make or buy a particular organizational function is also influenced by the powerful party in the innovation value chain.

From the case, it was found that the following five types of effort were made by the management level of different parties in order to generate the commitment of the related parties: (a) contractual effort, (b) persuasive effort, (c) coercive effort, (d) submissive effort, and (e) cooperative effort.

All the parties that played the role of the principal (in terms of a principal–agent relationship) put in contractual effort to get the commitment of their agents. The ACDPL JV used their contractual effort to get the commitment of their agents. The agents who engaged in contractual negotiations also put in their contractual effort to win the commitment of their principals. For instance, the contractor JV put in contractual effort to bargain with the ACDPL JV.

SB put persuasive effort into its attempt to win the commitment of the client and the other two JV partners. ACDPL JV was coercive while dealing with the main contractor JV after the crisis hit the project. At the same time, the main contractor JV was persuasive to the client. MBC and RMB were coercive to SB, whereas the latter was persuasive to the former parties.

The type of effort was found to be governed by the relative power and positions of the parties and the status of their working relationships. Each of the efforts was found to be directed at the project-level superordinate goals with the intention of fulfilling the organizational subordinate goals of the party putting in effort. MBC and RMB terminated their commitment because they did not expect that their short-term profit goal would be fulfilled, although they also had the long-term goal of nurturing their business relationship with CG – their capacity and tolerance level were not sufficient to pursue the long-term goal.

Therefore, in essence, each of the management-level efforts was, in a real sense, directed to win the commitment of the related party so that the party which was putting in effort could fulfill its organizational subordinate goals.

Proposition 6: The organization's management-level effort for the innovation would positively influence its project-level commitment to the innovation

Proposition 7: The organization's commitment to the innovation at the project level would increase its effort for the innovation at the project level

These two propositions state that the effort at the management level would mainly focus on making its project unit adequately resourceful, organized, and committed to implementing the innovation. Thus, the management-level effort would attempt to generate normative commitment within its project unit toward the organization's expected goals behind the innovation. The normative commitment would then be the motivational force for the project-level unit in putting effort into the innovation.

It was found that the project-level commitment and effort of the parties, such as MBC and RMB – which worked with weak and dwindling management-level commitment – was found to be weak and dwindling. On the part of SB, which worked with persistent management-level commitment despite the challenges, the project-level commitment and effort was found to be persistent as well. However, the project-level commitment and effort of the ACDPL was found to be weak and intermittent despite the persistent management-level commitment and effort of CG and AG. The project team of ACDPL was pressurized by the management level to deal with the main contractor's problems, which were beyond the project team's capacity to solve. Another reason for the intermittent project-level commitment was the decision structure in the management of the ACDPL. Although the function of the architectural and engineering consultant was internalized by AG, there was considerable delay in information processing and decision-making processes. The lack of integration within the decision structure of the ACDPL regarding the construction of the project made its project team frustrated, leading to a high turnover of key project personnel. This observation shows that internalizing organizational functions would not necessarily increase organizational integration, whereas, in order to effectively transfer management-level commitment and effort to the project level, internal organizational integration is highly important.

Another issue related to the relationship between management effort and project-level commitment is the extent of involvement of the management level at the project level. For instance, the management levels of all three partners of the main contractor JV were directly involved at their project levels. However, the management levels of CG and AG were not involved directly at the project levels; instead, they deployed their project teams. It was found that if the management level directly gets involved at the project level, the management-level effort would be readily reflected in the project-level commitment and efforts. Direct involvement of the management level at the project level is also a type of strong integration between the two levels. Therefore, once again, internal organizational integration is highly important to transfer the management-level effort into its project level.

Proposition 8: Challenges for the innovation at the project level would decrease the organization's project-level commitment to the innovation

There was a series of intense and serious project-level challenges for all the involved parties. There were changes in at least three sets of senior construction project teams of the ACDPL JV. The teams yielded under pressure and left the project. The overall commitment pattern of the ACDPL JV was therefore intermittent during the construction period. MBC and RMB of the main contractor JV showed weak and dwindling commitment, and they even left the project. SB, however, showed persistent commitment even though the client changed the amount and modality of the contract.

The above observations suggest that project-level challenges do not always adversely affect the project-level commitment. It was found that the more the parties are committed to their expected goals and the more the confined stake in the project, the more the parties would work harder and persistently to deal with the challenges that crop up in the project. However, even if a party is committed at the management level, if the internal organizational process is not firmly integrated, project-level challenges could adversely affect project-level commitment (cases of CG and AG). Moreover, if a party perceives lack of self-efficacy in dealing with the project-level challenges, there is a high possibility that the commitment of the party would turn weak and dwindling amid the challenges (cases of MBC and RMB).

Proposition 9: The organization's project-level effort for the innovation would enhance the overall effort consistency for the innovation

Effort consistency is a relative term. The efforts of relevant parties which are consistent, for example, in terms of the superordinate goal of the project, need not necessarily be so in terms of their individual subordinate goals. In this proposition, the overall effort consistency was assumed to be the effort consistency in terms of the superordinate goal of the project.

From the case, it was found that the effect of the project-level efforts of different parties on the overall effort consistency would depend on the extent to which the parties relate their project-level subordinate goals (that would fulfill their respective organizational-level subordinate goals) to the superordinate goal of the project. Therefore, it is obvious that the organization's project-level effort for the innovation would not always enhance the overall effort consistency for the innovation.

For the first six months, the superordinate goal of the project was satisfactorily in line with the project-level subordinate goals of the ACDPL and that of the main contractor JV. However, after six months, due to a series of challenges, the main contractor could not follow the construction schedule. As such, the project-level subordinate goals of the ACDPL and that of the main contractor JV (and its three partners) started getting misaligned with the superordinate goal. The unity of effort, or the effort consistency, therefore became very weak. The ACDPL then took charge of the construction and changed the contract modality into a labor contract. SB was appointed as the project manager under the new contract. The project-level subordinate goals of SB and that of the ACDPL then realigned toward the superordinate goal. The effort consistency improved, and it eventually helped to complete the construction of the project.

Proposition 10: The effort consistency for the innovation would positively influence the favorable results out of the innovation

As stated in Proposition 9, here again, the effort consistency and the favorable results have to be interpreted in terms of the superordinate goal of the project. The results, which are favorable in terms of the superordinate goal of the project, need not necessarily be so in terms of their individual subordinate goals. Nevertheless, in this proposition, the effort consistency and the favorable results were taken in terms of the superordinate goal of the project.

Because of the highly inconsistent efforts of the ACDPL and the main contractor JV, there was about a two-year construction delay and the interactions among the project participants were highly turbulent. Although the project was eventually completed, the lack of effort consistency did adversely affect the project-level superordinate goals.

Therefore, the observations in the case suggest that effort consistency for the innovation in terms of the project-level superordinate goal would positively influence the respective favorable results out of the innovation. Moreover, even if effort consistency is interpreted in terms of a subordinate goal of any relevant party, it would positively influence the respective favorable result for the party.

Proposition 11: The favorable results out of the innovation would positively influence the organization's project-level commitment to the innovation

The favorability of the results depends on the extent to which the expected goals are achieved. The hierarchy of the expected goals, as explained in Proposition 2, helps to explain the extent of achieved favorable results at different levels. Achievement of the project-level subordinate goals of an organization indicates favorable results for the organization at the project level. Similarly, the achievement of organizational subordinate goals indicates favorable results for the organization at the management level. Likewise, the achievement of project-level superordinate goals indicates favorable results for the client.

This review of the concept of different hierarchies of expected goals suggests that the favorable results at the project level not only positively influence the organization's commitment to the innovation at the project level, but they would also positively influence the organization's commitment at the management level. Proposition 11, therefore, needs to be rephrased as Proposition 11a.

Proposition 11a: The favorable results of the innovation would positively influence the organization's project- and management-level commitment to the innovation

The corresponding addition of the relational link between the favorable results and the management-level commitment of the involved organizations has been presented in the modified theoretical framework, as shown in Figure 8.2.

Another important aspect of the favorable result is its relative sense of interpretation by different parties. As the subordinate goals of innovation could be different for the involved organizations, the perception of an achieved favorable result would also be different for them. Therefore, an organization's perception of its favorable results depends on the judgment of the extent of achievement of its subordinate goal. If the superordinate goal of the project supports achieving the project-level subordinate goal (which would fulfill their respective organizational-level subordinate goals) of a party, or vice versa, the results that support the superordinate goal could be perceived as favorable results. If this is not the case, the fulfillment of individual subordinate goals would be taken as favorable results, which would be a case of weak effort consistency.

The observations in the case support that the favorable results out of innovation influence positively the organization's project- and management-level commitment to the innovation. Further, the perception and judgment of the favorable result affect the commitment and the subsequent efforts.

In this case, the progressive project results during construction were not satisfactory. The main troublesome issues were the construction delay and the escalation of project cost. Because of such unfavorable outcomes, the project teams of the ACDPL could not remain committed – there was a turnover of three key project personnel teams. However, as the prioritized organizational subordinate goals of the CG and AG were broader than the project-level subordinate goals, the management levels of the two groups were committed to completing the project. That is why they continued to put persistent effort into changing the project team, the main contractor, and even contract modality in order to contain further delays and cost escalation.

For the three partners of the main contractor JV, it was very difficult to withstand the unfavorable results during the construction of the project. Because of their small scale of operation, their prioritized project-level subordinate goal was to get sufficient cash inflow, which would supposedly fulfill their prioritized organizational subordinate goal of securing a profit. As the results were highly unfavorable compared with their expected goals, the two partners left their commitment to the project after the first six months of the construction stage. SB, however, remained persistently committed because it had to focus on its changed organizational subordinate goal of recovering the confined investment in the project. For SB, even remaining within the project was a favorable result in order to achieve its changed goal.

CONCLUSIONS

This chapter presented and explained the innovation case on the first group housing project in a developing country. The case data have been used to test the developed conceptual framework. The following are the conclusive points that can be summarized from the analysis:

a. Generally, the perceived constraints negatively influence the level of expected goals at the initiation stage of innovation. However, in the specific situation in which if any powerful or important party in the project wants to levy the risks pertaining to the perceived constraints in its profit, then more or less critical constraints might respectively induce a higher or lower expected profit goal.
b. The expected goals positively influence the management-level commitment toward innovation. However, it was found that the level of commitment would in real terms be influenced by the individual subordinate organizational goals. Therefore, the extent to which the organization could relate its subordinate organizational goals to the superordinate project-level goal would be an

important issue in the implementation process. At the same time, the nature and level of commitment would also be affected by the importance of the innovation for an organization and the consequent prioritized set of organizational expected goal(s).

c. It was found that generally an organization's management-level commitment to the innovation positively affects its management-level effort for the innovation. Specifically, the parties which are consistent and persistent in their commitment would continue their effort in the innovation, and the parties which have to be persistent to deal with the challenges would be more robust in putting in their effort.
d. It was found that the management- or project-level challenges do not always adversely affect the management-level commitment. The more priority that is given to the expected goals and the more the confined stake in the project, the more persistently the organization would work to deal with the challenges that crop up along the way. The management- or project-level challenges would adversely affect the management-level commitment if an organization does not give priority to the expected goals and if it has little stake in the innovation venture. Moreover, if an organization perceives a lack of self-efficacy in dealing with the management- or project-level challenges, there is a high possibility that its commitment would turn weak and dwindling amid the challenges.
e. It is not necessary that an organization's management-level effort for the innovation would influence positively the other related organization's management-level commitment to the innovation. The effectiveness of a party's management-level effort in winning the commitment of the other related party would depend on their power and position differentials. It was also found that the management-level effort, in a real sense, would be directed to win the commitment of the related party so that the party which was putting in effort could fulfill its organizational subordinate goals.
f. The organization's management-level effort for the innovation would not always positively influence its project-level commitment and thereby the effort toward the innovation. Internal organizational integration is highly important to transfer the management-level effort into its project level.
g. The project-level challenges do not always adversely affect the project-level commitment. Similarly, as in the case of the effect of management- or project-level challenges on the management-level commitment, it was found that the more the organization is committed to its expected goals and the more the confined stake in the project, the harder and more it would work to deal with the challenges that crop up in the project. However, even if the organization is committed at the management level, if the internal organizational process is not firmly integrated, project-level challenges could adversely affect the project-level commitment. Moreover, if an organization perceives a lack of self-efficacy in dealing with the project-level challenges, there is a high possibility that the commitment of the organization would turn weak and dwindling amid the challenges.

h. The efforts of relevant parties which are consistent, for example, in terms of the superordinate goal of the project, need not necessarily be so in terms of their individual subordinate goals. The effect of the project-level efforts of different parties on the overall effort consistency would depend on the extent to which the parties relate their project-level subordinate goals (that would fulfill their respective organizational-level subordinate goals) to the superordinate goal of the project. Therefore, the organization's project-level effort for the innovation would not always enhance the overall effort consistency for the innovation.
i. The effort consistency for the innovation in terms of the project-level superordinate goal would positively influence the respective favorable results out of the innovation. Even if the effort consistency is interpreted in terms of a subordinate goal of any relevant party, it would positively influence the respective favorable result for the party.
j. It was found that the favorable results of innovation influence positively the organization's project- and management-level commitment to the innovation. However, the perception and judgment of the favorable result affect the commitment and subsequent efforts.

REFERENCES

Aguilera, R. V., De Massis, A., Fini, R. and Vismara, S. (2023), Organizational Goals, Outcomes, and the Assessment of Performance: Reconceptualizing Success in Management Studies, *Journal of Management Studies*, doi.org/10.1111/joms.12994

Coccia, M. (2016), Sources of Technological Innovation: Radical and Incremental Innovation, *Technology Analysis and Strategic Management*, 29 (9), 1048–1061.

Gross, E. (1969), The Definition of Organizational Goals, *British Journal of Sociology*, 20, 277–294.

Katz, D. and Kahn, R. L. (1966), *The Social Psychology of Organizations*, John Wiley, New York.

Perrow, C. (1961), The Analysis of Goals in Complex Organizations, *American Sociological Review*, 26, 859–866.

Slaughter, S. (1998), Models of Construction Innovation, *Journal of Construction Engineering and Management*, 124 (3), 226–231.

Yin, R. K. (2018), *Case Study Research: Design and Methods*, Sage Publications, Thousand Oaks, CA.

PART III

INNOVATIONS IN CONSTRUCTION PROJECT PROCUREMENT

9. Policy-led innovations: the impact of Indigenous procurement policies on contractors' behaviour towards Indigenous people's participation in construction projects

George Denny-Smith, Martin Loosemore, Riza Yosia Sunindijo and Megan Williams

INTRODUCTION

Increasing inequities internationally require the construction industry to innovate new ways of working and engage with marginalised groups so it can achieve social sustainability goals such as those articulated in emerging social procurement policies and more broadly in the United Nations' (UN) Sustainable Development Goals (SDGs) (Raiden and King, 2023). These innovations involve project-level solutions that enable construction professionals to better cope with new 'social procurement' requirements where they must hire workers from disadvantaged and marginalised backgrounds to achieve social sustainability (Troje, 2023).

Therefore, this chapter investigates how policy-led innovations such as Indigenous procurement policies (IPPs) change contractor behaviour to achieve social sustainability. IPPs are used by local, state, and federal governments in Australia and other countries with colonised peoples to require contractors to increase Indigenous representation in their workforces and supply chains. The objective of IPPs is to increase Indigenous Australians' economic participation and close the socioeconomic 'gaps' existing between Indigenous and non-Indigenous people. We empirically focus on IPPs because of their wide application in Australia and their linkages to innovation and social sustainability, discussed below. The chapter progresses by first reviewing the concept of social sustainability and then moves to a discussion of IPPs as a policy-led innovation in construction. We discuss the importance, opportunities, and challenges of innovation in both contexts. We also outline the limitations of current understanding regarding how policy-led innovations affect construction companies to adopt innovative practice. Following this, we discuss the theoretical framework, Ngaa-bi-nya, used to understand how contractors' behaviour changes towards Indigenous participation and present the study methods. Finally, the findings are presented and discussed in relation to our research questions and the relevant literature.

SOCIAL SUSTAINABILITY

Social sustainability is defined as situations where people are not subject to social conditions that risk their health, where they can participate in social systems, have access to learning and development, are not exposed to adverse treatment, and can create their own forms of cultural, group, or individual expression (Missimer, Robèrt and Broman, 2017, p. 47). Relevant concepts of study in social sustainability usually concern issues such as health and social equity, human rights, labour rights, practices and decent working conditions, work health and safety, social responsibility and justice, community development and wellbeing, product responsibility, community resilience, and cultural competence (Balaman, 2019).

However, social sustainability is the least investigated of the three sustainability pillars of economic, environmental, and social (Torkayesh et al., 2021). Unfortunately, in construction, social sustainability is still often overlooked in favour of the economic and environmental aspects of sustainability (Troje, 2023). Yet as Sharma, Kumar and Singh (2023) note, social sustainability is critical for successful and ongoing socioeconomic development and impacts, and is influenced by stakeholders across the whole business ecosystem, like workers, consultants, suppliers, clients, and communities. As such, social sustainability is critical to the realisation of sustainable development agendas, particularly to limit the construction sector's adverse sociocultural impact. In fact, social sustainability is needed for realising environmental sustainability and to address issues associated with climate change, which seems to be the main sustainability agenda. How can sustainability objectives be met if the people needed to implement them experience social and financial dislocation and socioeconomic inequities that limit their ability to do so?

One approach to social sustainability is social procurement. Social procurement 'asks how a purchase can help address and solve the complex socio-economic issues faced by a community', thus determining the community impact of a project or purchasing decision and creating social value (LePage and Renaerts, 2023, p. 131). Social procurement is used by public and private sector clients who want their investments to create a social return in addition to the economic and infrastructure returns generated by construction. The construction industry is increasingly important to social procurement objectives to realise sustainable development outcomes because of its large contributions to countries' gross domestic product (GDP), its high employment, and its broad impact on the economic, social, and environmental pillars of sustainability, through the creation of new infrastructure, employment, and use of resources (Raiden and King, 2023). As LePage and Renaerts (2023) note, the aforementioned contributions of construction mean it can influence 12 of the 17 SDGs, and in particular: Good health and wellbeing (SDG3); Gender equality (SDG5); Decent work and economic growth (SDG8); Industry, innovation and infrastructure (SDG9); and Reduced inequalities (SDG10).

INNOVATION FOR SOCIAL SUSTAINABILITY

While social procurement is subject to an increasing amount of scholarly examination, a common theme in the literature is that social procurement is still challenging industry norms and ways of working. For example, Australian research on social procurement that addresses youth unemployment found that contractors are not yet prepared to adequately respond to and implement social procurement requirements imposed by social procurement policies (Loosemore et al., 2022). Loosemore et al. (2022) also found that social procurement forces contractors to adapt and develop new ways of working that effectively deliver social procurement, usually in the form of new cross-sectoral partnerships, but which could be supported by education campaigns about the need for, and how to implement, social procurement successfully.

Supporting the theme that the construction industry is unprepared or lacks the tools to effectively deliver social procurement, Troje's (2023) Swedish social procurement research found that contractors lack the time, resources, and training that enable effective social procurement. These insights supported prior research that found the three most common internal challenges to social procurement among 179 organisations, from multiple sectors, are a lack of organisational social procurement experience, competing organisational objectives, and resource commitments (Barraket et al., 2021). This shows there is a need for construction to innovate to deliver social procurement and, therefore, social sustainability. The construction industry has needed to adapt to social procurement requirements and there is little evidence of how social procurement is changing contractors' behaviour to promote social sustainability in their operations.

Therefore, this chapter aims to empirically document how IPPs, as one manifestation of social procurement policy, are changing contractors' behaviour to promote social sustainability by increasing Indigenous participation on construction projects. We examine IPPs as a policy-led innovation to require contractors to purchase from Indigenous businesses and employ Indigenous workers when delivering government projects. The research questions (RQs) this chapter addresses are:

(1) How do IPPs impact contractors' behaviours towards Indigenous people's participation on construction projects?
(2) What is the effect of contractors' changed behaviours due to IPPs on social sustainability outcomes in construction?
(3) What role do policy-led innovations like IPPs have in influencing sustainable attitudes in construction?

To begin answering these research questions, the following section reviews IPPs as a policy-led innovation in construction. The review finds that IPPs are a radical form of innovation that drives contractors to develop new approaches and results in cultural transformation in the industry.

IPPS AS POLICY INNOVATIONS IN CONSTRUCTION

Several scholars support the idea that policy has an important role to play in influencing innovation in construction. For example, Mullett (2022) argues that mandates – in the form of regulation or another authoritative source – that require contractors to meet sustainability expectations and measures are potential innovation drivers because they force clients, and therefore contractors, to take sustainability issues more seriously. Similarly, Zhang et al.'s (2023) innovation research concludes that governments have a critical role to optimise construction innovation because they can encourage innovation through policies and promoting innovative exchanges through technological or knowledge exchanges. Bailey et al.'s (2022) research on social innovation to create inclusive construction employment looks favourably on social procurement policies in Australia because they mandate the employment of people with a disability, who may not have been employed if the policies did not exist. As Māori auto-ethnographical insights on supplier diversity in New Zealand highlight, creating equity or social sustainability for Indigenous Māori peoples through procurement is a demand-side issue that needs policymakers to work with different stakeholders to support the growth and success of supplier diversity practices (Malcolm, Cairns and Pouwhare, 2023).

To this end, Australian governments now use IPPs as a policy innovation to stimulate demand for Indigenous workers and businesses. The aim of IPPs in Australia generally is to create a 'level playing field' that opens new networks and business avenues that have historically been closed to Indigenous Australians (Kinsela-Christie, 2019). IPPs open previously closed industry networks by requiring government agencies and contractors working on government projects to meet specified purchasing and employment targets. These work either as a percentage of the number of eligible contracts (currently 3%), or a percentage of the financial value of eligible contracts (currently 1.75%) that must be awarded to Indigenous businesses as a minimum (NIAA, 2023). These figures are based on the national IPP, the Commonwealth Indigenous Procurement Policy. Other state or local governments have their own targets based on the maturity of their IPP or its objectives. Targets are passed on through contracts, so a project worth $1,000,000 would have to allocate at least $17,500, as a working example of the Commonwealth IPP. Contractors who do not meet their IPP targets may have the corresponding difference withheld from their contract by government clients (NSW Government, 2022). The logic behind this innovation is that IPPs commit contractors to delivering social sustainability with Indigenous suppliers and employees. IPPs create more demand for their services, which leads to greater saturation in the market, which leads to socioeconomic advances through education, employment, and business experience.

PURPORTED SOCIAL AND ECONOMIC BENEFITS OF IPP POLICY INNOVATIONS

Advocates of these policy innovations propose that, over the long term, IPPs can address the socioeconomic inequities Indigenous Australians have experienced, compared with non-Indigenous Australians, since European invasion in 1788. For example, advocates such as Janke (2021) favour IPPs because they could create substantial employment and contracting work that will increase Indigenous employment and economic participation. Others concur, writing that IPPs can highlight the economic, social, and cultural contributions that Indigenous businesses can make to Australia when they are included in supply chains (see for example Evans et al., 2021).

The innovation of IPPs comes from the requirement that contractors must meet procurement and employment targets on government contracts. The key benefit of this is that Indigenous businesses can compete for and win work that they would not normally be considered for (Kinsela-Christie, 2019). The potential for this to deliver social sustainability is significant. For example, in addition to greater tax revenues, modelling has shown that there would be a $4.7 billion saving to governments if the employment gap between Indigenous and non-Indigenous Australians was closed by 2031 (Deloitte, 2014). The savings would be found in lower health and social security costs driven by greater wellbeing and wealth among Indigenous Australians. In addition to financial benefits, key social and cultural outcomes that IPPs can create include (Supply Nation, 2018):

- reconnecting with family and community
- more self-pride and autonomy
- enhanced sense of identity, belonging, safety, and pride in culture
- changing people's perceptions about Indigenous Australians and Indigenous businesses.

Other social benefits IPPs can create illustrate the potential of IPPs to deliver social sustainability. Benefits that can be realised by business owners and workers include (Kinsela-Christie, 2019):

- business owners and workers being role models for younger people to look up to
- self-determination and empowerment as people gain control over their career path and life
- reinvestment in businesses, families, and communities through strategic initiatives and community projects
- providing training that develops new skills and knowledge that support long-term career pathways
- the long-term benefits of employment as Indigenous businesses are more likely to employ and retain Indigenous people.

The points above show that IPPs can influence social sustainability outcomes. Key social sustainability concepts that IPPs can influence include social equity, labour outcomes and conditions, and community development and wellbeing (Balaman, 2019). Therefore, as an innovation, IPPs can have a significant influence on the construction industry and its influence on the social pillar of sustainability. However, as the following subsections discuss, implementing this policy-led innovation in an industry like construction faces barriers to their successful delivery. After discussing these barriers, we outline our approach to empirically researching how construction contractors are responding to IPPs, and how IPPs are affecting contractor behaviour for better or worse. We especially explain how we used an Aboriginal evaluation framework so that we could analyse the qualitative results of our study that aligned with the lived experiences of Indigenous contractors.

BARRIERS TO IMPLEMENTING IPPS AND OTHER SOCIAL PROCUREMENT POLICIES AS INNOVATIONS

The construction industry is notoriously slow at innovating. This is even more pronounced in the context of implementing innovations like IPPs and other social procurement policies. Risk aversion in construction usually deters innovation, as companies avoid it because contractors are worried about the uncertain nature of innovation or the possible increased costs of investing in it (Zhang et al., 2023). Further, the temporary and stop–start nature of construction projects can cause dislocation among teams, meaning learnings from innovation cannot be transferred from one project to another (Mullett, 2022).

In this context, IPPs require new approaches to implement, to which the industry is still learning how to respond (Troje, 2023). This makes them a radical innovation that results in social or organisation restructuring and cultural transformation (Bresnen and Marshall, 2001). This unfortunately creates significant barriers and resistance on projects when contractors are told they must deliver IPP and other social procurement requirements. For example, research on the use of construction social procurement in Australia found that subcontractors respond negatively to social procurement requirements for several reasons, including the extra costs to employ and supervise the groups social procurement targets, higher safety risks from less-experienced workers on site, and reduced productivity as new workers need to be trained in and acclimatised to construction roles (Loosemore et al., 2022).

Indeed, one barrier to implementing IPPs arises when contractors and third parties need to work together to deliver successful social procurement. The third parties that contractors must work for generally come from the social sector, employment agencies, or government departments requiring collaboration to identify, recruit, and retain the workers targeted by social procurement policies, in this case Indigenous workers (Troje, 2023).

Changes to how the industry operates, driven by IPPs, require novel ways of engaging with new stakeholders. But as research has found, tensions are created during social procurement delivery when multiple organisations must collaborate. Tensions are created because of competing objectives between the organisations working together, a lack of resources and training in how to deliver social procurement, and the traditional focus on time, cost, and quality in construction, which means social procurement may be considered as secondary to delivering a built form that meets the client's needs (Troje, 2023). These tensions create barriers to implementing social procurement policies like IPPs because they are imposed top-down on construction contractors who are not prepared for, or do not understand, the need for and methods to comply with social procurement requirements (Loosemore et al., 2021). Further barriers to enacting social innovations in construction can also include employment barriers experienced by the groups social procurement targets, biases and stigmas around those groups, and the challenges of informal norms and practices in construction which shape how the industry recruits, trains, and responds to minority groups (Bailey et al., 2022). In the case of IPPs, many Australians hold an inherent bias against Indigenous Australians (Shirodkar, 2019).

The above review shows possible barriers to implementing IPPs successfully. These barriers include:

- Projects are not resourced properly.
- Project budgets require them to seek lowest-cost options.
- Construction professionals do not know how IPPs can be delivered effectively.
- Traditional ways of working in construction limit how builders and suppliers work effectively with Indigenous businesses and workers.

However, contractors must still meet IPP targets, as they are a policy-led innovation. The potential penalties of not meeting IPP targets also necessitate that contractors find ways of complying with IPP requirements. Therefore, the following subsection reviews how contractors have been found to respond to IPPs and social procurement policies before outlining the theoretical framework that guided the study.

EFFECTS OF IPPS AND SOCIAL PROCUREMENT POLICY INNOVATIONS ON CONTRACTORS' BEHAVIOUR

Recent research has found IPPs and social procurement policies are challenging how contractors work to deliver the requirements placed on them. For example, Loosemore et al.'s (2022) Australian construction social procurement research found contractors generally only provide low-risk jobs to people employed through IPPs and social procurement policies. This is because contractors see the policies as an unfairly distributed risk that is placed on them. Research on the targeted employment of Māori peoples in New Zealand through social procurement supports this. That research found Māori construction workers can face precarious employment categorised by low-paid, low-skilled jobs with no training or development prospects (Hurt-Suwan and Mahler, 2021).

Other Australian research has also found that contractors respond to IPP requirements with a compliance mindset. This research found that in order to comply with IPP targets, contractors aim for short-term, menial jobs. For example, it has been reported that contractors employ short-term Indigenous labour and have instructed Indigenous workers to sit 'under a tree all day' as a tokenistic form of Indigenous employment (Denny-Smith, 2021, p. 640). This type of response to IPP requirements is damaging to Indigenous workers who are meant to benefit from the policies because it is harmful to the wellbeing and sense of self of Indigenous workers. This outcome is detrimental to IPPs being an innovation for social sustainability as it exposes people to adverse treatment and working conditions (Missimer, Robèrt and Broman, 2017; Balaman, 2019).

This review shows that while IPPs have potential as a social sustainability innovation, there are a variety of reasons why the construction industry may not deliver them effectively. Further, no previous research has looked at these reasons in the context of increasing the engagement and participation of Indigenous contractors and workers on construction projects. Understanding these responses using empirical observations is important to categorise them. This will help researchers and practitioners understand how these innovations can be effectively implemented on projects in the future. This will increase the industry's ability to implement these innovations and contribute to broader sustainability objectives. To do this, the following section outlines the Aboriginal evaluation framework used to analyse empirical results and provide these insights. After discussing the theoretical framework, we present the method used to collect data. The methods and analysis below can be used by innovation researchers in the future when working in new areas that are underexplored to show how innovations are impacting construction.

THEORETICAL FRAMEWORK

Ngaa-bi-nya (pronounced 'nar-binya' in Wiradjuri) was our theoretical framework to analyse how IPPs impact contractors' behaviour. Ngaa-bi-nya was developed by Williams (2018) in the field of Aboriginal and Torres Strait Islander health and social programmes as a tool to evaluate programmes that were culturally informed to Indigenous stakeholders. Ngaa-bi-nya has four domains with prompts that are relevant to improving the wellbeing and social and economic status of Aboriginal people. The domains are: the landscape in which policies and programmes exist; resources used and generated by programmes; ways of working by and with Aboriginal people; and learnings from programme and policy processes and outcomes. Subsets of statements in each domain serve as prompts for data collection, analysis, and interpretation to ensure it can understand and critique success and value from Indigenous peoples' perspectives.

In the context of empirically understanding how IPPs change contractors' behaviour, Ngaa-bi-nya is useful because it theorises actions based on Indigenous peoples' cultural knowledges, rights, needs, and aspirations, which IPP statements and other

Australian government policies often aim to meet. This also aligns with social sustainability objectives and the United Nations Declaration on the Rights of Indigenous Peoples (UNDRIP) (United Nations, 2007). For example, a subset of Ngaa-bi-nya asks how human, material, and in-kind resources and informal economies and relationships support Aboriginal and Torres Strait Islander programmes, in this case Indigenous businesses. Therefore, Ngaa-bi-nya prompts how contractors use or create these resources to work with Indigenous businesses to deliver IPPs. Table 9.1 shows the Ngaa-bi-nya domains and prompts relevant to this study investigating their impact on contractors' behaviour.

Table 9.1 indicates how the impact of IPPs on contractor behaviour was assessed. The Ngaa-bi-nya prompts above teach that, for innovations like IPPs to be better understood, they need to be investigated through the lens of relevant concepts. This can clarify whether the actions of contractors support or restrict social sustainability objectives relevant to Indigenous Australians.

Having outlined the theoretical approach to understand the impact of IPPs on social sustainability as a policy-led innovation, the following section outlines the participatory approach to apply Ngaa-bi-nya to this phenomenon. In line with social

Table 9.1 Ngaa-bi-nya Aboriginal evaluation framework domains and their use in studying the impact of this innovation on contractor behaviour

Ngaa-bi-nya domain	**Ngaa-bi-nya prompts to assess the impacts of IPPs on contractors**
Landscape	• The extent to which local Indigenous peoples have been engaged in identifying needs and setting priorities • Alignment between legislation and policies relating to the outcomes targeted by Indigenous procurement policies
Resources	• Resources (human, material, and in-kind) and informal economies and relationships supporting IPP implementation • Employment, Indigenous workforce development, and knowledge transfer to create long-term workforces • Networks that support construction employment opportunities created by Indigenous procurement policies
Ways of working	• Delivery of IPPs • How IPPs support self-determination in communities • How local stakeholders are engaged to plan for IPP implementation • How IPP implementation is supported by activities, relationships, and accountability processes
Learnings	• Assesses whether policy objectives have been met, what worked well, and what can be improved • Relates to the movement of ideas, actions, purpose, ways of being, and ways of relating

sustainability aims, our participatory approach is consistent with Indigenous ways of doing, where reciprocity is critical to the creation and sharing of knowledge (Williams, 2018). Following the methods used in this study, the results are presented and then discussed in relation to the research questions presented above.

METHODS

Reflexivity is essential in social sustainability research, as researchers may come from privileged positions that differ from the positions of others for whom social sustainability will benefit. This is even more important for Indigenous research topics where cultural backgrounds could influence research outcomes. Among the researchers and authors, for example, one belongs to the Wiradjuri nation in central New South Wales, and others are Anglo-Celtic and Indonesian – that is, non-Indigenous. Our cultural backgrounds led us to explore and discuss our training and disciplines of work, current practices, positionality and identity, and 'cultural assumptions, standpoints and biases' (Martin and Mirraboopa, 2003, p. 212). As a multicultural team researching a social innovation that involves more than one culture (i.e., non-Indigenous contractors working with Indigenous contractors), we agreed that this research must be co-designed with Indigenous stakeholders as research partners.

Therefore, this study followed Indigenous research principles supporting social and environmental wellbeing (Dudgeon et al., 2020) and sustainability-oriented innovation research that uses action research to understand how innovations are developed and implemented (Santa-Maria, Vermeulen and Baumgartner, 2022). Principles of community-based participatory research (CBPR) were used that emphasised participatory knowledge-making and actions, and therefore informed our methods.

COLLABORATIVE INNOVATION RESEARCH WITH INDIGENOUS STAKEHOLDERS

CBPR guides collaborative research processes that use the unique strengths of research stakeholders in an equitable way (Minkler, 2005). Research involving Indigenous stakeholders must be designed, conducted, and reported so that it also benefits Indigenous communities and research partners who have facilitated it (AIATSIS, 2020). Indeed, this research made sure to follow clear guidelines and adopted principles of reciprocity that meant the research could be communicated back to stakeholders for their use. For example, we had to ensure knowledge was not taken, but shared and given back, so that it was ethical, represented participants' experiences of IPPs, and overcame the ethical nuances of working with Indigenous people. Implementing these principles took three years of developing partnerships with Indigenous stakeholders. This is because research on social sustainability with Indigenous stakeholders is new in construction. Therefore, there are no prescribed methods or empirical concepts to readily use.

Mirroring the social equity and human rights principles of social sustainability, our partnership approach used Indigenous ways of generating and sharing knowledge. Indigenous scholars posit that research with Indigenous stakeholders should be community-based and informed by Indigenous ways of knowing and doing (Rossingh and Yunupingu, 2016). As such, 'collaborative yarning' was done with Indigenous stakeholders to reach shared understanding about the research and its aims, methods, and processes (Bessarab and Ng'andu, 2010). In this study, the lead author travelled to regional and remote New South Wales in Australia multiple times to meet with stakeholders. The author presented the research ideas and concepts and sought feedback on their alignment with the aims and objectives of those stakeholders. The research design was refined each time and reflected the changing context of the research. Flexibility was required to adapt research methods so the effects of IPPs on contractors' behaviour would enable empirical assessment of how the industry responds to policy-led innovations that aim for social sustainability. The research context was affected several times by, in no order, stakeholder commitments, project demands on stakeholder construction sites, COVID-19, limitations on collecting face-to-face data, and unexpected changes in availability.

Despite challenges faced in implementing CBPR, the authors developed an approach that reflects social equity and the rights of Indigenous stakeholders to determine how the research was designed and conducted. For example, after initially agreeing that yarning circles would be a culturally appropriate form of gathering qualitative data that reflects Indigenous cultural traditions of oral knowledge-creation (Bessarab and Ng'andu, 2010), semi-structured interviews were seen by Indigenous participants as culturally appropriate in a COVID-19 context because of the health risks that yarning circles would have imposed on highly vulnerable and Indigenous communities. Qualitative processes like semi-structured interviews empower research participants to explain social phenomena, especially in Indigenous contexts (Dudgeon et al., 2020). In an innovation context, semi-structured interviews can help explain how actors interact with each other to deliver sustainability innovations (Stål, Bengtsson and Manzhynski, 2022). Understanding how actors implement sustainability innovations is influential in determining the success of an innovation and whether actors can benefit from its outcomes (Lüdeke-Freund, 2020). Therefore, such approaches are critical to empirically understanding how IPPs affected contractor behaviour to achieve social sustainability.

INTERVIEW QUESTIONS

Semi-structured interview questions were developed based on the Ngaa-bi-nya evaluation framework in Table 9.1. Questions were initially based on the Ngaa-bi-nya prompts in Table 9.1 and informed further by the critical literature review earlier in this chapter and through the CBPR research process. For example, open conversations with community stakeholders were held to discuss interview questions and sense-check them so they aligned with the social equity aspirations of stakeholders.

Stakeholders advised that IPP implementation was also relevant to contractors' interactions with communities on current and future projects. This illustrated the importance of exploring how IPPs are implemented from a contractor–community perspective instead of a contractor–Indigenous business perspective exclusively.

The process above produced several interview questions grouped in Ngaa-bi-nya domains and tailored to the context of how IPPs impact contractor behaviour. The final interview questions are shown in Table 9.2.

Table 9.2 Interview questions to examine how IPPs impact contractor behaviour

Ngaa-bi-nya domain	Questions to examine the impact of IPPs on contractors' behaviour
Landscape	• Whether IPPs had promoted self-determination among local Indigenous communities • Whether IPPs had improved Indigenous peoples' and communities' socioeconomic position
Resources	• Whether a range of Indigenous businesses have worked on IPP projects on different work packages • The training and employment opportunities IPPs create • The financial outcomes experienced by Indigenous businesses and workers because of IPPs • Whether local Indigenous workers had been engaged on IPP projects and developed their skills and experiences
Ways of working	• Whether participants experienced that IPPs promote the social determinants of health and wellbeing • The extent to which IPPs promote a greater sense of cultural identity among Indigenous workers • Whether or not, and to what extent, local communities had been engaged during an IPP project to advise on business and employment opportunities and processes • Whether or not community concerns had been addressed on IPP projects and how their concerns were managed by project teams
Learnings	• The type and nature of challenges and setbacks that participants had experienced on IPP projects • Whether or not the challenges and setbacks had been overcome and the impact this had on relationships between participants and other contractors or local communities • How much trust, reciprocity, and sharing participants had experienced among project stakeholders, including communities, on IPP projects

SAMPLING, RECRUITMENT, AND ANALYSIS

Purposeful sampling using Indigenous Business Direct (IBD), an online database maintained by Supply Nation of registered and member businesses, was used to identify participants. Only Indigenous construction businesses were invited to participate in the study. Participants were invited and selected because they owned or were senior managers at Indigenous construction businesses and had worked on a government project subject to IPP requirements. Some snowball sampling (Noy, 2008) occurred where Indigenous participants recommended additional participants from non-Indigenous businesses. Additional non-Indigenous businesses were recommended because they had worked closely with Indigenous contractors to deliver projects previously. The participants who recommended them believed they had deep experience of IPP implementation that would be valuable to the study.

Most interviewees came from general or civil construction (n=17). Most participants are identified with their real names, respecting the principles of Indigenous data sovereignty (Janke, 2021) and ethical approval for this study. Only a small number of participants chose to use a pseudonym and most chose the option in the informed consent form to use their real name when writing research findings. Where two participants had the same name, a numerical identifier has been used to differentiate them. No context is given when using participants' names, to maintain their privacy: they own their data; we report on the findings.

Analysis followed an iterative process that recognised sustainability innovation research is still in an explanatory stage (Lüdeke-Freund, 2020). Analysis included several immersions between theory and data, and deductive and inductive coding, to develop an empirical model answering our research questions (Stål, Bengtsson and Manzhynski, 2022). Regular sense-checking with Ngaa-bi-nya and participants ensured the experiences of participants were reflected in findings. In addition to promoting social equity, this helped manage researcher positionality and subjectivity, and recognised ontological and epistemological influences on data collection and analysis (Braun and Clarke, 2006).

For example, after uploading transcribed interviews to NVivo, a round of structural coding was done to systematically categorise and sort data (Saldaña, 2021). Structural codes were based on Ngaa-bi-nya domains. A second round of coding was then done based on the questions in Table 9.2. This 'second cycle' of coding (Saldaña, 2021) identified data relevant to research questions and eliminated redundant data that were relevant to Ngaa-bi-nya but not necessarily relevant to the aims stated above. A third round of coding was done to ensure rigour in our analytical process. The third cycle of coding involved a final review of data to distil themes that explained how IPPs impact contractor behaviour.

Table 9.3 is an excerpt of the analytical process to reach the findings presented below.

Deep, collaborative, and rigorous analysis of empirical data is critical to move sustainability innovation research and practice from descriptive and explanatory to practical and evidence-based (Santa-Maria, Vermeulen and Baumgartner, 2022; Stål,

Table 9.3 *Data analysis and analytical codes from this qualitative research*

Data quotes	Structural code	Research question	Theme
And then the IPP has allowed us to be seen by them [tier one contractors]. Because if the IPP wasn't there, we wouldn't be seen.	Landscape	RQ2	Involvement in IPP delivery and governance
This is… procurement in action… We get work… we put on an apprentice… it's [the most positive impact of IPPs] the employment.	Landscape	RQ3	Economic, social, and cultural development
You've got to demonstrate that you have the ability, the financial capacity, and the skill set and quality requirements to actually deliver that job… it's given all of us here that opportunity to develop, grow and meet, overcome challenges.	Resources	RQ1	Opportunities realised by Indigenous businesses
And the same with using local skills… we would often identify those skills and let a contractor know that there are skills available in the community and that they should contact [them].	Resources	RQ1	Stronger connections in the construction industry
We actually take some contractors through a bit of a training programme and try and get their business up to a standard that's ready to do government work for [the] IPP.	Ways of working	RQ2	Shared strategies and solutions

Bengtsson and Manzhynski, 2022). In an IPP context, the analytical process above enabled the authors to develop a deep relationship with the data that explained the impact of IPPs on contractors to work to social sustainability. The findings of this new approach to analysing and explaining innovation in construction is presented below under the four Ngaa-bi-nya domains. Themes that report changes to contractor behaviour are presented under the relevant domain so our reporting is consistent with Indigenous ways of knowing and doing.

FINDINGS

This section presents the findings relevant to how IPPs change contractors' behaviour towards Indigenous participation.

Landscape

Involvement in IPP delivery and governance

Data coded to this theme referred to whether contractors involve Indigenous contractors in the delivery of IPPs. Data that supports this theme indicates that contractors are working more closely with Indigenous contractors to involve them in projects and identify how they can deliver projects. For example, one participant explained how IPP requirements increase exposure for their business to major infrastructure projects because contractors have to work with them. As Ash explained, that IPP targets require Indigenous contractors in infrastructure supply chains is a strategic opportunity they have used to deliver projects and generate revenue:

> And then the IPP has allowed us to be seen by them [tier one contractors]. Because if the IPP wasn't there, we wouldn't be seen. (Ash)

In addition to direct involvement with main contractors, other participants reported that they have actively been involved in how IPPs are administered. Participants explained that IPPs create an environment where non-Indigenous workers must work more closely and more frequently with Indigenous workers. Participants reported that this exposure is involving Indigenous people in the governance and delivery of IPPs. Therefore, contractors view Indigenous contractors and workers as more influential stakeholders who should be involved in day-to-day functions. The quote below explains one participant's experience of IPPs changing the attitudes and behaviours of non-Indigenous people. In the participant's experience, non-Indigenous staff had come to have a greater appreciation for Indigenous culture and contributed to viewing Indigenous staff and contractors as equal partners in the workplace:

> I think the staff, the non-Indigenous staff are seeing, obviously, a whole different side of Australia with the communities that we work in, and building a great appreciation of where we're at, and, you know, coming along with that journey with us. (Tim)

Resources

The strongest themes explaining how IPPs change contractor behaviour came from data coded under the Resources domain. These themes are the strongest because they explain how non-Indigenous contractors work closely with Indigenous contractors and the types of relationships this creates. The two themes relevant to our research questions were 'Opportunities realised by Indigenous businesses' and 'Stronger connections in the construction industry'.

Opportunities realised by Indigenous businesses

Findings under this theme indicate that IPPs create more contracting and employment roles that require Indigenous businesses and workers to fill them. These opportunities are being driven, at first, by the contractual requirements that main contractors must meet. Findings explain that IPPs may change behaviour because contractors want to avoid the financial penalties applied for non-compliance. As the following participant explained, forcing contractors to engage more with Indigenous businesses creates opportunities for them:

> And we won the job, which was great. So that's going to create a lot of opportunity because they have to meet their… government tender requirements on the IPP targets that they have in that. (John2)

Participants also explained that, once they secured initial work on projects, they still had to successfully deliver projects to ensure future work. Representing the effect of making non-Indigenous businesses consider them for work they would not normally be considered for (Kinsela-Christie, 2019), the following participant explained that they have experienced the changing perceptions of non-Indigenous contractors as they delivered work. This appears to be a significant impact of IPPs on contractor behaviour. First, IPPs change contractors' behaviour because they now have to work with Indigenous businesses. Second, Indigenous businesses who deliver work successfully come to be seen as a reliable delivery partner:

> You've got to demonstrate that you have the ability, the financial capacity, and the skill set and quality requirements to actually deliver that job… And then… under the IPP, I was procuring from [Tier one contractor]. (Ash)

Stronger connections in the construction industry

Findings indicate that another significant impact of IPPs on contractor behaviour comes from the stronger connections Indigenous businesses create in the construction industry. Similar to above, data indicated the prescriptive nature of IPPs mandating Indigenous suppliers drives non-Indigenous contractors to connect with Indigenous contractors. As Greg noted, 'mainstream contractor[s] will have certain social procurement targets to meet in the delivery of that project'. Greg's point is that IPPs make non-Indigenous contractors think about how they will meet IPP targets. This demonstrates how IPPs are a radical innovation (Bresnen and Marshall, 2001) because they actively make contractors rethink how they structure and engage their supply chains to deliver projects that include Indigenous contractors.

Other participants reported using this new environment strategically to develop their and their business's experience. For example, the following participant explained how they use their new connections to promote Indigenous engagement as much as possible while meeting procurement targets. The participant had a strategic goal of using their business to create as much Indigenous employment as possible in the construction industry through these connections.

> But we were definitely working with them [Tier one contractor] to try and encourage… as much Indigenous engagement as possible… with [their] targets. (Nathan1)

The following participant gave an operational example of how businesses are now working closely with non-Indigenous contractors to collaboratively identify where Indigenous businesses and workers can deliver projects. Demonstrating this changed behaviour supports the social equity as well as the community development and well-being elements of social sustainability. The participant describes their work with main contractors to raise their awareness of local Indigenous skills that could contribute to a project:

> And the same with using local skills… we would often identify those skills and let a contractor know that there are skills available in the community and that they should contact some to see if… those skills were going to be appropriate for the project. (Geoff)

IPPs therefore appear to create an environment where non-Indigenous contractors are required to connect with Indigenous contractors to engage them in their supply chains. Non-Indigenous contractors have reacted by working with Indigenous contractors to identify where they can deliver projects. Indigenous contractors use this to create more work for themselves, their staff, and their communities. This is a positive impact of IPPs on contractor behaviour as it opens industry networks that may have been previously closed to Indigenous businesses.

Ways of Working

The key theme under this domain referred to Indigenous and non-Indigenous contractors working together to share strategies and solutions. Findings explain how IPPs create opportunities for Indigenous and non-Indigenous companies to work together to develop new solutions to enable IPPs to succeed in delivering social sustainability.

Shared strategies and solutions

Data coded to this theme were noticeable for emphasising the importance participants placed on collaboration to help other Indigenous contractors grow and deliver projects. While the importance of collaborative partnerships has been discussed in a cross-sector social procurement context (Loosemore et al., 2021, 2022), little attention has thus far been given to intra-sector collaboration to deliver IPPs. This is important because delivering IPPs requires multiple relationships: main contractors, subcontractors, labour hire, and consultants, to name some examples. Findings under this theme highlight that the different actors now work together to deliver IPP outcomes, like, for example, understanding government contracts or delivery requirements on government projects.

Discussing these new forms of collaboration, the participant below explained how their business works with Indigenous contractors to upskill them on government requirements. The outcome of this is that their business will engage them as a subcontractor on a government project or support them to tender for a government contract in the future:

> So we actually take some contractors through a bit of a training programme and try and get their business up to a standard that's ready to do government work for [the] IPP... One thing that we guarantee at the end of that course is either a contract with [Company] or a contract with government through the IPP. (Nathan2)

In addition to collaboration for technical skills, as illustrated above, other participants explained how they actively foster collaboration with other Indigenous contractors to share contract opportunities. The participant explained that if they share procurement opportunities with other Indigenous contractors, they will increase interactions between Indigenous and non-Indigenous contractors. Their strategy for this was that it would lead to better IPP implementation and outcomes they could potentially create:

> So we want to encourage... networking and collaborative work with other Indigenous enterprises as well... I collaboratively work with them, saying, we'll advertise their services if we... are approached by a client that wants us to do that, (and) we'll just subbie it to them. (John1)

Other participants explained they actively sought to work with local communities on projects to identify opportunities for local businesses and workers. This is a deviation from the traditional construction view of communities as a risk that must be managed. It indicates that IPPs are changing behaviour in construction to work with Indigenous stakeholders to embed long-term outcomes in project delivery. For example, the participant below explained their approach of working with local communities to identify ongoing work opportunities. In the participant's experience, IPPs could be limited because of the short-term nature of project work. Therefore, to support effective IPP delivery and outcomes, they aimed to achieve good standing in the community by identifying construction and maintenance opportunities for local businesses and workers:

> As an Indigenous business, we actually partnered with [the Contractor], because we only had two and a half months to deliver the project. You know that that's the whole piece of getting involved with the community in [Location]. Not just goin' up there, doin' a job and leaving, right. (Tim)

DISCUSSION

The findings above provide evidence that IPPs have an effect on contractors' behaviour towards Indigenous participation on construction projects. These changes typically involve new motivations to work with Indigenous contractors and the level of involvement Indigenous contractors can have in these new relationships. This section discusses the importance of these findings in relation to the research questions above and the relevant innovation literature.

HOW DO IPPS IMPACT CONTRACTORS' BEHAVIOUR REGARDING INDIGENOUS PROCUREMENT AND PARTICIPATION ON CONSTRUCTION PROJECTS?

Overall, the findings above offer new insights into how contractors respond to innovations like IPPs to deliver them. Interfacing with research that discusses uncertainty avoidance (Mullett, 2022) as a barrier to innovation in construction, the findings explain that contractors actively take steps to reduce uncertainties related to implementing IPP deliverables. Delivering IPPs creates uncertainties because many contractors have not worked with Indigenous contractors before and this is challenging their business-as-usual operations. In response, contractors now work closely with Indigenous contractors to include them in their supply chains. The tangible effects of this can be seen in the increasing number of Indigenous engagement managers that large contractors now employ to deliver Indigenous participation and procurement in their business. The findings also indicate that contractors realise they need to be better connected with Indigenous businesses. They therefore change how they do business by actively seeking out and working with Indigenous contractors, and this is strong evidence of how IPPs impact contractors' behaviour to deliver Indigenous participation and procurement.

The above point highlights how the findings corroborate Australian and international research on social procurement implementation. For example, the new connections IPPs influence seem to drive new relationships between Indigenous and non-Indigenous contractors. These new relationships are being used to enhance the outcomes IPPs create through long-term supply partnerships and collaborations. This finding reinforces other research that found social procurement requires new skills and knowledge in construction that challenge how contractors do business (Loosemore et al., 2021; Troje, 2023). The findings above build on prior research and show that contractors adapt to these new circumstances by finding new ways (i.e., innovating) to ensure a pipeline of work, which in this case is derived from how they work constructively with Indigenous contractors. As research shows that Indigenous contractors create a range of outcomes that can contribute to social sustainability, such as economic, social, and cultural development and self-determination (Denny-Smith et al., 2024), these findings are a working example of how policy-led innovations (IPPs) can influence behavioural changes that contribute to social sustainability outcomes.

WHAT IS THE EFFECT OF CONTRACTORS' CHANGED BEHAVIOURS DUE TO IPPS ON SOCIAL SUSTAINABILITY OUTCOMES IN CONSTRUCTION?

The findings also give new insight into how the behaviours influenced by IPPs can contribute to social sustainability in construction. While Loosemore et al.'s (2021) research cautioned that contractors may cynically 'game' social procurement requirements as pushback against policy targets being imposed in a top-down fashion, our findings illustrate that Indigenous contractors had experienced constructive partnerships with non-Indigenous contractors in response to IPPs. IPPs have altered contractors' behaviour to collaborate regularly with Indigenous contractors. The effects of collaboration reported above are that Indigenous contractors use it to develop their own experience and create more work opportunities. This empowers them to invest in staff and their business, which can reinforce social sustainability determinants like social equity and socioeconomic development.

Regarding how stakeholders interact to deliver IPPs and social sustainability, our findings also support innovation research on the tensions and opportunities when implementing sustainable innovations. While determining 'success' in a sustainability innovation context can be subjective and hard to define in general because different stakeholders' definition of success can vary (Lüdeke-Freund, 2020), the findings indicate that IPPs can support social sustainability in construction because they define what good outcomes look like (i.e., success) and how contractors can achieve them. In such instances, the effect of IPPs on contractors' behaviour has been to coordinate what social sustainability can look like in construction and to create space for contractors to innovate in their business operations and culture to deliver IPPs. The other contribution of this insight is to explain how IPPs influence collaboration that addresses power imbalances in construction to find mutually beneficial solutions for sustainability innovations (Stål, Bengtsson and Manzhynski, 2022). Because contractors change their behaviour to implement IPPs, they work more closely with Indigenous contractors. They then find optimal partnerships, where Indigenous contractors continue to develop their experience and ensure continuity of work and non-Indigenous contractors continue to deliver their IPP requirements.

WHAT ROLE DO POLICY-LED INNOVATIONS LIKE IPPS HAVE IN INFLUENCING SUSTAINABLE ATTITUDES IN CONSTRUCTION?

Given how IPPs change contractors' behaviour, and the effects the changes could have on social sustainability outcomes in construction, our findings indicate that policy innovations are crucial in driving future sustainability activities in construction. Indeed, this finding reinforces sustainability innovation research that found public policies support the implementation of sustainable innovations by providing a level of security that counters perceptions of financial risk (Lüdeke-Freund, 2020). Our findings explained that non-Indigenous contractors may engage more enthusiastically with Indigenous contractors once they are more familiar with them and their capabilities. This is important because it can reinforce knowledge to implement these initiatives and the required resources, something lacking from social sustainability practice in construction (Troje, 2023).

Further, our findings indicate that policy innovations do have a critical role in balancing power to implement sustainable activities. Empirically evidencing that IPPs do create a level playing field and open closed industry networks (Kinsela-Christie, 2019), policy innovations like IPPs can actively change contractors' behaviour so they begin considering how they can run their businesses to meet policy requirements and therefore sustainability outcomes. This addresses power imbalances that have closed industry networks to Indigenous contractors previously and shows how collaborative partnerships can overcome uncertainties of working with new partners to progress social sustainability (Stål, Bengtsson and Manzhynski, 2022). As such, the role of IPPs is to prescribe actions between Indigenous and non-Indigenous contractors to collaborate and develop economic outcomes and social sustainability progress.

LIMITATIONS

The findings in this chapter have minor limitations that may affect their generalisability. First, findings are based on interviews with 24 participants. While the participants had deep experience of the impact of IPPs, our sample size is unlikely to cover the full breadth and types of examples of the impact of IPPs, for better or worse. Second, we have excluded 'Learnings' themes when presenting the findings. This is because during analysis only one theme was found under the 'Learnings' domain. The theme (Self-determination as an outcome of IPPs) was left out of reported findings as it was not relevant to the impact of IPPs on contractors' behaviour. A larger sample may find relevant 'Learnings' themes that give strong evidence of the usefulness of policy-led innovations to change industry attitudes to social and other sustainability objectives. Nevertheless, these limitations have a minor effect on exploratory studies like this that can be refined and tested with more empirical work.

CONCLUSION

In this chapter we sought to explore how IPPs influence contractors' behaviour towards Indigenous peoples' participation in construction projects.

Our findings indicate that policy-led innovations may be crucial to driving behavioural change in an industry that seeks to avoid risks. Policy-led innovations create deterrents and incentives that motivate contractors to embrace these innovations and consequently adapt how they work to effectively deliver them. In an IPP context, main deterrents include financial penalties and reputational damage for not meeting Indigenous procurement and employment targets. Incentives for contractors include brand awareness for being a good delivery partner to government clients and Indigenous subcontractors, maximising chances of successful future tenders, meeting internal Reconciliation Action Plan targets, and leaving a positive social legacy in the communities where they build.

These findings make a significant contribution to international efforts to address socioeconomic inequities and better include marginalised communities in the workforce. First, we have shown the value of policy-led innovations to address the aforementioned global challenges. Policy-led innovations can have a significant effect on the behaviour of contractors to better include the groups that benefit from socially sustainable outcomes. We have shown policies like IPPs influence innovation in construction companies to better engage with Indigenous-owned businesses and workers. The second significant contribution of this research is to address the lack of social sustainability research in large industries like construction. Indeed, our findings indicate that construction can innovate to achieve social sustainability in response to policy drivers. This is a significant insight for companies and governments aiming for social sustainability objectives. At the macro level, policies like IPPs are valuable for directing the outcomes a government wants to achieve and indicates to an industry what to do to achieve those outcomes – in this case, Indigenous procurement. At the micro level, our research also provides insights for forward-thinking contractors who want to improve Indigenous participation in their business and supply chains. We have found that greater collaboration and opening up industry networks can positively influence this. In our experience, this drives innovation because it opens contractors to new ideas and new ways of working – a potential solution to the inhibitors of social sustainability in construction discussed earlier.

Internationally, our findings are instructive for other governments with social sustainability objectives. Our research shows that policy-led innovations can drive behavioural change for social sustainability outcomes in construction; this means that other countries may adopt our findings to inform their own policies to drive country-specific outcomes. While colonised countries like Canada and New Zealand have or are developing Indigenous procurement policies, our findings could be used to review those policies for improved development. The findings above could also be transplanted in developed countries that experience struggles adopting refugees and migrants into the population; governments could use our insights. Governments aiming to influence industry behaviour to accept new members could develop their own prescriptive procurement policies that require contractors to engage with migrants more often. In this sense, our research is also a possible avenue to address the social dislocation caused by large numbers of refugees and undocumented migrants internationally.

In conclusion, policy-led innovations may become more crucial as construction and other industries transition to a sustainable future. The benefit of policy-led innovations is that they can be hard to avoid and are subject to public auditing. They create an environment where contractors must act to improve their performance and therefore are useful to move the dial towards more sustainability (economic, social, or environmental) innovations in the future. While more empirical work is needed to identify and mitigate the risks that contractors could undermine IPP implementation through unscrupulous behaviours, the findings in this chapter have established the value of innovation policies in influencing a more socially sustainable society.

REFERENCES

AIATSIS. (2020) *AIATSIS Code of Ethics for Aboriginal and Torres Strait Islander Research*. Canberra: Australian Institute for Aboriginal and Torres Strait Islander Research.

Bailey, S., Carnemolla, P., Loosemore, M., Darcy, S. and Sankaran, S. (2022) 'A critical scoping review of disability employment research in the construction industry: Driving social innovation through more inclusive pathways to employment opportunity', *Buildings*, 12(12), 2196.

Balaman, Ş.Y. (2019) *Decision-making for biomass-based production chains*. Amsterdam: Elsevier.

Barraket, J., Hiruy, K., Walshe, J. and Moussa, B. (2021) *The state of social procurement in Australia and New Zealand*. St Marys: IPA Personnel Services and Chartered Institute of Procurement and Supply.

Bessarab, D. and Ng'andu, B. (2010) 'Yarning about yarning as a Legitimate method in Indigenous research', *International Journal of Critical Indigenous Studies*, 3(1), 37–50.

Braun, V. and Clarke, V. (2006) 'Using thematic analysis in psychology', *Qualitative Research in Psychology*, 3(2), 77–101.

Bresnen, M. and Marshall, N. (2001) 'Understanding the diffusion and application of new management ideas in construction', *Engineering, Construction and Architectural Management*, 8(5/6), 335–345.

Deloitte (2014) *Economic benefits of closing the gap in Indigenous employment outcomes. Report for Reconciliation Australia*. Sydney: Deloitte.

Denny-Smith, G. (2021) 'How compliance imperatives stifle the impact of social procurement', in L. Scott and C. Neilson (eds) *Proceedings of the 37th Annual ARCOM Conference*. UK: Association of Researchers in Construction Management, 634–643.

Denny-Smith, G., Williams, M., Loosemore, M., Sunindijo, R.Y. and Piggott, L. (2024) 'What social value do Indigenous contractors create?', *Construction Management and Economics*, 42(1), 16–34.

Dudgeon, P., Bray, A., Darlaston-Jones, D. and Walker, R. (2020) *Aboriginal participatory action research: An Indigenous research methodology strengthening decolonisation and social and emotional wellbeing*. Melbourne: The Lowitja Institute.

Evans, M., Polidano, C., Moschion, J., Langton, M., Storey, M., Jensen, P. and Kurland, S. (2021) *Indigenous business sector snapshot 1.0*. Melbourne: Indigenous Business Research Group, University of Melbourne.

Hurt-Suwan, C.J.P. and Mahler, M.L. (2021) 'Social procurement to reduce precarious employment for Māori and Pasifika workers in the construction industry', *Kōtuitui: New Zealand Journal of Social Sciences Online*, 16(1), 100–115.

Janke, T. (2021) *True Tracks: Respecting Indigenous knowledge and culture*. Sydney: NewSouth Publishing.

Kinsela-Christie, K. (2019) *Supplier diversity how: A 5 step guide to Indigenous business procurement*. Bunbury, WA: IPS Management Consultants.

LePage, D. and Renaerts, E. (2023) 'Achieving the UN SDGs through the integration of social procurement in construction projects', in T. Walker, C. Cucuzzella, S. Goubran and R. Geith (eds) *The role of design, construction and real estate in advancing the Sustainable Development Goals*. Cham: Palgrave Macmillan, 129–147.

Loosemore, M., Denny-Smith, G., Barraket, J., Keast, R., Chamberlain, D., Muir, K., Powell, A., Higgon, D. and Osborne, J. (2021) 'Optimising social procurement policy outcomes through cross-sector collaboration in the Australian construction industry', *Engineering, Construction and Architectural Management*, 28(7), 1908–1928.

Loosemore, M., Keast, R., Barraket, J. and Denny-Smith, G. (2022) 'The risks and opportunities of social procurement in construction projects: A cross-sector collaboration perspective', *International Journal of Managing Projects in Business*, 15(5), 793–815.

Lüdeke-Freund, F. (2020) 'Sustainable entrepreneurship, innovation, and business models: Integrative framework and propositions for future research', *Business Strategy and the Environment*, 29(2), 665–681.

Malcolm, K., Cairns, F. and Pouwhare, T. (2023) 'Empowering indigenous enterprise through supplier diversity – The case for Puna Awarau in Aotearoa (New Zealand)', *Equality, Diversity and Inclusion: An International Journal*, In-Press. Available at: https://doi.org/10.1108/EDI-10-2022-0301

Martin, K. and Mirraboopa, B. (2003) 'Ways of knowing, being and doing: A theoretical framework and methods for Indigenous and Indigenist research', *Journal of Australian Studies*, 27(76), 203–214.

Minkler, M. (2005) 'Community-based research partnerships: Challenges and opportunities', *Journal of Urban Health*, 82(Supp. 2), ii3–ii12.

Missimer, M., Robèrt, K.-H. and Broman, G. (2017) 'A strategic approach to social sustainability – Part 2: A principle-based definition', *Journal of Cleaner Production*, 140, 42–52.

Mullett, P. (2022) 'Challenges to innovation in construction', in S.H. Ghaffar, P. Mullett, E. Pei and J. Roberts (eds) *Innovation in Construction*. Cham: Springer International Publishing, 89–115.

NIAA. (2023) *Indigenous Procurement Policy*. Canberra: National Indigenous Australians Agency. Available at: https://www.niaa.gov.au/indigenous-affairs/economic-development/indigenous-procurement-policy-ipp (Accessed: 28 August 2022).

Noy, C. (2008) 'Sampling knowledge: The hermeneutics of snowball sampling in qualitative research', *International Journal of Social Research Methodology*, 11(4), 327–344.

NSW Government (2022) *Aboriginal Procurement Policy*. Sydney: NSW Government.

Raiden, A. and King, A. (2023) 'Added value and numerical measurement of social value: A critical enquiry', *Buildings and Cities*, 4(1), 767–782.

Rossingh, B. and Yunupingu, Y. (2016) 'Evaluating as an outsider or an insider: A two-way approach guided by the knowers of culture', *Evaluation Journal of Australasia*, 16(3), 5–14.

Saldaña, J. (2021) *The coding manual for qualitative researchers*. 4th edn. London: Sage.

Santa-Maria, T., Vermeulen, W.J.V. and Baumgartner, R.J. (2022) 'The Circular Sprint: Circular business model innovation through design thinking', *Journal of Cleaner Production*, 362, 132323.

Sharma, D., Kumar, P. and Singh, R.K. (2023) 'Empirical study of integrating social sustainability factors: An organizational perspective', *Process Integration and Optimization for Sustainability*, 7(4), 901–919.

Shirodkar, S. (2019) 'Bias against Indigenous Australians: Implicit association test results for Australia', *Journal of Australian Indigenous Issues*, 22(3–4), 3–34.

Stål, H.I., Bengtsson, M. and Manzhynski, S. (2022) 'Cross-sectoral collaboration in business model innovation for sustainable development: Tensions and compromises', *Business Strategy and the Environment*, 31(1), 445–463.

Supply Nation (2018) *The sleeping giant: A social return on investment report on Supply Nation certified suppliers*. Sydney: Supply Nation.

Torkayesh, A.E., Ecer, F., Pamucar, D. and Karamaşa, Ç. (2021) 'Comparative assessment of social sustainability performance: Integrated data-driven weighting system and CoCoSo model', *Sustainable Cities and Society*, 71, 102975.

Troje, D. (2023) 'Social sustainability in projects: Using social procurement to create employment in the Swedish construction sector', *Project Management Journal*, 54(1), 52–69.

United Nations (2007) *United Nations Declaration on the Rights of Indigenous Peoples*. New York: United Nations.

Williams, M. (2018) 'Ngaa-bi-nya Aboriginal and Torres Strait Islander program evaluation framework', *Evaluation Journal of Australasia*, 18(1), 6–20.

Zhang, R., Chen, T., Zhang, Y. and Wang, Z. (2023) 'Exploring differences in level of construction innovation: An empirical analysis in China', *Environmental Science and Pollution Research*, 30(16), 48179–48200.

10. Multi-stakeholders' perspectives in public procurement of innovation in construction: analysing benefits, beneficiaries and actors of innovation

Per Erik Eriksson, Johan Larsson and Malena Havenvid

INTRODUCTION TO PUBLIC PROCUREMENT OF INNOVATION IN CONSTRUCTION

Public Procurement of Innovation

In the European Union, public demand has been increasingly emphasised as a major potential driver of innovation (Edler & Georghiou, 2007; Obwegeser & Dueholm Müller, 2018). As noted by Uyarra and Flanagan (2010: 123), 'the notion that public procurement can be used to promote innovation is high on the agenda of European policy-makers at all levels'. Hence, both policymakers and scholars recognise the potential demand-side power of public clients' procurement strategies to drive innovation (Uyarra et al., 2014; Suhonen et al., 2019; Rainville, 2021). When public procurement is utilised to stimulate innovation in a broad sense, it is commonly called 'public procurement of innovation' (Kundu et al., 2020; Uyarra et al., 2020; Mwesiumo et al., 2021).

Previous authors have argued that public procurement is an especially important policy tool to facilitate innovation related to grand societal challenges and so-called 'wicked problems', such as climate change, public health, and security (Edquist & Zabala-Iturriagagoitia, 2015; Uyarra et al., 2020; Mwesiumo et al., 2021). Arguably, the Sustainable Development Goals (SDGs) in the United Nations' Agenda 2030 constitute such grand societal challenges that public clients need to address through innovation (Ika & Munro, 2022; George et al., 2024). To do so, 'the public sector may be willing to pay premium cost or bear some efficiency losses if it wants to encourage certain policy goals and societal needs, such as sustainability, social inclusion, etc.' (Uyarra & Flanagan, 2010: 130).

Possibilities to inform public procurement strategies meaningfully in practice have, however, been limited by failure to fully acknowledge the highly varied and complex nature of both innovation and public procurement in recent debates and literature (Uyarra & Flanagan, 2010). Hence, both the innovation and procurement concepts need more comprehensive, detailed, and context-specific treatment to inform public clients. Furthermore, adjustments to traditional public procurement procedures may be needed to facilitate innovation (Suhonen et al., 2019), as well as developing public clients' in-house competences and capabilities. Accordingly, low competences of public procurers and lack of organisational support and resources are major barriers to innovation through public procurement (Uyarra et al., 2014; Ntsondé & Aggeri, 2021; Mwesiumo et al., 2021; Rainville, 2021).

Public Procurement of Innovation in Construction

Public procurement occurs in diverse industrial sectors, but it is especially prominent in the construction, health, and transport sectors (Georghiou et al., 2014), where large public clients may use procurement to shape their markets. The project-based construction sector is a particularly relevant empirical context to address for two main reasons. First, many scholars and policy initiatives emphasise the importance of the client's role in stimulating innovation within the construction industry and built environment (e.g., Wolstenholme, 2009; Barbosa et al., 2017; Järvenpää et al., 2023). More specifically, many studies have shown that construction clients' procurement strategies play crucial roles in the facilitation (or hindrance) of innovation in construction projects (Dulaimi et al., 2002; Rose & Manley, 2014; Eriksson, 2017; Hedborg Bengtsson et al., 2018; Larsson et al., 2022). Nevertheless, the client's innovation needs are seldom considered when choosing procurement strategies (Eriksson et al., 2017). Second, the lack of client competence is particularly clear in the procurement of complex goods and services, such as construction projects (Uyarra et al., 2014). A major reported reason for this lack of client competence regarding suitable procurement practices is insufficient demand articulation (Uyarra et al., 2020). Thus, in order to procure innovation, public construction clients need to improve their understanding of innovation needs, that is, 'the characteristics of what is to be procured in terms of needs and requirements' (Lenderink et al., 2022: 668).

In construction contexts, most innovation processes occur at the business project level (i.e., in construction projects), as construction work is customised to deliver one-off solutions that meet client requirements (Ivory, 2005; Ozorhon, 2013; Havenvid et al., 2016). Consequently, innovation efforts are generally focused on traditional short-term project benefits, such as reductions in time and costs, or increases in quality-related variables for the project actors (Loosemore, 2015; Ozorhon et al., 2016). Hence, they may miss opportunities to address longer-term benefits, related to sustainable development, for broader ranges of external stakeholders and the whole society (Ivory, 2005; Ozorhon et al., 2016; Larsson & Larsson, 2018). As public construction clients have increased their attention to long-term sustainability effects during the last decade (Rundquist et al., 2013), it is essential to assess the benefits and beneficiaries of innovation more broadly. Furthermore, different outcomes have different benefits for different stakeholders, both in terms of project actors and external stakeholders, so their interests must be considered when deciding whether to pursue a certain innovation need or not (Gil et al., 2012). Arguably, public clients have an especially important, but challenging, role to play by managing complex networks of project actors as well as external stakeholders and the pressures to satisfy diverse, and often conflicting, priorities in terms of both benefits and beneficiaries of innovation.

To facilitate the formulation and implementation of urgently required place- and case-sensitive public procurement procedures (Uyarra & Flanagan, 2010; Uyarra et al., 2020), we argue that client organisations must analyse and understand their innovation needs before designing appropriate strategies. Hence, in this chapter, we discuss central features of clients' innovation needs, in terms of benefits and beneficiaries of innovation outcomes and the actors of innovation processes to enable more systematic analysis of the multi-dimensional innovation needs in the project-based construction industry. The logic underpinning this discussion is that clients must first analyse and understand their innovation needs to optimise decisions regarding the design and implementation of procurement strategies that enhance innovation processes and the achievement of targeted innovation outcomes. In addition to reviewing relevant literature, our discussion is based on a complementary interview study conducted to illuminate more clearly the benefits, beneficiaries, and actors of innovation in a specific public construction procurement context (the Swedish infrastructure sector). In total, 19 respondents were interviewed, representing the three key sets of project actors: client, contractors, and design consultants.

ANALYSING THE BENEFITS OF INNOVATION

Analysis of the benefits of an innovation involves assessment of its intended purposes and effects (Bygballe & Ingemansson, 2014; Brockmann et al., 2016). In the innovation management literature, the purpose (or benefit) of innovation is not extensively discussed. Arguably, this is because the implicit or explicit purpose generally is to improve the performance and competitive advantage of the innovating firm (Crossan & Apaydin, 2010), whereas societal effects are largely neglected (Meissner et al., 2024). Similarly, Hartmann et al. (2008: 439) note that private construction clients' innovation behaviour 'mainly depends on the contribution of the constructed facility to the competitive positioning of the organization'. However, this general and straightforward benefit of competitive advantage is not applicable to public client organisations, because 'competition does not primarily underlie strategic choices and innovation activities' (Hartmann et al., 2008: 439). Public clients' long-term societal concerns and short-term concerns for each temporary project may sometimes coincide, but often conflict, which complicates analysis of the benefits of innovation.

It is therefore common to differentiate benefits into the two overarching categories of short-term and long-term outcomes (Ozorhon et al., 2016; Bandé et al., 2024). Similarly, in an analysis of Heathrow Airport's Terminal 5 project, Gil et al. (2012: 457) concluded that 'assessments of benefits could vary dramatically since some new technologies aimed to bring project efficiencies (benefiting directly the sub-project team that owned the budget), whereas others aimed to improve operations after project handover'. We believe that acknowledgement of this division and potential associated tensions is especially crucial for public clients in the project-based construction industry, since they must consider both traditional *short-term benefits* of specific projects and *long-term benefits* related to sustainable development.

Short-term Benefits

Short-term benefits are connected to traditional project performance measures, such as the iron triangle of time, cost, and quality (Ozorhon & Oral, 2017; Bandé et al., 2024). Construction innovation efforts have mainly focused on cost and time reduction, since the most important aims of actors in a construction project generally include completion within budget and on time (Gann & Salter, 2000; Ozorhon, 2013; Ozorhon et al., 2014). These two success criteria often go hand in hand, as construction firms often implement innovations to increase efficiency (Goodrum & Haas, 2000). Accordingly, prior research has found that investments in innovation must pay off in the specific project in which they occur (Eriksson & Szentes, 2017), implying a strong focus on short-term benefits. Another contributor to the focus on short-term benefits is that innovation in construction projects is mainly driven by efforts in the design and production phases to overcome problems encountered in a specific project and associated hindrances (Bygballe & Ingemansson, 2014). Thus, as innovation in construction mainly occurs within the timeframe of individual projects and generally targets project-specific issues, the benefits tend to be short term and highly project-related.

Long-term Benefits

Long-term benefits involve broader and more permanent effects of innovation, including improvements in variables related to environmental and/or societal aspects of sustainability, beyond a temporary project (Arif et al., 2013; Eriksson et al., 2014; Vosman et al., 2023). A systematic bibliometric review of 257 construction innovation studies has shown that interest in sustainability-related innovation has exploded in recent years (Wang et al., 2023). Public procurement initiatives intended to stimulate long-term environmental effects are often referred to as sustainable, circular, or green (Ntsondé & Aggeri, 2021; Krieger & Zipperer, 2022). However, at an overarching level, construction innovations focusing on efficiency may also have significant long-term societal benefits (Arif et al., 2013). For example, when short-term construction costs are reduced, the properties produced may become more affordable, and hence more accessible to a greater proportion of the population, thereby facilitating heterogeneity and social sustainability. Initiatives that are intended to improve the short-term efficiency of construction may consequently have long-term societal consequences for end-users. The SDGs of Agenda 2030 are examples of long-term benefits that are regarded as crucial for the whole society (Ika & Munro, 2022). However, long-term benefits may also include organisational matters such as improving involved firms' images, experience and knowledge, future business collaborations, and market growth (Slaughter, 1998; Ozorhon et al., 2016).

ANALYSING THE BENEFICIARIES OF INNOVATION

According to the innovation management literature, innovation outcomes are generally intended to benefit the private sector firm undertaking the innovation effort (e.g., Jansen et al., 2006; Mueller et al., 2013; Meissner et al., 2024). However, this general focus on internal benefits is insufficient for public clients because their innovation outcomes may create public value affecting diverse stakeholders (Chen et al., 2020). In the project-based construction industry, innovations are implemented in inter-organisational projects, thereby affecting multiple project actors in the supply chain, that is, stakeholders internal to the project (Widén et al., 2014; Larsson & Larsson, 2018). However, innovations may also affect external stakeholders connected to society as a whole (Uyarra et al., 2020; Meissner et al., 2024). As specific innovations may have varying effects (positive and potentially negative) for different stakeholders, both internal and external to the project supply chain (Bygballe & Ingemansson, 2014; Larsson et al., 2022), benefits of innovation should be analysed through the eye of the beholder. Accordingly, when analysing the benefits of an innovation, it is also important to reflect on for whom (project actors and/or external stakeholders) the benefits are created, and/or who will be affected by the implementation (i.e., beneficiaries). This is because 'the project-based decision to adopt (or not) a new technology is a collective outcome involving all key actors that would see their activities impacted by the innovation, either during project delivery or after hand-over to operations' (Gil et al., 2012: 453). Due to the strong interconnections between the benefits and beneficiaries of innovation outcomes, these two features must be analysed together. Beneficiaries of innovation may be divided into two categories: *project actor(s)* and *external stakeholder(s)* (Bandé et al., 2024), as outlined in the following text.

Project Actor(s)

Many innovations are meant to benefit one or multiple firms or actors within a project-based supply chain. These benefits may involve both short-term and long-term effects for project actors (Ozorhon et al., 2016). The most common benefits are related to improvements in project performance, especially reductions in project costs and time schedules, which may be beneficial for several project actors, such as the client and contractors. However, project actors may also benefit from more long-term effects, such as improvements in image and reputation (Goodrum & Haas, 2000), acquired experience and capability (Ozorhon et al., 2014), and future business collaboration opportunities (Ozorhon, 2013).

External Stakeholder(s)

Innovations may also affect external stakeholders outside the project supply chain, for example, end-users, neighbours, taxpayers, and the surrounding society (Bandé et al., 2024). In the innovation management literature, external stakeholders in the surrounding society are largely neglected (Chen et al., 2020; Meissner et al., 2024), but they have received much more attention in the literature on public procurement of innovation. This is because public clients may procure innovation related to 'wicked problems' and grand societal challenges (Edquist & Zabala-Iturriagagoitia, 2015; Uyarra et al., 2020; Mwesiumo et al., 2021; George et al., 2024). Construction innovation literature increasingly acknowledges that the implementation of innovation may not always be restricted to affected project-level actors. Public infrastructure clients, for example, often demand innovations that benefit an entire transport system or utilities and surrounding society. The effects of innovation implementation then become more difficult to control and constrain, as they extend beyond the influence of a single organisation or individual (Harty, 2008). The effects on external stakeholders are arguably becoming stronger and more widespread, as the construction industry intensifies efforts to develop services and innovations that meet broader societal and environmental concerns in the global business environment (Rundquist et al., 2013). Consequently, the motivation to implement sustainability-related innovations significantly affects both the benefits and beneficiaries of innovation outcomes.

External stakeholders have different interests and expectations, which may lead to some supporting certain construction projects and others opposing them (Chan & Oppong, 2017). Furthermore, as effects on innovation outcomes for different stakeholders may substantially differ (Gil et al., 2012), it is crucial to identify potential external stakeholders that are adversely affected by an innovation in advance in order to reduce their incentives to hinder implementation and diffusion (Widén et al., 2014).

ANALYSING THE ACTORS OF INNOVATION

As described by Chesbrough (2003), an innovation process may be 'closed' (handled in-house by one organisation) or 'open' (involving multiple collaborating actors). The analysis of (project) actors of innovation is arguably strongly connected to the analysis of the beneficiaries. This is because systemic innovations (e.g., innovations tackling grand challenges) with outcomes that affect many stakeholders (beneficiaries) are often considered complex and require strong inter-organisational collaboration during the innovation process (Chen et al., 2020; Vosman et al., 2023; George et al., 2024). Collaborative relationships may therefore facilitate innovation (Dulaimi et al., 2002; Harty, 2008; Rutten et al., 2009; Ozorhon, 2013), whereas adversarial relationships among key (project) actors may be major obstacles to innovation (Kumaraswamy & Dulaimi, 2001; Rose & Manley, 2014). In construction management, prior research has shown that key stakeholders affected by the outcome should be involved and engaged in the innovation process to increase the chances of successful innovation diffusion (Widén et al., 2014). Innovation in construction projects may involve diverse project actors, but the main project actors are often the client, contractor, and design consultant, as discussed below.

Client

Numerous studies have shown that customers (i.e., clients) play an important role in either driving or supporting innovation through requirements, customisation demands, and use over time (e.g., von Hippel, 1986). In the construction innovation literature, clients have been identified as important actors in initiating innovation (Blayse & Manley, 2004; Hartmann et al., 2008). Construction clients can be defined as the actors who 'initiate, commission, and pay for a construction project and formulate and communicate the requirements of a construction project to be accomplished for the intended usage of the facility' (Hartmann et al., 2008: 437). Hence, clients govern projects by setting specifications and requirements for the project organisation, thereby strongly influencing particular project outcomes, such as innovation (Kometa et al., 1994; Gann & Salter, 2000; Newcombe, 2003). In construction, innovation must be implemented in inter-organisational projects and thus is a collective effort that requires the interaction of several project actors, in which the client can act as an 'innovation supporter' (Nam & Tatum, 1997). Some scholars also call for greater participation of public clients in innovation processes through co-creation (Järvenpää et al., 2023).

Determining the client's role in, and ability to drive, innovation is, however, not trivial, as it depends not only on the client's internal 'innovation capability' (Nam & Tatum, 1997; Manley, 2006), but also on how it may trigger innovative behaviour of other project actors, such as contractors (Winch, 1998; Håkansson & Ingemansson, 2013). For professional clients, who can generally 'mobilise greater resources to acquire and assimilate innovations' than infrequent clients (Hartmann et al., 2008: 440), their overall innovation capability is a complex function of the expertise of specialised units and individuals, routines and skills in the internal organisation, and a set of network relationships (Manley, 2006). This means that a professional client's innovation capability partly depends on resources and activities external to its own organisation, and capacities to mobilise these resources and activities through relationships (Havenvid et al., 2016). Thus, the types of relationships clients have and can potentially deploy with other actors, especially main contractors, are important components of clients' abilities to trigger innovative behaviour in project organisations. Both formal contractual arrangements and informal social sentiments supporting interaction processes, such as in partnering projects, have reported the ability to foster innovation from both a short- and long-term perspective (e.g., Kumaraswamy & Dulaimi, 2001; Bygballe et al., 2010; Crespin-Mazet et al., 2015; Havenvid et al., 2016).

Contractor

In the CoPS (complex product systems) model, Winch (1998) identifies clients, regulators, and professional institutions as the 'innovation superstructure' – those who demand innovation – and trade contractors, specialist consultants, and component suppliers as the 'innovation infrastructure' – those who implement it. As an executor of client requirements and organiser of inputs of other trades, the main contractor serves as the 'systems integrator' or mediator in the production phase. Hence, contractors have a key mediating role between members of the superstructure, who stimulate searches for new ideas, and those in the infrastructure, who execute production and manage the implementation of the ideas (Rutten et al., 2009). One consequence of this mediating role is that unless contractors are positively affected by (beneficiaries of) a new idea and have the skills to develop it, innovation will not occur or will be slow. Therefore, the likelihood of innovation occurring in construction projects strongly depends on the contractors' activities and relationships (Winch, 1998; Dulaimi, 2022).

Rutten et al. (2009) argue that a 'true' systems integrator must govern the design as well as the construction phase. This is the case in design-build contracts, which provide incentives for contractors to minimise costs to succeed in the tendering phase and therefore to generate cost-reducing innovations (Hartmann et al., 2008; Larsson et al., 2022; Järvenpää et al., 2023). Recent studies have also shown that contractors may innovate for long-term benefits (e.g., reducing lifecycle costs) for the client when they have appropriate incentives in design-build-maintain contracts (Larsson et al., 2022). In contrast, a split contract (e.g., design-bid-build) dividing the system integrator role between consultant and contractor may hinder innovation in construction (Winch, 1998; Rutten et al., 2009; Eriksson & Westerberg, 2011).

Consultant

According to the innovation management literature, consultants are sources of expert knowledge that can be used for various innovative purposes, acting, for example, as 'cross-pollinators' of experience and knowledge between firms in different fields and locations, and as intermediaries between clients and other specialist firms (Bessant & Rush, 1995). It has been argued that consultancy firms' ability to transfer technological knowledge is based on their social relationships (Gans & Stern, 2003). From a social network perspective, they fill 'structural holes' (Burt, 1992) by building and maintaining extended social networks. They have also been recently identified as important actors in 'open innovation' through the ability to provide R&D services that have traditionally been carried out in-house (Bianchi et al., 2016).

In construction, design consultants play a key role in providing innovative design solutions to clients and acting as brokers of technologies from other consultants to the client (Ivory, 2005). According to the CoPS model, if a contract involves a main design consultant, this actor will act as a systems integrator in the design phase (Winch, 1998). This means that project-level innovation is highly dependent on the relationship between the consultant and client in early design phases (Ivory, 2005) and the client's ability to stimulate innovative joint performance (Eriksson & Kadefors, 2015). However, consultants may also play important roles in supporting the contractor's innovation in design-build contracts (Larsson et al., 2022).

EMPIRICAL METHOD

To complement the presented literature review and empirically illustrate the benefits, beneficiaries, and actors of innovation, we interviewed key actors in a relevant public construction procurement context: the Swedish infrastructure sector. Respondents with pertinent knowledge and experience of the Swedish infrastructure sector were selected in collaboration with the Swedish Transport Administration (STA) – the largest public procurer of infrastructure in Sweden, representing the client perspective in this study – as follows. Several suitable respondents were first identified by the Swedish Transport Administration's representatives and the researchers individually, then jointly evaluated to identify a suitable number and configuration of respondents with various kinds of experience. To obtain a representative group of respondents with the required diversity of experience and knowledge, they were selected to include:

1. representatives of all three key actors (client, contractor, and engineering consultant)
2. representatives with experience in diverse types of projects (road and railway), and projects in various phases of infrastructure lifecycle, including planning, design, production, and operations and maintenance (O&M)
3. representatives of diverse hierarchical levels, functions, and divisions in the participating organisations, such as division/department managers, project managers, and innovation managers.

The final sample consisted of 19 respondents, including 9 representatives of the client organisation STA (Trafikverket in Swedish) and 10 representatives from various kinds of suppliers (three consultancy firms, five construction contractors, and two O&M contractors). See Table 10.1 for a summary of the respondents and the interviews. About half of the interviews were conducted digitally (via Zoom) and half through in-real-life (IRL) meetings.

To collect data consistently and coherently, we applied an interview guide covering relevant background information about the respondent, more general innovation aspects, as well as particular questions about each of the three key focal features of innovation (benefits, beneficiaries, and actors). More specifically related to the empirical material presented here, the respondents were asked to reflect on innovation in the Swedish infrastructure sector, for example, in terms of benefits, beneficiaries, and actors of innovation, based upon their professional roles and experiences. The durations of the semi-structured interviews ranged between 70 and 135 minutes, depending on the respondent's scope of experience and role. In total, the interviews lasted 26 hours and 45 minutes. Deviation from the initial questions was allowed to enable the pursuit of relevant insights raised during the interviews.

In addition, a one-day workshop was held, in which the actors (representatives of the client, contractors, and consultants) were asked to reflect on and discuss

Table 10.1 Respondents and interviews

Respondent no.	Actor type	Respondent role	Length [min]	Type
R1	Construction contractor	R&D/innovation	90	IRL
R2	Construction client	Project director R&D/innovation	100	IRL
R3	Construction contractor	Project director	110 (same as R4)	IRL
R4	Construction contractor	Business manager	110 (same as R3)	IRL
R5	Engineering consultant	Business manager	90	IRL
R6	O&M client	Project manager	85	Zoom
R7	Construction client	District manager (construction projects)	85	IRL
R8	Construction client	Business developer	75	IRL
R9	Engineering consultant	Project manager	70	IRL
R10	O&M client	District manager (O&M)	80	IRL
R11	O&M contractor	Area manager (O&M)	75	Zoom
R12	Engineering consultant	Head of innovation	75	IRL
R13	Planning client	Senior role, planning national level	90	Zoom
R14	O&M contractor	Project manager	90	Zoom
R15	Construction client	Business developer	135	Zoom
R16	Construction contractor	Site manager	90	Zoom
R17	Construction contractor	Project manager (design)	90	Zoom
R18	Construction client	Division manager	85	Zoom
R19	Construction client	Project manager, large projects	90	Zoom

preliminary research results, thereby providing additional data. While the interviews provided perspectives of individual actors, the workshop provided an arena for discussion and thus opportunities to document reflections among the actors regarding their different perspectives.

EMPIRICAL FINDINGS

Benefits of Innovation

Almost all respondents emphasised the importance of reflecting on and discussing the benefits of innovation to analyse and understand why it is needed. One respondent stated: 'I am a "why" person, I can't work if I don't understand why I work' [R15]. Another expressed similar reasoning, saying: 'If we want to get our employees and others involved in a transition, the purpose is always important' [R7]. Another major reason noted for the importance of benefit and value is that it is a critical element of the very definition of innovation, since 'innovations must create value' [R1]. To facilitate such value-creation, one must first understand 'what is the goal of innovation' [R3]. Some respondents also emphasised that it is the client's task to prioritise benefits and articulate which is most important in each project to direct innovation efforts towards the selected goal.

Due to the project-based nature of the infrastructure sector, many respondents noted that the purpose of most innovations is to benefit the project at hand by providing short-term advantages during its execution. Many innovations involve improving efficiency in a current project, through time and cost reductions. One consultant respondent highlighted this as follows: 'In the projects that we do work innovatively, and there are not that many, we focus on improving efficiency to reduce cost' [R5]. Accordingly, an innovation 'must make business sense' [R1], otherwise it will never be developed and implemented. The focus on short-term benefits was emphasised most strongly by project-level respondents, who perceived every interference as a disturbance of the schedule, whereas firm-level professionals more frequently also appreciated the importance of long-term effects. In addition, several respondents stated that when innovation is considered in projects, it 'becomes a burden' [R13] that adds to pressures associated with meeting time, cost, and quality requirements, rather than a promoter of efficiency.

Many respondents indicated that a stronger focus on more long-term sustainability-related effects would be beneficial, in line with recent increases in interest in these effects, such as reductions in CO_2 emissions and other adverse environmental impacts, or, for example, increases in recycling. These long-term societal-level effects are demanded by politicians, putting pressure on the infrastructure sector to change and innovate. Hence, many respondents see the increased focus on sustainability in society as an opportunity to legitimise innovation: 'There are often requirements to reduce the climate impact of various things, which can give rise to a need to come up with innovations' [R9]. In this manner, specific regulatory demands to use sustainable materials and production methods are influential, as stated by a client respondent involved in a large project: 'We would not have spent time on making climate calculations to figure out which production processes and products have the lowest climate impact if we were not directed "from above"' [R19].

However, due to the ubiquitous focus on short-term benefits, innovation cannot be implemented at the expense of cost control. Many respondents, therefore, emphasised that the focus on short-term benefits hinders innovation that offers potential long-term benefits. Thus, several respondents expressed beliefs that striving to enhance sustainability is difficult, since it may cause goal conflicts and require project actors to balance short- and long-term benefits. Some also discussed examples of innovations with sustainability-related goals that were successfully developed but not implemented because of the required increases in investment. Such cases with adverse effects of conflicting goals clearly highlight the need for prioritisation of different goals and benefits. One client respondent stated: 'You may need to prioritise, in terms of control and what to work with' [R8], to enable work with long-term benefits. Another respondent emphasised the shift in focus needed to address such goal conflict: 'We need to be ahead of the curve and that can sometimes be associated with higher costs, but in the long run, it means much greater benefit, although it may cost a little more initially' [R18]. Otherwise, short-term benefits such as time and cost reductions will always come first. Furthermore, one client respondent also argued that goal conflicts may be resolved by applying appropriate timeframes: 'I'm not sure that reducing climate impact needs to cost more money, perhaps in the short term, but not the long term' [R2]. To elaborate on this idea, some innovations may increase costs in the project(s) where they are initially implemented, but in the longer run, their broader-scale implementation may not be more costly. Another client respondent argued that in large projects, there may not necessarily be a trade-off between short- and long-term benefits because, during their long (multiple-year) execution, many new processes and methods are tested.

Due to potential goal conflicts arising from the differences between their positive and negative effects in the short term and long term, public clients with large project portfolios may pursue different innovation benefits in different projects. A client respondent discussed how clients may establish a more systematic approach to fostering innovation by utilising the entire project portfolio: 'A large governmental authority like Trafikverket, with such a large [project] portfolio, needs to point strategically at some desired directions' [R19].

Beneficiaries of Innovation

Many respondents emphasised that most innovations mainly benefit the specific project and its involved project actors. In fact, benefits are mainly restricted to the specific supplier that initiates the innovation, for example, short-term reductions in the contractor's costs or improvement in site safety during the project. In addition, consultants pursue innovations that provide them short-term benefits during the project at hand. A common denominator of these innovations is that they are often based on incremental process improvements that provide quick results.

Some respondents discussed how innovations may also benefit external stakeholders, mainly taxpayers, who exacerbate the focus on short-term benefits, even making it highly logical in the infrastructure sector due to the perceived importance of avoiding wasting taxpayers' money. One client respondent stated: 'It's the taxpayers' money we're working with, and we must use it in the best possible way' [R2].

Another short-term benefit for external stakeholders lies in reducing disturbances for third parties during projects, by minimising traffic shutdowns and improving traffic diversions. One client respondent argued that the focus on this short-term societal benefit may diminish focus on other, more efficiency-related benefits: 'Sometimes we should consider shutting down traffic to improve production efficiency' [R18]. This may reduce costs and time in the ongoing project but also release resources for use in other projects. However, some respondents also mentioned that innovations may create long-term benefits for other external stakeholders, especially end-users (e.g., travellers and road users). A client respondent emphasised that: 'Satisfied customers, e.g., the bus companies. What are we [the client] really for: the business [project] is there to create value for citizens' [R10].

Some respondents discussed the connections between benefits and beneficiaries by highlighting the differences between stakeholders and their perceived innovation benefits. A construction contractor emphasised that some innovations may be worthy of pursuing, even if they increase costs for the ongoing project: 'If the client considered the societal value, more money could be allocated to innovation' [R4]. A client respondent similarly argued that public clients may strive to achieve societal-level benefits, whereas suppliers, such as contractors, mainly strive to obtain short-term benefits: 'Trafikverket has a societal responsibility; we can't ask contractors to have that in a project' [R2]. Similarly, a contractor respondent stated: 'As a contractor, our drive is to find more work that we can charge money for; this is also a way for us to make improvements' [R14].

Several respondents emphasised the possibility of obtaining benefits for both project actors and external stakeholders: 'You, I and everyone else who gets the most bang for the buck, but also the contractor who gets to develop his actions and practices' [R6]. Other examples are safety improvements, since innovations that reduce construction site work accidents will benefit the project actors and have positive societal effects. However, innovations that are mainly intended to provide long-term societal benefits, such as reducing climate impacts, may also be beneficial for the project actors if incentives and contracts are appropriately designed.

Examples of new solutions having unexpected negative effects on the lives of citizens and end-users (external stakeholders) were also raised. For instance, when a contractor decided to use a new substance for binding dust on roads to reduce dust in the air and surrounding surfaces, it was later discovered that it had negative effects on car paint. Thus, introducing new solutions in infrastructure can have direct positive and negative effects on citizens and end-users. Hence, many respondents called for more focus on the interests of external stakeholders, such as road users and commuters or travellers, when formulating requirements in tendering documents and executing projects.

Actors of Innovation

Contractor respondents argued that innovation processes are mostly executed by contractors during the tendering phase of projects, and thus mostly involve in-house competences, without much involvement of external actors. One contractor respondent explicitly regarded the generation of innovation as an important element of dealing with competition: 'We generate our own solutions; we don't invite others to participate in those processes' [R14]. A drawback is that during the tendering process, contractors mostly focus on innovations that can provide time and cost benefits in the ongoing project, in order to win the contract. Before the contract is won, the contractor does not want to pay other actors to contribute to innovation efforts, so subcontractors and material suppliers are rarely involved in innovation processes during the tendering phase. Another problem with the strong focus on innovation in the tendering process is that this phase is too short, as highlighted by a contractor respondent: 'This innovation process is rather useless. We spend four weeks thinking about something smart, then five years to produce it' [R4]. A client respondent agreed that this isolated and short-term focus has major limitations and argued that pure design-build contracts have become somewhat outdated and lead to 'siloed work' [R15] by the contractor.

Another drawback with the heavy reliance on contractors as innovators lies in their dependence on others in the supply chain to carry out the innovation work in their role as integrators between superstructure and infrastructure, as well as limited resources for external monitoring (benchmarking) to facilitate the adoption of innovations generated in other contexts. Accordingly, it is suggested that contractor organisations should devote more central resources to R&D: 'To achieve the Agenda 2030 goals we must allocate resources for it, but we don't do that. Lack of resources is a big problem, both in terms of money to spend and the right people to involve' [R1]. Another contractor respondent compared the Swedish contractors with foreign contractors that allocate more resources to innovation by specifically commissioned units: 'If we should strive for more product innovation, we need to establish a group of technical experts within the firm. Some foreign competitors have that' [R4].

However, some client respondents suggested that the client is often involved in the innovation process, together with other project actors: 'The project members themselves, the closest ones, both with us [client] and on the supplier side' [R10] are engaged. The same respondent also stated: 'If you're more connected to us through a [business] transaction, it will be easier. If you're someone completely from the outside, it's a huge obstacle' [R10]. Accordingly, actors that have contractual relationships with the client are considered to be more involved in innovation processes. There may also be positive aspects of collaboration from a risk allocation perspective, especially for contractors. If the responsibility for innovation is allocated to a single contractor, the risk may deter some trials of new innovative solutions: 'Risk allocation is important: if the risk is too big, then you don't dare to take a chance with innovation' [R3].

There seems to be somewhat more involvement of the client in consultants' innovation processes conducted in early project phases than in later phases such as production. However, from a consultant's perspective, Trafikverket is still involved too little in innovation efforts during planning and design phases, and it would be particularly valuable to involve actors with technical O&M expertise to strengthen the long-term asset management perspective: 'The asset management and maintenance part of the client should be involved more so that we can find smarter solutions that affect the maintenance' [R5]. A representative of the client's O&M division expressed similar reasoning: 'It could be even better to take lessons from operation and maintenance into new projects to avoid repeating the same mistakes again' [R7]. However, the same respondent highlighted difficulties in involving O&M resources in investment projects since they have their own O&M contracts that they need to prioritise.

Some respondents called for wider involvement of different actors in more collaborative innovation processes to benefit from complementary resources and competences: 'A combination of different competences is required' [R15]. Concerning stakeholder involvement, a client expressed a need to move away from single actors' needs and instead focus on developing mutual goals in collaboration with various community actors in very early phases:

> We're getting better at setting a direction together with others that in turn can generate innovative and effective measures, which contribute to fulfilling the transition we need. If we initiate a dialogue in this [early] phase, we'll get wider participation and understanding once we get into the investment projects, then we can also coordinate other ongoing activities better and more effectively. [R13]

This indicates that more collaborative innovation processes may be important for broadening the input of competences, spreading risks, and improving acceptance of the innovation outcomes.

CONCLUDING DISCUSSION

To conclude this chapter, we highlight the importance for public construction clients to identify innovation needs by analysing benefits, beneficiaries, and actors of innovation in order to enable more accurate demand articulation. Below we propose ways in which public clients can embark on this analysis as well as ways in which they can utilise their client role and public procurement strategies to facilitate more innovation in efforts to tackle grand societal challenges.

Propositions Regarding the Analysis of Innovation Needs

The empirical findings support recent suggestions in the literature on public procurement of innovation that public clients need to improve their competence and routines related to innovation demand articulation (Uyarra et al., 2020; Lenderink et al., 2022). More specifically, we argue that analysis of the benefits of innovation is crucial to improve client organisations' understanding of innovation needs and enable clearer demand articulation. The findings illustrate that a clear statement of the benefits of innovation is important to motivate people on both the client and supplier sides to engage in the innovation process. Thus, we propose:

- Public clients should improve their understanding of innovation needs by analysing the intended benefits of innovation and addressing the question, *Why is innovation needed?*

The empirical findings support conclusions in prior construction management studies that specific innovations may have varying effects (positive and potentially negative) for different stakeholders (Gil et al., 2012; Bygballe & Ingemansson, 2014; Larsson et al., 2022). It is thus crucial that benefits of innovation are analysed through the eye of the beholder. Accordingly, it is important to reflect on for whom (project actors and/or external stakeholders) benefits are created, identify beneficiaries that are affected by an innovation, and determine how the effects are likely to motivate people to engage (or deter them from engaging) in innovation processes. Thus, we propose:

- Public clients should improve their understanding of innovation needs by analysing the beneficiaries of innovation and addressing the question, *For whom is the innovation intended?*

The empirical findings highlight that most innovation processes are initiated and pursued by a specific supplier (contractor or consultant) that will benefit from the outcome. However, the findings also emphasise the importance of greater collaboration, especially to tackle grand challenges (Vosman et al., 2023; George et al., 2024). In line with prior construction management research findings (Harty, 2008; Rutten et al., 2009; Ozorhon, 2013), our respondents' expressions suggest that inter-organisational collaboration in construction innovation processes is crucial for more systemic innovation outcomes. Hence, key stakeholders (beneficiaries) that are expected to be affected by the outcome (benefits) should be involved as actors in the innovation process to increase the chances for successful innovation diffusion. Thus, we propose:

- Public clients should improve their understanding of innovation needs by analysing potentially suitable actors of innovation and addressing the question, *Who should initiate and participate in the innovation process?*

While the empirical findings and earlier research show that it is impossible to fully predict the spectrum of various effects of a specific innovation (positive and negative) across time and space (i.e., who will be affected, how, and when), this should not discourage attempts to identify beneficiaries and the actors who should be involved. On the contrary, analysis of innovation needs in terms of benefits, beneficiaries, and actors of innovation is important for reducing gaps between foreseen and unforeseen contingencies by improving understanding of these key aspects and the strategic targeting of public procurement of innovation. This is especially relevant for public clients who have large project portfolios and thus can strongly influence their supply markets (Wolstenholme, 2009; Barbosa et al., 2017; Järvenpää et al., 2023).

Propositions Regarding How to Facilitate More Innovation

Meissner et al. (2024) argue that classical innovation management theories and practices have poor sustainability as they focus on obtaining short-term benefits related to monetary gains. In contrast, sustainable innovation management needs to balance anticipated short- and long-term impacts (ibid.). Our empirical results related to the two overarching categories of benefits – short term and long term (Gil et al., 2012; Ozorhon et al., 2016) – indicate that a clear challenge is to systematically balance short-term gains in individual projects with desired long-term outcomes. Furthermore, the findings highlight that innovation processes in many construction settings are adversely affected by goal conflicts, as different stakeholders may have radically differing perceptions of innovation benefits. This may create conflicts between short-term and long-term benefits of innovations, and hence tensions between stakeholders that may support and resist their implementation. Thus, we propose:

- Public clients need to balance and combine short-term gains for suppliers with long-term societal goals.

To address the tensions arising from discrepancies between short- and long-term benefits, public clients need to identify appropriate short-term drivers or incentives for suppliers to innovate in ways that also help efforts to meet long-term societal goals. Such incentives must address the increases in costs and risks associated with innovation and their distribution among the actors of innovation. If done appropriately, and in the early phases of a project, this will provide firmer foundations for collaboration based on mutual trust (Crespin-Mazet et al., 2015; Havenvid et al., 2016). As concluded in prior construction management studies (Eriksson & Szentes, 2017; Larsson & Larsson, 2022), procurement strategies should include the facilitation of communication and discussion about joint goals and co-creation even during the planning and design phases to enhance the balance of short-term and long-term innovation efforts. Thus, we propose:

- Public clients should consider using procurement strategies to promote intense long-term collaboration from the early phases of projects onwards.

More internal R&D investment by contractor and consultancy companies is required to improve innovations of products or procedures with long-term benefits before their implementation in inter-organisational construction projects. The empirical findings highlight that such investments are less frequent in project-based construction contexts than in many other settings since suppliers need to invest resources before they even know that there is a business opportunity. Business opportunities do not occur before the client needs are known and communicated to the suppliers through tendering documents. Thus, finally, we propose:

- Public clients that strive for more systemic innovations should communicate their long-term ambitions and innovation needs in a strategic manner.

REFERENCES

Arif, M., Syal, M., Florez, L., Castro, D. & Irizarry, J. (2013). Measuring sustainability perceptions of construction materials. *Construction Innovation*, 13(2), 217–234.

Bandé, A., Ika, L. & Ouédraogo, S. (2024). Beneficiary participation is an imperative, not an option, but does it really work in international development projects? *International Journal of Project Management*, 42, 1–13.

Barbosa, F., Woetzel, J. & Mischke, J. (2017). *Reinventing Construction: A Route of Higher Productivity.* McKinsey Global Institute, Houston.

Bessant, J. & Rush, H. (1995). Building bridges for innovation: The role of consultants in technology transfer. *Research Policy*, 24, 97–114.

Bianchi, M., Croce, A., Dell'Era, C., Di Benedetto, A. & Frattini, F. (2016). Organizing for inbound open innovation: How external consultants and a dedicated R&D unit influence product innovation performance. *Journal of Product Innovation Management*, 33(4), 492–510.

Blayse, A. M. & Manley, K. (2004). Key influences on construction innovation. *Construction Innovation*, 4(3), 143–154.

Brockmann, C., Brezinski, H. & Erbe, A. (2016). Innovation in construction megaprojects. *Journal of Construction Engineering and Management*, 142(11), 04016059.

Burt, R. (1992). The network structure of social capital. *Research in Organizational Behavior*, 22, 345–423.

Bygballe, L. E. & Ingemansson, M. (2014). The logic of innovation in construction. *Industrial Marketing Management*, 43(3), 512–524.

Bygballe, L., Jahre, M. & Swärd, A. (2010) Partnering relationships in construction: A literature review. *Journal of Purchasing and Supply Management*, 16(4), 239–253.

Chan, A. P. & Oppong, G. D. (2017). Managing the expectations of external stakeholders in construction projects. *Engineering, Construction and Architectural Management*, 24(5), 736–756.

Chen, J., Walker, R. M., & Sawhney, M. (2020). Public service innovation: A typology. *Public Management Review*, 22(11), 1674–1695.

Chesbrough, H. (2003). The era of open innovation. *MIT Sloan Management Review*, 44(3), 35–41.

Crespin-Mazet, F., Ingemansson Havenvid, M. & Linné, Å. (2015). Antecedents of project partnering in the construction industry: The impact of relationship history. *Industrial Marketing Management*, 50, 4–15.

Crossan, M. & Apaydin, M. (2010). A multi-dimensional framework of organizational innovation: A systematic review of the literature. *Journal of Management Studies*, 47(6), 1154–1191.

Dulaimi, M. F. (2022). The climate of innovation in the UAE and its construction industry. *Engineering, Construction and Architectural Management*, 29(1), 141–164.

Dulaimi, M. F., Ling, F., Ofori, G. & De Silva, N. (2002). Enhancing integration and innovation in construction. *Building Research & Information*, 30(4), 237–247.
Edler, J. & Georghiou, L. (2007). Public procurement and innovation: Resurrecting the demand side. *Research Policy*, 36(7), 949–963.
Edquist, C. & Zabala-Iturriagagoitia, J. M. (2015). Pre-commercial procurement: A demand or supply policy instrument in relation to innovation? *R&D Management*, 45(2), 147–160.
Eriksson, P. E. (2017). Procurement strategies for enhancing exploration and exploitation in construction projects. *Journal of Financial Management of Property and Construction*, 22(2), 211–230.
Eriksson, P. E. & Szentes, H. (2017). Managing the tensions between exploration and exploitation in large construction projects. *Construction Innovation*, 17(4), 492–510.
Eriksson, P. E. & Westerberg, M. (2011). Effects of cooperative procurement procedures on construction project performance: A conceptual framework. *International Journal of Project Management*, 29(2), 197–208.
Eriksson, P. E., Olander, S., Szentes, H. & Widén, K. (2014). Managing short-term efficiency and long-term development through industrialized construction. *Construction Management and Economics*, 32(1–2), 97–108.
Eriksson, P. E., Lingegård, S., Borg, L. & Nyström, J. (2017). Procurement of railway infrastructure projects: A European benchmarking study. *Civil Engineering Journal*, 3(4), 199–213.
Eriksson, T. & Kadefors, A. (2015). Designing and implementing incentives for engineering consultants: Encouraging cooperation and innovation in a large infrastructure project. *Engineering Project Organization Journal*, 5(4), 146–159.
Gann, D. M. & Salter, A. J. (2000). Innovation in project based, service-enhanced firms: The construction of complex products and systems. *Research Policy*, 29, 955–972.
Gans, J. A. & Stern, S. (2003). The product market and the market for ideas: Commercialization strategies for technology entrepreneurs. *Research Policy*, 32(2), 333–350.
George, G., Fewer, T., Lazzarini, S., McGahan, A. & Puranam, P. (2024). Partnering for grand challenges: A review of organizational design considerations in public–private collaborations. *Journal of Management*, 50(1), 10–40.
Georghiou, L., Edler, J., Uyarra, E. & Yeow, J. (2014). Policy instruments for public procurement of innovation: Choice, design and assessment. *Technological Forecasting & Social Change*, 86(1), 1–12.
Gil, N., Miozzo, M. & Massini, S. (2012). The innovation potential of new infrastructure development: An empirical study of Heathrow Airport's T5 project. *Research Policy*, 41, 452–466.
Goodrum, P. M. & Haas, C. T. (2000). Variables affecting innovations in the US construction industry. In *Construction Congress VI: Building Together for a Better Tomorrow in an Increasingly Complex World* (pp. 525–533).
Håkansson, H. & Ingemansson, M. (2013). Industrial renewal within the construction network. *Construction Management and Economics*, 31(1), 40–61.
Hartmann, A., Reymen, I. & Van Oosterom, G. (2008). Factors constituting the innovation adoption environment of public clients. *Building Research & Information*, 36(5), 436–449.
Harty, C. F. (2008). Implementing innovation in construction: Contexts, relative boundedness and actor-network theory. *Construction Management and Economics*, 26(10), 1029–1041.
Havenvid, M. I., Hulthén, K., Linné, Å. & Sundquist, V., (2016). Renewal in construction projects: Tracing effects of client requirements. *Construction Management and Economics*, 34(11), 790–807.
Hedborg Bengtsson, S., Karrbom Gustavsson, T. & Eriksson, P. E. (2018). Users' influence on inter-organizational innovation: Mapping the receptive context. *Construction Innovation*, 18(4), 488–504.

Ika, L. & Munro, L. (2022). Tackling grand challenges with projects: Five insights and a research agenda for project management theory and practice. *International Journal of Project Management*, 40(6), 601–607.

Ivory, C. (2005). The cult of customer responsiveness: Is design innovation the price of a client-focused construction industry? *Construction Management and Economics*, 23(8), 861–870.

Jansen, J., Tempelaar, M., van Den Bosch, F. & Volberda, H. (2006). Exploratory innovation, exploitative innovation, and performance: Effects of organizational antecedents and environmental moderators. *Management Science*, 52(11), 1661–1674.

Järvenpää, A-T., Larsson, J. & Eriksson, P. E. (2023). How public client's control systems affect contractors' innovation possibilities. *Construction Innovation*, 23(in press).

Kometa, S. T., Olomolaiye, P. O. & Harris, F. C. (1994). Attributes of UK construction clients influencing project consultants' performance. *Construction Management and Economics*, 12(5), 433–443.

Krieger, B. & Zipperer, V. (2022). Does green public procurement trigger environmental innovations? *Research Policy*, 51(6), 1–27.

Kumaraswamy, M. & Dulaimi, M. F. (2001). Empowering innovative improvements through creative construction procurement. *Engineering, Construction and Architectural Management*, 8(5/6), 325–334.

Kundu, O., James, A. D. & Rigby, J. (2020). Public procurement and innovation: A systematic literature review. *Science and Public Policy*, 47(4), 490–502.

Larsson, J., Eriksson, P. E., Lingegård, S. & Järvenpää, A. T. (2022). Innovation outcomes and processes in infrastructure projects: A comparative study of Design-Build and Design-Build-Maintenance contracts. *Construction Management and Economics*, 40(2), 142–156.

Larsson, L. & Larsson, J. (2018). Sustainable development in project-based industries: Supporting the realization of explorative innovation. *Sustainability*, 10(3), 683.

Lenderink, B., Halman, J. & Voordijk, H. (2022). Innovation and public procurement: From fragmentation to synthesis on concepts, rationales and approaches. *Innovation: The European Journal of Social Science Research*, 35(4), 650–674.

Loosemore, M. (2015). Construction innovation: Fifth generation perspective. *Journal of Management in Engineering*, 31(6), 04015012.

Manley, K. (2006). The innovation competence of repeat public sector clients in the Australian construction industry. *Construction Management and Economics*, 24(12), 1295–1304.

Meissner, D., Harms, R., Kratzer, J. & Zhou, J. (2024). The dead end of classical innovation management and unsustainable innovation. *Technovation*, 124, 1–5.

Mueller, V., Rosenbusch, N. & Bausch, A. (2013). Success patterns of exploratory and exploitative innovation: A meta-analysis of the influence of institutional factors. *Journal of Management*, 39(6), 1606–1636.

Mwesiumo, D., Glavee-Geo, R., Olsen, K. M. & Svenning, G. A. (2021). Improving public purchaser attitudes towards public procurement of innovations. *Technovation*, 101, 1–13.

Nam, C. H. & Tatum, C. B. (1997). Leaders and champions for construction innovation. *Construction Management and Economics*, 15(4), 259–270.

Newcombe, R. (2003). From client to project stakeholders: A stakeholder mapping approach. *Construction Management and Economics*, 21(8), 841–848.

Ntsondé, J. & Aggeri, F. (2021). Stimulating innovation and creating new markets: The potential of circular public procurement. *Journal of Cleaner Production*, 308, 1–8.

Obwegeser, N. & Dueholm Müller, S. (2018). Innovation and public procurement: Terminology, concepts, and applications. *Technovation*, 74–75(1), 1–17.

Ozorhon, B. (2013). Analysis of construction innovation process at project level. *Journal of Management in Engineering*, 29(4), 455–463.

Ozorhon, B. & Oral, K. (2017). Drivers of innovation in construction projects. *Journal of Construction Engineering and Management*, 143(4), 04016118.

Ozorhon, B., Abbott, C. & Aouad, G. (2014). Integration and leadership as enablers of innovation in construction: Case study. *Journal of Management in Engineering*, 30(2), 256–263.

Ozorhon, B., Oral, K. & Demirkesen, S. (2016). Investigating the components of innovation in construction projects. *Journal of Management in Engineering*, 32(3), 04015052.

Rainville, A. (2021). Stimulating a more circular economy through public procurement: Roles and dynamics of intermediation. *Research Policy*, 50(4), 1–11.

Rose, T. M. & Manley, K. (2014). Revisiting the adoption of innovative products on Australian road infrastructure projects. *Construction Management and Economics*, 32(9), 904–917.

Rundquist, J., Emmitt, S., Halila, F., Hjort, B. & Larsson, B. (2013). Construction innovation: Addressing the project–product gap in the Swedish construction sector. *International Journal of Innovation Science*, 5(1), 1–10.

Rutten, M., Dorée, A. & Halman, J. (2009). Innovation and interorganizational cooperation: A synthesis of literature. *Construction Innovation*, 9(3), 285–297.

Slaughter, E. S. (1998). Models of construction innovation. *Journal of Construction Engineering and Management*, 124(3), 226–231.

Suhonen, N., Tammi, T., Sastamoinen, J., Pesu, J. & Turtiainen, M. (2019). Incentives and risk-sharing in public procurement of innovations. *Journal of Public Procurement*, 19(2), 129–145.

Uyarra, E. & Flanagan, K. (2010). Understanding the innovation impacts of public procurement. *European Planning Studies*, 18(1), 123–143.

Uyarra, E., Edler, J., Garcia-Estevez, J., Georghiou, L. & Yeow, J. (2014). Barriers to innovation through public procurement: A supplier perspective. *Technovation*, 34(10), 631–645.

Uyarra, E., Zabala-Iturriagagoitia, J. M., Flanagan, K. & Magro, E. (2020). Public procurement, innovation and industrial policy: Rationales, roles, capabilities and implementation. *Research Policy*, 49(1), 1–11.

von Hippel, E. (1986). Lead users: A source of novel product concepts. *Management Science*, 32(7), 791–805.

Vosman, L., Coenen, T., Volker, L. & Visscher, K. (2023). Collaboration and innovation beyond project boundaries: Exploring the potential of an ecosystem perspective in the infrastructure sector. *Construction Management and Economics*, 41(6), 457–474.

Wang, N., Xu, Z. & Liu, Z. (2023). Innovation in the construction sector: Bibliometric analysis and research agenda. *Journal of Engineering and Technology Management*, 68, 1–14.

Widén, K., Olander, S. & Atkin, B. (2014). Links between successful innovation diffusion and stakeholder management. *Journal of Management in Engineering*, 30(5), 1–7.

Winch, G. (1998). Zephyrs of creative destruction: Understanding the management of innovation in construction. *Building Research & Information*, 26(5), 268–279.

Wolstenholme, A. (2009). *Never Waste a Good Crisis: A Review of Progress since Rethinking Construction and Thoughts for Our Future*. London: Constructing Excellence.

11. Creating a fertile learning ecosystem for innovation: an integrated programme of alliancing projects approach

Derek H.T. Walker and Peter E.D. Love

INTRODUCTION

Traditional construction project delivery forms such as design-bid-build (DBB) have been widely criticised for failing to adequately support innovation in construction materials or delivery systems (Latham, 1994; Egan, 1998). While some change has been evident over the past two decades, a vital driver of improvement has been in project procurement approaches for supply-chain team integration and collaboration (Egan, 2002). More emphasis has been laid on project owner leadership capabilities to meaningfully engage with the supply chain by adopting a hands-on rather than hands-off relationship with design, delivery, and facility operation teams.

The emergence of integrated project delivery (IPD) contract forms has facilitated new ways of achieving this. Heathrow Airport's Terminal 5 in the UK provided a vanguard collaborative, integrated team project that demonstrated how project owner capabilities combined with team collaborative integration could develop superior project team capabilities that were further enhanced. Examples include Crossrail, the Thames Tideway Project, the London Olympics, and other projects undertaken using NEC3/4 contract forms (Davies and Brady, 2015). The USA health provider Sutter Health, responding to a need for rapid innovative construction in California, adopted lean construction principles and the alliancing concept with its IPD projects (Lichtig, 2005; Alves and Lichtig, 2020). In Australia and New Zealand, the alliancing form of project delivery has triggered many innovative construction practices (Australian Constructors Association, 1999; Davis and Love, 2011; Lloyd-Walker, Walker and Mills, 2014; Ibrahim, Costello and Wilkinson, 2018). Similar conclusions have been drawn about alliance projects in Finland (Hietajärvi, 2017; Lahdenperä, 2017; Aaltonen and Turkulainen, 2018) and other parts of Europe, notably the Netherlands (Rutten, Dorée and Halman, 2009). The Level Crossing Removal Project (LXRP) is an unusual example of a programme of alliance projects.

The central research question posed in this chapter is: *How does the LXRP's collaborative alliancing approach set it apart from more traditional approaches to procurement and enable the facilitation and realisation of innovative practice?* From the 1970s onwards, design and construct (D&C), partnering, and team integration through public–private partnerships (PPP) have been touted by many academics and industry representatives as encouraging innovative practices. These project delivery forms may support innovative practice, but they lack the intense collaborative integration that alliancing and IPD effectively use to support dialogue between project delivery team participants (Walker, Love and Matthews, 2022). Part of the answer to the 'how does alliancing support innovative practices?' question is about the project delivery choice, part is about a project owner's well-articulated innovation strategy developed for Crossrail (Davies et al., 2014), part is about owner capabilities that ensure enlightened engagement with the project team (Gulino, Sergeeva and Winch, 2020), and part is about governance mechanisms that ensure innovation strategy realisation. Effective incentive arrangements (Department of Infrastructure and Regional Development, 2015) and an effective supportive platform are required to make innovative practices happen (Love et al., 2024).

This chapter focuses on the LXRP, a 15-year programme with five alliance project teams delivering an AU$19.8 billion programme. This case study can be regarded as an exemplar through its strategising for innovation and continuous improvement and making that strategy work to achieve its innovation aspirations. The chapter briefly provides the LXRP context to acquaint readers with its background relevance to this chapter's aim, and several key elements of the LXRP procurement choice and governance illustrate how innovation strategy was turned into action.

WHY STUDY THE LXRP AS A CASE STUDY EXEMPLAR?

Eisenhardt and Graebner (2007) identify several design challenges when undertaking case study research, particularly on a single case study. Essentially, a single case study, for example, the LXRP, provides a unique opportunity to study a programme of projects to understand better *how* and *why* questions and *what* questions are investigated from a context perspective. The chosen research approach (single case study) should be justified, and the research questions stated provide a sound rationale for their formation and why they may be best answered using a single case study. The research techniques used should also make sense and be plausible. The conclusion from such research contributes to theory by unpacking complexity where the interface of several theoretical lenses is used to analyse data. Results may not necessarily be generalisable, but they should explain the *how* and *why* of a specific case in its context (the *what*) so that they may be adapted to different contexts. The theoretical contribution is delivered through gaining a more nuanced understanding of the case study and how it may be adapted to other relevant situations.

In this case, our research question was chosen because the LXRP demonstrates, through strategy and actions, how it successfully pursues one of its key strategic objectives: to enable and realise continuous improvement and innovation diffusion. Both authors had previously researched the LXRP and were intrigued by how the LXRP operates. Choosing this case study falls into Davis' (1971) rationale for an interesting topic: most projects fail to capitalise on learning and fall victim to the 'eternal beginner syndrome' by constantly 'reinventing the wheel' (Flyvbjerg,

Budzier and Lunn, 2021). Understanding the LXRP's approach could be adapted elsewhere.

Our research design (explained in more depth later) is focused on *how* and *why* questions; therefore, our research design included reviewing public and private organisational documents and, importantly, content analysis of open-ended question interviews with 24 highly experienced key LXRP experts who have been intimately involved with the LXRP programme and were willing to share insights and observations about how the programme pursued its continuous improvement and innovation diffusion strategy. Broad interview questions were asked, some about specific aspects of the programme's strategy and delivery and how the enacted strategy delivers continuous improvement and innovation. Other questions related to the organisation's workplace culture, leadership style, and governance mechanisms operated in practice that motivate and reward innovation-sharing and how it *felt* to work in the LXRP.

One reason for selecting this case study was partially opportunistic, because we already had contacts with organisational LXRP leaders through our previous research who were willing for us to study their programme of projects. The LXRP is a megaproject comprising an integrated programme of projects undertaken as alliance work packages over 15 years. It provided insight into opportunities for LXRP organisational learning to realise substantial performance and social benefits. As such, it is a rare example of what may be achieved in construction megaprojects. According to the literature, megaprojects are plagued by cost and time overruns (Flyvbjerg, 2011), wasteful rework, and poorly captured knowledge and lessons that might fruitfully be used to obviate repeated problems typical of megaprojects (Love et al., 2022). The LXRP has delivered an impressive performance so far. According to mid-2022 internal LXRP documents, seven years into the 15-year programme, it is significantly ahead of schedule; its actual outturn costs (AOC) are around 0.5% below its target outturn cost (TOC), with estimated costs at around 2.5% below similar projects. It has also delivered substantial social benefits, gaining national and international award-winning recognition[1] for stakeholder engagement. It is a remarkable case study example in organisational learning terms, having successfully developed a novel continuous improvement system. This positions it as an ideal case study.

WHAT IS UNIQUE ABOUT THE LXRP WARRANTING A CONTINUOUS IMPROVEMENT AND INNOVATION CASE STUDY?

We first elaborate on the LXRP scope and context, which justifies its categorisation as special and relevant to construction management learning organisational theory.

LXRP Scope

The LXRP is a programme that provides alliance project work packages. The Labor Party pledged in its 2014 Victorian State election campaign to remove 50 of the most dangerous and congested rail crossings and rebuild and revitalise station precincts. The programme started in May 2015 as the LXRA, the Level Crossing Removal Authority, with 20 of the 50 crossings pledged to be removed by 2018. A further 25 crossing removals were promised during the 2018 Victorian State election. In 2021, that commitment grew by 10 more crossing removals. In 2022, more crossing removals were announced, bringing the programme scope to 110 crossing removals, 51 new stations and their precinct urban renewal transformation by 2030. The LXRP is delivered as an AU$19.8 billion programme of projects. The project delivery choice was alliancing, with each initial alliance work package being delivered through five alliance syndicates procured through a competitive TOC process (Walker et al., 2024).

The TOC process is argued to deliver a more reliable cost/time and delivery plan than conventional project procurement methods (Walker, Vaz Serra and Love, 2022), partly because of the composition of alliance syndicates. These include a design team, contractor team, facility operator team, and the project owner representative team. At stage 1 of the process, the design team and contractor team (with teams often a joint venture of two or more organisations) develop a design and TOC over a six- to nine-month period involving deep collaboration with intensive input from the facilities operator (rail and road) and project owner representative. This provides cross-disciplinary perspectives and risk and uncertainty advice through the involvement of the facility operator team and project owner representatives that led to numerous innovations on many of its projects, including an example on Alliance Package 1, where use of a novel steel sheet piling system proposed by the rail operator team resulted in substantial time savings (Walker, Matinheikki and Maqsood, 2018). These initial alliances' TOCs were subject to reference class benchmarking and external cost/time evaluation to maintain competitive tension. With scope expansion from 50 to 110 crossing removals and associated station and precinct works, the procurement system was changed to a single TOC process, where one of the five initial alliance work package syndicates was invited to bid for additional alliance work packages (AWPs) where the TOC was developed and validated against an increasingly more sophisticated performance and design benchmark database. As-built work designs were reused and adapted for specific crossing/station contexts, facilitating modularisation innovations, greater knowledge-sharing across the programme, and systems development for continuous improvement and innovation diffusion. AWPs were first invited to syndicates that had proven themselves and where the work package best fitted geographically across the Melbourne rail corridors.[2] The evolving programme scope from 50 to 110 crossings, associated works, and the 15-year horizon offered opportunities to package works into manageable chunks, with TOCs of hundreds of millions of dollars. The prospect of a continued pipeline of AWPs allowed alliance syndicates and the LXRP to capitalise on innovations from one project to others within the programme across the 15-year duration. Team participant retention was improved through this strategy. As a result, this enabled the LXRP's 'corporate knowledge' to be retained. Succession and career path planning were enhanced as a learning organisation support mechanism. Programme scope linked to highly effective project alliance package integration was a high-impact factor on motivation and the workplace stability environment through job security and work continuity.

Complex Context

The LXRP is a highly complex programme that can be undertaken from numerous perspectives. The political context was complicated. For decades, successive state governments shunned addressing the growing problems associated with dangerous and congested rail crossings and the dilapidated and 'tired' looking rail stations. Developing a programme delivery strategy that addressed political complexity and public support was vital.

The scale complexity of 110 crossing removals, 51 new stations, and station precinct urban renewal transformation required a step-change in skills and competence of the construction industry's entire supply chain to be capable of delivering a programme of this scale. Clearly, resource constraint complexity was a significant challenge, particularly in training and upskilling the workforce to cope with anticipated management demands.

Technically, much of the work involves known technologies. However, interfaces between old and new signalling systems pose complex challenges. The pace of construction needed to complete the works also posed complexity due to a need to improve delivery performance and increase productivity. Indeed, modularisation, continuous improvement, and skills development should also be central to any project delivery strategy.

A further complexity involved interfacing the construction works with a live working rail network and active road system. Thus, the programme design and delivery teams should include expert advice from rail and road authorities regarding technical matters and knowledge about permits, administrative procedures, and other management interface issues.

The LXRP Response to the Programme Complexity

Other complexity issues were identified. Considering the above, it became evident that the project owner, the Victorian Government and its Department of Transport and Planning, needed to take a hands-on leading role at the programme's front end and throughout its delivery. This suggested taking an alliancing approach, where the programme team(s) comprise an integrated alliance between design teams, key construction contractors, the facilities (train and roads) operators, and a highly competent project owner representative team. The scope and scale indicated breaking up the programme into a series of alliance work packages because no single design or construction contractor could undertake the 15-year programme. Complexity-addressing logic further suggests an alliancing approach because, while there are many known-knowns at numerous levels, there are also numerous unknowns and unknowables that demand cross-discipline, multiple-perspective-taking (Snowden and Rancati, 2021). Achieving optimal modularisation and enabling continuous improvement across projects within the programme requires a high level of project team integration led by a competent project owner leadership team. Therefore, it became evident that a collaborative and highly integrated form of project delivery was necessary. This approach required additional innovative practices triggered by the extent of project–programme integration and cross-project and within-project cross-discipline collaboration.

HOW DO WE MAKE SENSE OF THE LXRP RESPONSE TO PROGRAMME COMPLEXITY?

This section briefly outlines our research approach and theoretical frame of reference. The single case study research approach has been justified. The context of the case study explained above positions our research paradigm as *critical realism*. Somerville explains this as:

> characterised as a combination of a realist ontology (or theory of being, of how the world is) with a fallibilist epistemology (or theory of knowledge, of how the world is known). Realism here means the belief that reality exists (in this case, the way the project is organised and delivered) independently of the human mind, with human consciousness being typically understood as perception (the observer), cognition (the knower), or thought (the thinker). Fallibilism is the belief that our knowledge of reality can always be proved wrong. That is, our knowledge has no sure foundations (2012, p. 291).

We therefore report on what we see occurring through tangible evidential forms (e.g., authoritative documents) and evidence transmitted through interviews (where we analyse transcripts and make sense of what we were told). Triangulation was sought through internal documents made available to us and from our observations on several separate research projects undertaken on the LXRP by us. We adopted a grounded theory approach using extant theory to abductively analyse and make sense of the data. This entailed moving cyclically from identifying isolated quotes as data points, linking their meaning and relevance to theoretical concepts, consolidating these into themes, and finally grouping themes into categories, also triangulating these data points with internal documentation and our observations from separate research projects on the LXRP. We used the analytical tool NVivo to code the data. We interviewed 24 experts engaged in the LXRP programme, each with extensive experience in alliance work. Many were cross-disciplinary and had worked for design, contractor, and government organisations. Further details of the research approach and other findings may be found in Walker et al. (2024). The chapter authors have been researching alliance construction projects and programmes in general for over two decades from a learning organisational perspective.

RESULTS AND FINDINGS

Data was coded into relevant core organisational learning nodes, enabling a structure of nodes to emerge following Gioia, Corley and Hamilton (2013). Identified first-order concepts could be logically grouped into continuous improvement and innovation diffusion, and second-order themes suggest aggregate dimensions that help explain the case study data's relevance to Szulanski's (1996) sticky knowledge theory. Knowledge transfer, such as continuous improvement from lessons learned (and retained) and innovation diffusion (Rogers, 2003), requires a fertile ecosystem for learning and effectively sharing insights and knowledge influenced by governance mechanisms and workplace culture.

The 24 interviewee transcripts and notes comprised more than 300 pages of interviewee quotes. We present a few quotes to illustrate how we analysed the data into the three aggregate dimensions illustrated in Figure 11.1.

> the ability to iterate and learn and enable a learning culture and a system thinking around learning culture, which we have done – it's only because we've had that opportunity that has given us the ability to take that advantage and apply it.
>
> …is definitely the framework of the alliance, and having five alliances, and giving them certainty of pipeline so they're committed; they've got a culture of looking at the horizon and investing in those discretionary things and not having that project lifecycle of a start, a middle and a finish, and people tailing off and disappearing into the sector again and looking for the next thing. We've got people that have hung around and gone from one to the next.

This quote provides data on several first-order concepts, for example, organisational drivers (*the PAA*) and a programme mindset that facilitates sharing knowledge and innovation across projects within the programme, with a pipeline of continuous work over 15 years. It also illustrates the encouragement of critical thinking that relates to participant drivers. Each quote was analysed to make sense of the interviewees' insights and messages and how that told a bigger story of objects – such as the joint coordination committee (JCC), innovation key performance indicators (KPIs), people's curiosity and willingness to explore ideas through dialogue with others – helped shape more general themes that in turn help form dimensions that may explain how the LXRP reduced the stickiness of knowledge transfer and innovation diffusion.

Figure 11.1 illustrates how continuous improvement is impacted by organisational enablers, drivers, and team participant drivers. Love et al. (2024) investigated continuous improvement and innovation diffusion in the LXRP and compared mechanisms and techniques used by the Crossrail megaproject that were further adapted for other UK megaprojects. Many of the first-order concepts and second-order themes are discussed in full in that book chapter.

The key to fully appreciating the fertile ecosystem context for reducing the stickiness of knowledge transfer is understanding how an organisation enables, empowers, and guides people to share knowledge. That ecosystem needs support mechanisms that drive and facilitate organisational learning and knowledge-sharing as well as

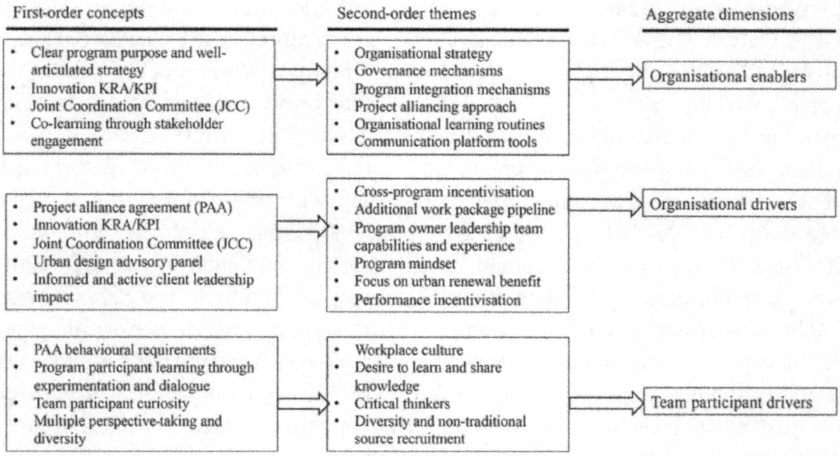

Figure 11.1 *Sticky knowledge reduction analysis*

people with critical thinking skills, who rely upon or benefit from that knowledge, having the capacity, motivation, and ability to share knowledge and engage in dialogue to adopt/adapt that knowledge in different or similar contexts.

Organisational Knowledge Transfer Enablers

Numerous first-order coded data items emerged from data analysis for this dimension. The project's purpose was clearly articulated and reflected in its business case (Victorian State Government, 2017). The purpose extends beyond separating road and rail network grades at critical junctions and improving the road and rail networks. The programme was explicitly articulated as transformational to leave a positive legacy through revitalising and improving the urban environment surrounding the rebuilt rail station precincts and to improve integrating land use along rail corridors to create vibrant community hubs. The social transformational purpose also extended to how the project would be procured and resourced. Its purpose extended transformation, actively facilitating and enabling recruitment of individuals from diverse backgrounds that have been largely ignored as a latent skills resource. This purpose included increasing the employment of many people with transferable and relevant deployable skills, including females, First Nations people, and recent refugee immigrants. It also extended to procure locally and encourage small local business enterprises to supply goods and services to the programme (Magree, Scott-Young and Lloyd-Walker, 2024).[3] Articulation of the LXRP strategic intent was consolidated in the publication of the programme's '5-Greats': (1) great network; (2) great places; (3) great partnerships; (4) great engagement; and (5) great people. These greats were explained in detail in the internally distributed LXRP blueprint one-page poster in

2023. Additionally, the blueprint highlighted the LXRP vision and values described as the '5-Cares'. These '5-Cares' articulated the programme's values, comprising: (1) creativity; (2) accountability; (3) relationships; (4) empowerment; and (5) safety. The blueprint provides a potent enabler that clearly and simply communicates what needs to be achieved and the programme team characteristics required.

Vision and strategy provide an answer to what needs to be achieved but also require a coherent system of governance that is consistent with ensuring that the strategy is followed. Müller (2017) argues that a governance style may exhibit inherent assumptions about the purpose of designed governance mechanisms. These differentiate between a shareholder and stakeholder orientation, and outcome and behaviour control. The underlying assumption accepted in designing governance arrangements affects how project participants respond to leadership styles and the behavioural norms expected of them. A shareholder behaviour-control approach tends towards a highly conformist orientation, whereas a stakeholder outcome-control orientation suits a 'versatile artist' orientation. The orientation influences how innovation diffusion and sharing lessons learned may be conducted. In a project-related study of governmentality, Müller et al. (2016) introduced three general dimensions of governmentality, which illustrate a project organisation's governance profile. Governmentality may lie on a continuum from an authoritarian approach to liberal and neoliberal, which reflects power asymmetry between the leadership hierarchy and coalface workers. Governmentality is assumption-based and designed to fit a particular purpose. So, it is also impacted by the governance system designers and may be measured on a process, project-specific, or values continuum. Governance mechanisms may range across a control, balanced, and trusting continuum. Adherence to governance institutions reflects complexity in meeting external governance standards, ranging from low (few) to high (many). For example, the LXRP case could include the Victorian Auditor General Office and a range of government departmental guidelines. In an early alliancing project study, Clegg et al. (2002) focused on how specific project alliance agreement (PAA) requirements influenced the oversight mechanism and how expected project participant behaviours were framed. Data analysed from the LXRP case study illustrated governance mechanisms designed to encourage, require, and monitor continuous improvement measures through key result areas (KRAs) and KPIs designed explicitly for continuous improvement and innovation diffusion.

Achieving close and effective team effort integration within AWPs and interfaces between AWPs within a programme require specific organisational enablers. A common team integration approach is the use of steering or coordination committees. The LXRP created an entity they term the joint coordination committee (JCC). Still, its remit extends beyond gathering the AWP directors and key executive teams to coordinate and integrate each project's contribution to the programme.

Walker, Love and Betts (2024, Chapters 3 and 9) describe the JCC, presenting an LXRP internal document illustrating the JCC's role as coordinating programme partners and governance through collaboration to realise the '5-Greats' reproduced in Figure 11.2. It also shows the central role of the JCC from a strategic, tactical, and operational perspective. At the strategic level, it coordinates project directors,

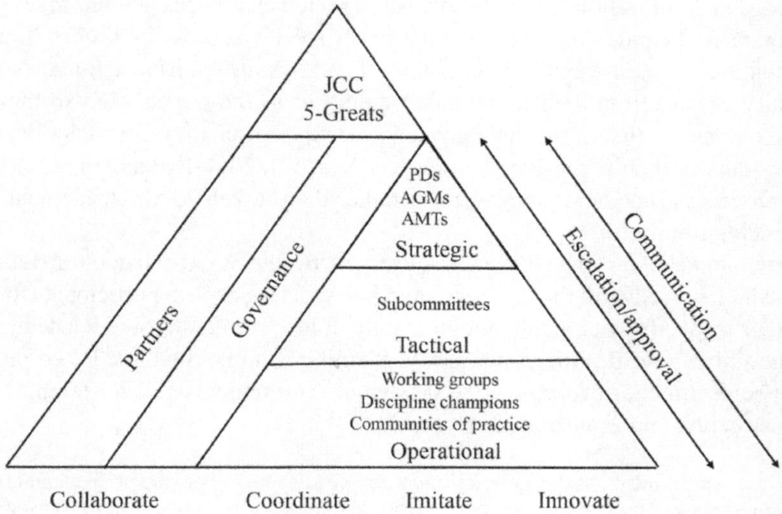

Source: LXRP internal document with permission.

Figure 11.2 JCC model

alliance general managers, and alliance management team participants. At a tactical level, the JCC has a set of discipline-based specialist subcommittees established for cross-fertilisation of ideas and innovations. At an operational level, there are working groups, discipline champions, and communities of practice (CoPs). This structure is illustrated as aiming to coordinate, imitate, and innovate through sharing innovative ideas and seeking and using JCC/LXRP-approved funding to conduct experiments that lead to continuous improvement. Innovation may be the design and development of materials, construction techniques, or process improvement innovations.

The JCC serves a pivotal role in programme integration mechanisms. The above has led to the development of organisational learning routines, mainly through the JCC as a CoP similar to that described in the CoP literature (Orr, 1990; Wenger, McDermott and Snyder, 2002). The LXRP developed an extranet communication platform to integrate all programme participants and enable them to collaborate, submit, review, and access innovation proposals. The JCC extranet is a significant enabler of innovation diffusion and continuous improvement.

Organisational Knowledge Transfer Drivers

A significant organisational driver is the PAA. Its design directly influences alliance governance requirements. Accountability and responsibility, for example, are based on the whole *project's* performance and not on a particular discipline team's performance. It promotes a 'we all sink or swim together' project mindset that drives knowledge and innovation to flow to solve unexpected challenges and improves planning. Team participants all share a stake in an AWP's success and are willing to share continuous improvement intelligence. The PAA also shifts each team's fiduciary duty of care from their home-based employer to the project, and so they are relieved of concern that in sharing knowledge or innovation, they are breaching their fiduciary duty to their home-based employer (Agarwal, 2014; Lydenberg, 2014). The PAA encourages stakeholder engagement rather than stakeholder management with an alliancing mindset.

Internal stakeholder engagement is governed by the PAA behavioural requirements, which fit well with the '5-Cares', and by expecting project participants to treat each other respectfully and collaborate readily. It also results in low knowledge and exercising hierarchical power asymmetry, enabling genuine dialogue based on participant subject-matter expertise. The behaviour requirements call for (Department of Infrastructure and Transport, 2011, p. 12):

> risk and opportunity sharing; commitment to 'no disputes'; best-for-project unanimous decision-making processes; 'no fault – no blame' culture; good faith; transparency expressed as open book documentation and reporting; and a joint management structure.

These requirements drive collaboration, prompt genuine dialogue, and trigger innovation and experimentation.

External stakeholder engagement is predicated on similar values. So, the LXRP established a rich co-learning environment between AWP stakeholder engagement team members, traders, and the public when designing disruption and disturbance minimisation plans and actions.

The LXRP PAAs have specifically designed KRAs/KPIs to shape and drive innovation specifically with both a project and programme focus through a series of programme-focused measures. KPI1.1 has a reward for the AWP team developing an approved, validated innovation submission. When AWP teams believe they have developed a genuine innovation for continuous improvement, they can apply the JCC to assess and judge the proposal's merits collegially. A one-page specific form and a process for an improvement or innovation idea exists. It is clearly and briefly articulated, and the problem or opportunity is explained in terms of the object (problem or opportunity), its extent, the nature of the defect being addressed, and its impact. The idea is articulated as aligned to the programme objectives and '5-Greats', benefits are listed, accountabilities are itemised, and next steps are proposed. The form also has space for an attached visual presentation. Achieving this KPI by convincing the JCC of its merit is incentivised.

This process has two main innovation-driving features. First, AWP teams are motivated to improve, innovate, and, most importantly, share ideas. Second, the JCC and submission process through the CoPs and JCC subcommittees drives innovation at a programme level because the JCC process involves all five programme alliance teams. Additionally, KPI1.2 is designed to reinforce this knowledge diffusion. That KPI rewards AWP teams for adapting/adopting an approved KPI1.1. It makes its

KPI1.2 submission that presents evidence of savings and improvements that result in *both* the KPI1.1 originators and the KPI1.2 applicant team sharing an incentive for the improvement. This process shares some similarities with a system devised for Crossrail in the UK (Davies et al., 2014). However, the LXRP system appears more sophisticated, using its JCC extranet as a communication platform to drive accessibility to knowledge about innovations and improvements.

The LXRP external stakeholder engagement and its diversity and inclusion strategy drive co-learning from multiple perspectives. An LXRP-appointed urban design and advisory panel that includes urban planning experts, LXRP staff, and local representatives drives its transformation strategy. The aim is to identify opportunities for improving a precinct along the rail corridors associated with particular AWPs. The panel explores and reviews ideas about transforming the precincts and advises on proposals for the least intrusive way work can proceed while minimising disruption to traders/businesses and the public. Numerous innovations and improvements are driven by these governance mechanisms, which are shared across the programme.

A critical driver of innovation and organisational learning is the LXRP's leadership team's qualities of being an informed and active client. This is enhanced at the programme level, where its leadership team drives project integration and collaboration to trigger learning and knowledge diffusion. The lack of effective project integration within programmes can resemble the general problem faced by the construction project industry. Teams come together and may develop some great ideas and innovations, but they quickly disband, and much of this organisational knowledge becomes corporate amnesia; the 'eternal beginner syndrome' prevails (Flyvbjerg, Budzier and Lunn, 2021). The LXRP avoids this through its procurement choice of alliancing based on rigorous performance by the teams and cross-referencing to substantial benchmark data, as well as through its strategy of releasing AWPs over time to guarantee a lengthy pipeline of work over 15 years. The LXRP is also an informed, active, hands-on client. Its senior leadership team and many AWP team participants have extensive large-scale infrastructure project experience, many of them having past cross-disciplinary roles as designers, contractors, and in government as client representatives. Many team members also had direct experience in alliancing and, more specifically, the Regional Rail Link (RRL) programme in the early 2000s (Victorian Auditor-General's Office, 2018, p. 32). The LXRP can, therefore, be classed as having high-level owner capabilities that are a project delivery success factor (Winch and Leiringer, 2016).

In summary, organisational learning drivers include cross-programme incentivisation through KPIs, the long pipeline of project package work and innovation performance incentives. LXRP leadership qualities also drive learning through adhering to a well-articulated purpose and strategy for the programme of effectively integrated projects. The leadership team had recent experience with a similar large-scale megaproject, the RRL, and many of the team participants had also worked on projects such as Crossrail in the UK. The client was, therefore, capable of driving innovation effectively.

Team Participant Drivers

Organisational knowledge integration is an important institutional driver for continuous improvement and innovation diffusion that directly involves teams of people. The most valuable knowledge is tacit, which needs socialisation to reap meaningful organisational benefits (Nonaka and Takeuchi, 1995). We can refer to early theory to understand how people make a difference in driving knowledge dissemination. How does knowledge from an individual make its way through an organisation? It happens through organisational supportive processes, individual attitudes, motivation, and the ability to learn.

Szulanski (1996) researched the nature and causes of 'sticky knowledge' and found that much of the problem rests with people and their workplace environment not encouraging, sometimes inhibiting, knowledge exchange. Among these causes, he lists the knowledge-giver as reticent or unable to explain ideas clearly, knowledge recipients as unable to comprehend the knowledge for various reasons, or poor relationships binding the provider–receiver. The workplace culture and environment are vitally important, and Szulanski (1996) refers to them as having a barren versus fertile organisational context.

The above sections highlighted organisational enablers and drivers that shape the knowledge transfer context. The LXRP can be seen as a fertile organisational learning context with a supportive workplace culture guided by highly competent leaders and, importantly, followers taking a shared leadership approach where personal expertise is respected and valued (Drescher et al., 2014; Scott-Young, Georgy and Grisinger, 2019).

The LXRP case study demonstrates interview data that referred to the PAA as driving positive learning behaviours. Teams were exposed to multiple perspectives through the diversity and inclusion recruitment strategy, with several interviewees stating that they were surprised by unexpected ideas prompting innovation from these diverse perspectives. Curiosity about the unexpected, rather than hiding or ignoring what is perceived as bad news or potential problems, was driven by the PAA behavioural requirements and the evident desire of many AWP team participants to experiment and exhibit readiness to improvise. It was also apparent from interview transcript data that the workplace culture actively drove continuous improvement through innovation. Assumptions were being routinely questioned. Having diverse and traditionally ignored team participants speak up and ask for assumptions to be explained and clarified helps find innovative solutions to challenges through effective dialogue. Often, when we are forced to explain assumptions, it becomes evident that we may have misunderstood or misinterpreted the challenge's context. So, fresh perspectives help clarify the challenge and lead to discovering its resolution, which lies at the heart of how individual and organisation-wide learning can flourish.

Crossan, Lane and White (1999) explained how organisational learning starts with individual intuition when confronting difficulties and challenges by devising 'workarounds' to existing routines. These can be shared through group socialising when genuine dialogue occurs between individuals and throughout the group.

Dialogue is important because it is not an activity where one party 'wins' a debate; instead, it is an open process of exchanging insights and perspectives so that a more holistic resolution is agreed upon (Bohm and Nichol, 1996; Womack, 2011). Group-refined knowledge can then become shared through organisational integration as cross-group exchanges extend the dialogue between groups until the 'workarounds' become institutionalised routines and processes. In the LXRP case, team integration occurs within projects and dynamically across projects within the programme.

From a person's perspective, intuition is triggered by curiosity. They wonder why and how what was intended may produce an unexpected outcome. Curiosity requires system interaction and critical thinking skills. Kolb's (1984) learning theories explain that learning is a process that starts with a concrete experience (such as facing a difficulty or challenge), critically reflecting on that (requiring analytical skills), abstracting the reflection through the use of existing knowledge about relevant theories or heuristics (critical thinking) that supports a hypothesis, then actively experimenting to validate or refute the hypothesis (being open-minded).

All this does not happen automatically. Creating a fertile ecosystem requires a clear and consistent strategic intent to promote curiosity dialogue and absorb new knowledge. The organisation needs to provide proactive leadership and workplace culture through enacting its strategy. In this situation, a learning organisation reduces sticky knowledge for continuous improvement and innovation diffusion.

DISCUSSION

The above sections have explained organisational learning from a project delivery mode choice perspective. Our research question is: *How does the LXRP's collaborative alliancing approach set it apart from more traditional approaches to procurement and enable the facilitation and realisation of innovative practice?*

The Organisational Learning Impact of a Collaborative IPD Project Delivery Form

The introduction briefly discusses how an integrated collaborative project delivery form such as the LXRP differs from traditional construction delivery approaches such as DBB and D&C or PPP. A key defining difference between alliancing and traditional approaches is contractual and psychological. Contractually, the PAA structure integrates disciplinary participants into an integrated single team that includes the project owner representative and often, as is the case with the LXRP, operator participants such as the rail service operator and the road authority. From an organisational learning perspective, the LXRP structure immediately widens available knowledge, insights, and perspectives into one coherent alliance project team. Also, loyalty and duty of care are transformed through the PAA. Participants accept a best-for-project outcome, and their fiduciary duty is to the project and not their home organisation's employer (Agarwal, 2014; Lydenberg, 2014). Performance is, therefore, squarely on the entire project. Sharing knowledge and innovation is not seen as betraying confidence but as being an authentic project–corporate citizen. Knowledge and learning transfer across the project become naturally part of the agreement.

The LXRP case study is exemplary because it demonstrates taking the project allancing approach to a higher level. Many programmes or projects are managed with less well-integrated processes, leadership motivation, and outcomes than the LXRP exhibits. The five initial alliance work packages were integrated into the programme. Still, research undertaken on one of these (Walker, Matinheikki and Maqsood, 2018) found that the initial programme mindset of interviewed individuals was far less mature than interviews undertaken during 2023 (Walker, Love and Betts, 2024). This may be explained by a shedding of the 'eternal beginner syndrome' (Flyvbjerg, Budzier and Lunn, 2021) as the LXRP progressed towards the AWP approach with its incentivisation of shared learning through innovation diffusion and the development and maturation of mechanisms such as the JCC, its platform, and the clear explication of strategy and purpose concisely provided in the 2023 blueprint. Therefore, the programme of projects delivery form, with its psychological leveraging of loyalty to the AWP *and* the programme, extended fiduciary duty to the programme, making organisational learning more conducive to benefiting the programme *and* individual AWPs.

The Organisational Learning Impact of Governance Incentive Design

Traditional project delivery forms such as DBB, D&C, and PPP constrain incentives due to a win–lose potential. PPPs usually deliver the physical infrastructure object as a D&C contract, so PPP delivery and D&C are grouped for this reason. Each project is fresh, so the 'eternal beginner syndrome' generally applies. There is little incentive for the D&C participants to establish an organisational learning system such as the LXRP's. The time span of a single project and the departure of many project participants to new projects disburses learning, and while there may be knowledge repositories that may hold information, that knowledge is notoriously difficult to capture because it is highly context- and perception-based and often stored and curated in non-standardised forms. For effective organisational learning, it is essential to maintain the reference source *and* access to those who generated the knowledge and lessons learned. The LXRP JCC and extranet platform provide a learning repository in a standardised, consistent form, and the strategy of AWPs and the long pipeline of work helps retain key knowledge brokers.

Several interviewees stated that the LXRP governance and AWP procurement strategy also encourage career development. Many participants can consider their LXRP involvement as a career. They absorb and co-create knowledge in a fertile organisational learning environment, can shadow those they report to, and temporarily act in these roles occasionally. They also have a large measure of employment stability. People can grow their careers through their evolving roles, providing a potent incentive for individuals, as several interviewees specifically mentioned.

KPI1.1 and KPI1.2 are particularly potent governance incentives because they reward the AWP team that generates an improvement or innovation and reward other AWP teams for adapting or adopting the innovation. The JCC and its extranet platform make the process transparent. The AWP TOC process also adopts the reuse of as-built designs to be adapted for a new AWP context. The LXRP leadership group actively questions which innovations and learnings from other projects in the programme are considered, and this further incentivises the use of the innovations and learning repository facility and prompts cross-project dialogue between innovation generators and the TOC during the TOC process.

The Organisational Learning Impact of People and the Workplace Culture

Interviewee data revealed numerous references to the PAA behavioural requirements, either specifically or describing what working on the LXRP programme is like. Prominent among the data were alliance participants (and non-alliance subcontractors) learning through experimentation and dialogue within a no-blame and curiosity workplace cultural setting. Low power and information asymmetry encouraged multiple-perspective-taking that supported genuine dialogue. These first-order concepts supported the identification of second-order themes of the desire and motivation of project participants to learn and share knowledge, experiences, and insights. It became evident that project participants enjoyed confronting challenges that tested their critical thinking ability. The data also suggests that diversity and inclusion of participants who would normally be ignored, sidelined, or not recruited were also part of the *feeling* of the workplace culture.

The impact of the above can be explained from the extant literature. Walker and Lloyd-Walker (2014) researched the nature of the project alliance workplace environment. Quoting from one of their 22 interviewed project alliance participants, they framed the workplace in terms of its *ambience*. Their coded data suggested that the *feel* of the alliance atmosphere created by human interactions within a project alliance tended to be markedly different from that found on most traditional DBB or D&C-delivered construction projects. This ambience went beyond being best described as culture, atmosphere, or environment because interviewees described an almost palpable sense of collaboration and harmony directed at a best-for-project outcome. Importantly, they found alignment between the espoused values articulated through the PAA and senior leadership teams and the culture in use. They revealed that positive cultural enablers included: participant levels of knowledge, skills, attitude, and experience; recruitment, staffing, and professional development opportunities; and the project alliance's community citizenship behaviours. They also identified cultural drivers: (1) the project owner representative's risk appetite; (2) participant teams' legacy experience of collaborative project work; (3) risk/conflict management practices; and (4) the balance of competitive and collaborative tension.

The construction industry is notorious for its toxic culture, with an exception being alliance and IPD projects that embrace inclusiveness and diversity (Clegg et al., 2023). The LXRP '5-Cares', PAA behavioural requirements, and the LXRP's diversity and inclusion strategy help counter workplace toxicity and create a positive *ambience* that supports learning and sharing knowledge.

Hackman and Oldham (1976) provide an intrinsic motivation model of worker psychological states. Maslow's (1943) higher-level self-actualisation needs are well accepted, and Hartmann (2006) found that improved innovation diffusion and continuous improvement was facilitated by giving immediate feedback, providing communication channels for implicit knowledge, allowing for autonomous work and task identity, initiating innovation projects and using a comprehensive reward and incentive system are appropriate managerial actions in this regard. This supports Szulanski's (1996) 'sticky knowledge' findings that motivation to offer and/or receive insights and knowledge critically impacted reducing knowledge transfer stickiness. The three relevant psychological states are (Hackman and Oldham, 1976): (1) task autonomy; (2) interpreted work responsibility (by the participant); and (3) interpreted feedback quality.

The LXRP data indicates that their workplace engagement strategy placed respect for expertise and critical thinking at a high level; people were expected to ask questions if they did not understand something critical to decision-making or action and question assumptions they felt uncomfortable about. The '5-Cares' clarifies that accountability and responsibility are core values espoused and enacted throughout the programme. Governance mechanisms such as KRAs/KPIs clearly articulate what is important to projects, and the programme and teams routinely share feedback. The JCC communication platform and the KPI1.1 and KPI1.2 processes also provide motivational feedback and peer recognition of innovations and improvements shared. This suggests that it feeds a virtuous motivation circle. Interviewees commented on how proud they felt about having their ideas and innovations recognised by both peers and the LXRP leadership team.

The Organisational Learning Impact Summary

Organisational learning needs to be strategised, clearly communicated, and nurtured, requiring the organisation to articulate its purpose, design mechanisms to drive it, and create a workplace culture that embraces it. The LXRP data analysis demonstrates that this occurred both within projects and across the programme of the project, which reduced the stickiness of knowledge transfer. A noteworthy enabler is that the LXRP's strategic approach to continuous improvement, such as lowering costs and enhancing performance, was complemented by its determination to leave positive industry-level skills and workplace culture legacy operationalised by a legacy KPI. This is an important issue because not only is continuous improvement and innovation diffusion expected to occur at the programme level, but also, programme participants are encouraged to disseminate any intellectual property associated with innovations across their home-based organisations and generally into the construction industry through presentations, networking, and other communication channels. The LXRP impact will be a long-term legacy for project delivery approaches and mechanisms.

CONCLUSIONS

This chapter's guiding research question was: *How does the LXRP's collaborative alliancing approach set it apart from more traditional approaches to procurement and enable the facilitation and realisation of innovative practice?*

We justified our choice of a single case study to help answer the research question as being useful to understand a unique or highly unusual case with a class of similar contexts. The chapter focuses on the megaproject infrastructure projects and programmes of projects context. Although findings cannot be generalised, they provide a straw-man model that other project owners may adopt when designing a project/programme delivery approach that can centrally feature organisational learning to reap associated benefits from that focus. Numerous interviewees from this case study stressed the limitations of blindly adopting or adapting the model without the necessary understanding of the context of this case study.

We suggest that three broad categories of entities need to develop the requisite knowledge and skills to deliver complex infrastructure projects, in general, using a collaborative and integrated procurement approach: (1) the client; (2) the supply chain; and (3) the programme participants.

1. The Client Representative Category

The client organisation needs to have highly skilled and experienced representatives. The term LXRP refers to clients needing an 'informed and capable team' (Gulino, Sergeeva and Winch, 2020). The LXRP had the advantage of its participants having experience on the RRL and other alliance projects over the past two or more decades. The project owner entity must also have the requisite strategic design skills to understand the programme/project purpose, craft a governance system to suit its context, and create a workplace ambience that supports the strategy. The strategic procurement choice of alliance projects was informed by RRL and other past experiences, and the opportunity presented by the extended scope and programme timeline allowed the programme to be broken down into manageable chunks or AWPs of several hundred million dollars up to a billion-plus. This provided a motivationally enhanced pipeline of work for design and delivery teams over 15 years.

2. The Supply Chain Category

The supply chain includes the AWP participant organisations and many suppliers, subcontractors, and service providers. Project alliance teams comprise a group of one or more design organisations, a contractor group of one or more construction contractors, the project owner representative, and the service operator(s). The LXRP had Metro Trains Melbourne (MTM) and VicRoads operator teams. Each participating organisation must have the corporate capabilities, expertise, and capacity to deliver complex infrastructure projects that span hundreds of millions of dollars.

The initial LXRP project procurement approach was a competitive alliance TOC process. Project alliancing has a decades-long history in Australia, and there is no shortage of potential alliance syndicate partners for alliances. However, the scale of the LXRP and its long duration placed strains on the marketplace. The process for syndicate recruitment and alliance team selection criteria is well established and documented in case studies such as the National Museum of Australia alliance project (Auditor-General of the Australian National Audit Office, 2000; Walker, Hampson and Peters, 2002). The Victorian Government's infrastructure alliancing experience was vital, and the RRL provided opportunities for relevant alliancing experience for Australian tier 1 contractors, global design organisations with a significant presence in Australia, and many LXRP participants. MTM and VicRoads also had alliancing experience spanning RRL and other Australian alliance projects. Shortly after the first five alliances were bid and work commenced, the LXRP changed its procurement strategy from competitive TOCs, where at least two bidding syndicates compete, to a single TOC approach, as outlined earlier in the chapter. This change reduced the strain on resource availability to develop TOCs from all participants, including the project owner representative group, and the growing substantial reference benchmarked performance data, knowledge bank of innovation, and availability of as-built designs' locked-in competitive tension ensured that AWPs represented value for money. Ensuring that this part of the supply chain was available and prepared was strategically managed by the LXRP.

The wider supply chain of suppliers, subcontractors, and service providers is extensive. The scale of spending is enormous, but it was strategically packaged into chunks that enabled global, national, and local organisations to compete. The repetitive nature of parts of the works resulted in substantial modularisation with a large pipeline of work ahead to facilitate the supply chain to adapt to the alliancing mindset and behaviours. LXRP interviewees mention the modular capacity to share as-built documentation as building information models (BIMs). These provide advantages and pose challenges to the supply chain. The LXRP strategy and programme's purpose, which is to facilitate local industry growth as a legacy, played an important role in expanding supply chain capabilities. Stakeholder engagement requires competent project/programme team participants similar to the stakeholders. For example, the urban planning expert panel members were responsible for co-creating learning and value, requiring them to develop collaborative and critical thinking skills.

3. The Programme Participant Category

Part of the TOC process and project and programme-level team participant recruitment requires knowledge, skills, attributes, and experience (KSAE) conducive to collaboration and co-development of organisational learning (Walker and Lloyd-Walker, 2020). Shortcomings and gaps can be, and were, rectified and bridged through a strategy for people development. The LXRP had a comprehensive team-participant training and development programme aligned with its 'great people' part of the '5-Greats'. While team participants may not have all the expected and desired alliance-related KSAE capabilities, they were enabled to attain these on the job. For example, the diversity and inclusion strategy catered for a graduate recruitment programme linked to university providers and training and development for people with potentially transferable skills, such as recent asylum seekers, recent immigrants, and motivated and able local people who had traditionally been ignored or not considered suitable to work on infrastructure projects. The wide range of skills and expertise required of the LXRP total workforce requires a step-change in thinking about resourcing people for such megaprojects, because the purpose and scope of these projects transcend construction or design technical competencies to encompass a host of roles associated with such areas as sustainability, community support and stakeholder engagement, local government administration understanding, urban planning, landscaping – the list is enormous. Programmes like the LXRP require entirely different skills and capabilities than standard construction-only projects.

Concluding Key Points

Innovation diffusion through KPI1.1 and KPI1.2 critically enhanced the programme's learning organisational outcome. Several governance mechanisms enabled and drove the LXRP as a learning organisation (e.g., the PAA as an overall framework and JCC as a key driver of learning and innovation diffusion). Interviewees stressed the importance of the AWP programme and the long pipeline of work as key shapers of the LXRP programme. The people aspect, including internal and external stakeholders, is also a major part of the LXRP example.

The differences between the IPD alliancing approach of a programme of AWPs by the LXRP set it fundamentally apart from traditional DBB and D&C. The level of sophistication of the project owner and supply chain is also a major factor. The participants and what they offer and bring to the programme drive the realisation of a learning organisation driven by a carefully designed project/programme organisation governance structure and workplace culture.

Interviewees commented that they would recommend the LXRP alliance *programme* model for other infrastructure facilities such as schools, prisons, hospital upgrades, and rebuilds. However, the model must be adapted to suit the particular context.

NOTES

1. For example, one award was for the LXRP Strategic Communications entity winning the International Association for Public Participation (IAP2) *2021 Australasia Organisation of the Year Award*, for its industry-leading approach to stakeholder engagement. This is an innovation that may prove globally groundbreaking for megaprojects.

2. South-Eastern Program Alliance, Metropolitan Roads Program Alliance, North-Western Program Alliance, Western Program Alliance and Southern Program Alliance.
3. Refer also to https://www.globalrailwayreview.com/article/63582/melbournes-huge-level-crossing-removal-project-saving-lives-improving-journeys/ for further examples of its innovative social value programme.

REFERENCES

Aaltonen, K. and Turkulainen, V. (2018). "Creating relational capital through socialization in project alliances." *International Journal of Operations & Production Management.* **38** (6): 1387–1421.

Agarwal, A.K. (2014). "Corporate governance: Changing trends in interpreting fiduciary duty." *Vikalpa: The Journal for Decision Makers.* **39** (3): 1–12.

Alves, T.d.C.L. and Lichtig, W. (2020). Integrated Project Delivery – Basic tenets and recommendations. In *Lean Construction: Core Concepts and New Frontiers.* Tzortzopoulos P., M. Kagioglou and L. Koskela (eds). Milton: CRC Press LLC, pp. 162–185.

Auditor-General of the Australian National Audit Office. (2000). *Construction of the National Museum of Australia and the Australian Institute of Aboriginal and Torres Strait Islander Studies.* Canberra, Australia.

Australian Constructors Association. (1999). *Relationship Contracting – Optimising Project Outcomes.* Sydney.

Bohm, D. and Nichol, L. (1996) *On Dialogue.* London: Taylor & Francis Group.

Clegg, S.R., Loosemore, M., Walker, D.H.T., van Marrewijk, A.H. and Sankaran, S. (2023). Construction cultures: Sources, signs and solutions of toxicity. In *Construction Project Organising.* Addyman S. and H. Smyth (eds). Chichester: John Wiley & Sons, pp. 3–16.

Clegg, S.R., Pitsis, T.S., Rura-Polley, T. and Marosszeky, M. (2002). "Governmentality matters: Designing an alliance culture of inter-organizational collaboration for managing projects." *Organization Studies.* **23** (3): 317–337.

Crossan, M.M., Lane, H.W. and White, R.E. (1999). "An organizational learning framework: From intuition to institution." *Academy of Management Review.* **24** (3): 522–537.

Davies, A. and Brady, T. (2015). "Explicating the dynamics of project capabilities." *International Journal of Project Management.* **34** (2): 314–327.

Davies, A., MacAulay, S., DeBarro, T. and Thurston, M. (2014). "Making innovation happen in a megaproject: London's Crossrail Suburban Railway System." *Project Management Journal.* **45** (6): 25–37.

Davis, M.S. (1971). "That's interesting!: Towards a phenomenology of sociology and a sociology of phenomenology." *Philosophy of the Social Sciences.* **1** (2): 309–344.

Davis, P.R. and Love, P.E.D. (2011). "Alliance contracting: Adding value through relationship development." *Engineering Construction & Architectural Management.* **18** (5): 444–461.

Department of Infrastructure and Regional Development (2015). National Alliance Contracting Guidelines Guidance Note No 5: Developing the Target Outturn Cost in Alliance Contracting. Department of Infrastructure and Regional Development A.C.G. Canberra, Commonwealth of Australia, p. 90.

Department of Infrastructure and Transport (2011). National Alliance Contracting Guidelines Guide to Alliance Contracting. Department of Infrastructure and Transport A. C. G. Canberra, Commonwealth of Australia, p. 168.

Drescher, M.A., Korsgaard, M.A., Welpe, I.M., Picot, A. and Wigand, R.T. (2014). "The dynamics of shared leadership: Building trust and enhancing performance." *Journal of Applied Psychology.* **99** (5): 771–783.

Egan, J. (1998). *Rethinking Construction – The Report of Construction Task Force.* London, p. 38.

Egan, J. (2002). *Accelerating Change. Rethinking Construction.* London: Construction Industry Council, p. 44.

Eisenhardt, K.M. and Graebner, M.E. (2007). "Theory building from cases: Opportunities and challenges." *Academy of Management Journal.* **50** (1): 25–32.

Flyvbjerg, B. (2011). Over budget, over time, over and over again. In *The Oxford Handbook of Project Management.* Morris P.W.G., J. Pinto and J. Söderlund (eds). Oxford: Oxford University Press, pp. 321–344.

Flyvbjerg, B., Budzier, A. and Lunn, D. (2021). "Regression to the tail: Why the Olympics blow up." *Environment and Planning A: Economy and Space.* **53** (2): 233–260.

Gioia, D.A., Corley, K.G. and Hamilton, A.L. (2013). "Seeking qualitative rigor in inductive research." *Organizational Research Methods.* **16** (1): 15–31.

Gulino, M.L., Sergeeva, N. and Winch, G. (2020). "Owner capabilities in social infrastructure projects: Towards an expansion of the dynamic capabilities' framework." *International Journal of Managing Projects in Business.* **13** (6): 1263–1282.

Hackman, J.R. and Oldham, G.R. (1976). "Motivation through the design of work: Test of a theory." *Organizational Behavior and Human Performance.* **16** (2): 250–279.

Hartmann, A. (2006). "The role of organizational culture in motivating innovative behaviour in construction firms." *Construction Innovation: Information, Process, Management.* **6** (3): 159–172.

Hietajärvi, A.-M. (2017). Capabilities for managing project alliances. Thesis Monograph, *International Journal of Managing Projects in Business.* Oolu, Finland.

Ibrahim, C.K.I., Costello, S.B. and Wilkinson, S. (2018). "Making sense of team integration practice through the 'lived experience' of alliance project teams." *Engineering, Construction and Architectural Management.* **25** (5): 598–622.

Kolb, D.A. (1984) *Experiential Learning: Experience as the Source of Learning and Development.* Englewood Cliffs, NJ: Prentice Hall.

Lahdenperä, P. (2017). "Towards a coherent theory of project alliancing: Discovering the systems: Complex mechanisms yielding value for money." *Construction Economics and Building.* **17** (2): 41–61.

Latham, M. (1994). *Constructing the Team.* London.

Lichtig, W.A. (2005). "Sutter Health: Developing a contracting model to support lean project delivery." *Lean Construction Journal.* **2** (1): 105–112.

Lloyd-Walker, B.M., Walker, D.H.T. and Mills, A. (2014). "Enabling construction innovation – the role of a no-blame culture as a collaboration behavioural driver in project alliances." *Construction Management and Economics.* **32** (3): 229–245.

Love, P.E.D., Davies, A., Walker, D.H.T. and Betts, M. (2024). LXRP continuous improvement and innovation diffusion. In *Strategic Integrated Program Delivery: Learning from the Level Crossing Removal Project.* Walker D.H.T., P.E.D. Love and M. Betts (eds). Abingdon: Routledge.

Love, P.E.D., Matthews, J., Ika, L.A. and Fang, W. (2022). "Error culture and its impact on rework: An exploration of norms and practices in a transport mega-project." *Developments in the Built Environment.* **10**: 14.

Lydenberg, S. (2014). "Reason, rationality, and fiduciary duty." *Journal of Business Ethics.* **119** (3): 365–380.

Magree, J., Scott-Young, C. and Lloyd-Walker, B.M. (2024). LXRP human resource development. In *Strategic Integrated Program Delivery: Learning from the Level Crossing Removal Project.* Walker D.H.T., P.E.D. Love and M. Betts (eds). Abingdon: Routledge.

Maslow, A.H. (1943). "A theory of human motivation." *Psychology Review.* **50** (4): 370–396.

Müller, R. (2017). Governance models and paradigms. In *Governance and Governmentality for Projects: Enablers, Practices, and Consequences*. Müller M. (ed.). London: Taylor & Francis Group, pp. 36–50.

Müller, R., Zhai, L., Wang, A. and Shao, J. (2016). "A framework for governance of projects: Governmentality, governance structure and projectification." *International Journal of Project Management*. **34** (6): 957–969.

Nonaka, I. and Takeuchi, H. (1995). *The Knowledge-Creating Company*. Oxford: Oxford University Press.

Orr, J. (1990). *Talking About Machines: An Ethnography of a Modern Job*. PhD thesis. Ithaca, NY: Cornell University.

Rogers, E.M. (2003). *Diffusion of Innovation*. New York: The Free Press.

Rutten, M.E.J., Dorée, A.G. and Halman, J.I.M. (2009). "Innovation and interorganizational cooperation: A synthesis of literature." *Construction Innovation*. **9** (3): 285–297.

Scott-Young, C.M., Georgy, M. and Grisinger, A. (2019). "Shared leadership in project teams: An integrative multi-level conceptual model and research agenda." *International Journal of Project Management*. **37** (4): 565–581.

Snowden, D. and Rancati, A. (2021). *Managing Complexity (and Chaos) in Times of Crisis. A Field Guide for Decision Makers Inspired by the Cynefin Framework*. Luxembourg: Publications Office of the European Union.

Somerville, P. (2012). Critical realism. In *International Encyclopedia of Housing and Home*. Smith S.J. (ed.). San Diego: Elsevier, pp. 291–295.

Szulanski, G. (1996). "Exploring internal stickiness: Impediments to the transfer of best practice within the firm." *Strategic Management Journal*. **17** (Winter Special Issue): 27–43.

Victorian Auditor-General's Office. (2018). *Assessing Benefits from the Regional Rail Link Project*. Printer V.G. Melbourne.

Victorian State Government. (2017). *Level Crossing Removal Project – Program Business Case*. Government V. Melbourne.

Walker, D.H.T., Hampson, K.D. and Peters, R.J. (2002). "Project alliancing vs project partnering: A case study of the Australian National Museum Project." *Supply Chain Management: An International Journal*. **7** (2): 83–91.

Walker, D.H.T. and Lloyd-Walker, B.M. (2014). "The ambience of a project alliance in Australia." *Engineering Project Organization Journal*. **4** (1): 2–16.

Walker, D.H.T. and Lloyd-Walker, B.M. (2020). Knowledge, Skills, Attributes and Experience (KSAE) for IPD Alliancing task motivation. In *The Routledge Handbook of Integrated Project Delivery*. Walker D.H.T. and S. Rowlinson (eds). Abingdon: Routledge, pp. 219–244.

Walker, D.H.T., Love, P.E.D. and Betts, M. (2024). *Strategic Integrated Program Delivery: Learning from the Level Crossing Removal Project*. Abingdon: Routledge.

Walker, D.H.T., Love, P.E.D., Betts, M. and Jacobsson, M. (2024). LXRP delivery choices. In *Strategic Integrated Program Delivery: Learning from the Level Crossing Removal Project*. Walker D.H.T., P.E.D. Love and M. Betts (eds). Abingdon: Routledge.

Walker, D.H.T., Love, P.E.D. and Matthews, J. (2022). The value of dialogue in alliancing projects. In *The 9th International Conference on Innovative Production and Construction*. Melbourne: CIB. https://virtual.oxfordabstracts.com/#/event/public/1970/program?session=35425&s=600

Walker, D.H.T., Matinheikki, J. and Maqsood, T. (2018). *Level Crossing Removal Authority Package 1 Case Study*. RMIT University Melbourne, Australia.

Walker, D.H.T., Vaz Serra, P. and Love, P.E.D. (2022). "Improved reliability in planning large-scale infrastructure project delivery through Alliancing." *International Journal of Managing Projects in Business*. **15** (8): 721–741.

Wenger, E.C., McDermott, R. and Snyder, W.M. (2002) *Cultivating Communities of Practice.* Boston: Harvard Business School Press.
Winch, G. and Leiringer, R. (2016). "Owner project capabilities for infrastructure development: A review and development of the 'strong owner' concept." *International Journal of Project Management*. **34** (2): 271–281.
Womack, P. (2011). *Dialogue.* London: Taylor & Francis.

12. Recent and nascent innovations in the evolution of construction project procurement: with examples from Australia

Hemanta Doloi and Mohan Kumaraswamy

INTRODUCTION

Procurement is well worth top management prioritisation, being potentially one of the most high-value-adding areas in construction projects, whereas inappropriate procurement can lead to disasters. Risks in procurement management increase with the size and complexity of the project. Procurement in projects can be taken to occur at two levels: one at the project strategy formulation level, that is, upstream, and the other at the project construction or implementation level, that is, downstream, whereas the latter stage is also called 'delivery'. Project delivery processes essentially transform projects from concepts and planning to operationalisation and 'getting it done'. Also, meaningful commercial and societal priorities should drive project strategy, which, in turn, should be formulated and deployed in procurement and delivery at both upstream and downstream stages. These priorities should therefore be harnessed in generating useful innovations that boost overall project performance.

While traditional procurement may simply refer to a process of buying goods and services, or a product based on the itemised cost including the profit margins of the contracting parties, modern projects that involve a multitude of stakeholders and global supply chains involve significant complexities in procurement and delivery processes. Upstream procurement strategies in traditional design-bid-build scenarios were rarely linked to overall project performance and were somewhat distant from the built-asset delivery, while the commercial and societal drivers of the project itself hardly received the attention they deserved. However, resource-intensive construction projects, especially high-value mega projects to deliver important public infrastructure, lead to changes in the characteristics of the built environment (PMBOK, 2021), and demand careful planning and implementation of procurement strategies focusing on both upstream and downstream stages. For example, climate change and global warming have significant positive correlations with the construction industry in general. Traditional construction procurement may not be seen as responsible for dealing with climate challenges, including global warming. But with the emerging need for environmentally sustainable construction practices, construction procurement that factors in issues such as smart and sustainable project design, material types and specifications, and responsible supply chains has become highly significant for creating a climate-positive outcome. For instance, the Victorian Government under the Social Procurement Framework promotes sustainable practices over and beyond the mere compliance requirements. The framework ensures the adaptation of sustainability principles across public projects to minimise adverse environmental impact and increase positive climate outcomes (Victoria Government, 2018). On the social sustainability front, the framework incorporates a range of measures such as local materials, local workmanship, local job creation, and engagement of social enterprises, etc. These measures not only entice context-specific and need-based design and project specifications, but also result in significant reductions of the overall carbon footprints in those projects. Therefore, innovation in procurement practices is needed, focusing on a whole-of-life approach to address the challenges across multiple fronts such as sustainable design, construction, and operations, including the reduction of carbon footprints, especially in large-scale infrastructure projects.

As many countries are currently signatories to the Paris Climate Agreement, their essential, albeit ambitious, contributions to global climate improvements must merge with their public procurement agendas, given the development needs of most countries. The need of the day is therefore for innovative procurement and smart delivery policies that align and synergise these potentially contradictory agendas, rather than bring them into conflict. Considering the OECD countries alone, the volume of public procurement in these countries is over 13% of GDP. The volume of public procurement of non-member countries such as India accounts for over 30% of GDP. Redefinition of goals with innovative ideas and creative thought processes in formulating procurement strategies for fostering sustainable growth and development is thus critically needed across all countries. The UN's Sustainable Development Goals (SDGs) provide a sound platform for realigning public procurement to reposition long-term societal needs, where the community is at the core of decision-making criteria and processes in development projects.

An innovative procurement policy fostering out-of-the-box project procurement strategies is key for delivering public infrastructure more sustainably. To achieve an overall sustainability outcome, the innovations in procurement need not necessarily focus on the construction industry alone but also need to embrace all interrelated sectors such as transport, water, energy, etc. Procurement of construction projects involves multiple parties, from the concept to planning, tendering, execution, and operation stages. Due to the high value and high volume of financial transactions involved in construction procurement, the processes are subject to a vast number of standards, regulations, and compliance requirements, especially when it comes to procuring mega-infrastructure. Alongside the procurement processes, the roles, responsibilities, and functions of the multiple stakeholders spanning upstream and downstream stages are highly significant, not only in the innovative planning of procurement strategies but also in delivering the infrastructure and services to the end-users. Taking, therefore, a whole-of-life perspective of the project procurement processes and drawing from numerous examples of current and creative procurement practices, especially in the context of Australian projects, this chapter puts forward a novel research-based approach for targeting value-adding innovations in construction procurement. The authors do not claim this novel approach to be a panacea for addressing all, or even many, of the ever-increasing climate challenges. But these innovative approaches should help to boost value for money and align processes and outcomes with relevant societal values in the high-value construction projects that shape the built environment landscape for life. References to the SDGs in the proposed procurement framework highlight the significance of context-specific and fit-for-purpose construction, which eventually also targets holistic sustainability outcomes.

The first part of this chapter focuses on the significance of procurement, its evolution, and the opportunities for adding value in procurement innovations. Then, a framework for innovative procurement and delivery is introduced, highlighting a step-by-step articulation of procurement processes and the meeting of targets. Finally, the implementation strategy of the innovative procurement functions, along with the required stakeholders' roles, responsibilities, and coordination, is discussed with a hypothetical research-based example. The need for a holistic approach integrating multi-dimensional considerations, especially the social aspects, is highlighted. The discussion also briefly touches on the UN SDGs as underpinning the way forward towards meeting the targets, with innovative procurement and delivery that harnesses, leverages, and benefits from the underlying drivers as well.

CONSTRUCTION PROJECT PROCUREMENT AND ITS BASIC EVOLUTION IN GENERAL

Construction project procurement includes the procurement of design, construction, and project management services, as well as materials and equipment. It has been defined by the W092 Working Commission on Procurement Systems of the CIB (International Council for Research and Innovation in Building and Construction) as 'the framework within which construction is brought about, acquired or obtained'. Jefferies and Rowlinson (2016) also cited a broader definition by Lenard and Moshini at the 1998 CIB W92 Conference in Montreal as 'a strategy to satisfy client's development and/or operational needs with respect to the provision of constructed facilities for a discrete life-cycle' (p2). This more goal-focused and longer-term procurement perspective resonates with current sustainability imperatives and provides a more appropriate backdrop for this chapter.

Indeed, the evolution of construction project procurement is not a new phenomenon, as humans have always needed shelter, which required some form of construction, thereby involving procurement too. By virtue of every construction project being unique, some degree of innovation from the contractual arrangements, subcontractors' engagement to procuring goods and services emerges in every project, aiming at efficiencies in resource usage and maximising profitability in privately procured projects. Many possible construction procurement routes are widely published in the mainstream literature. A starting point in summarising their evolution, as intended in this chapter, would be the master builders of yore who were commissioned by emperors, kings, pharaohs, high priests, and the like, to 'design and build' palaces, irrigation systems, hilltop temples, cathedrals, mosques, pyramids, and other mega-structures. This single master builder model evolved along with a decline in 'royal edicts' to the 'separated Design-Bid-Build' model that dominated the construction project landscape in the 19th and 20th centuries (Harp, 1991, Battersby, 2003). Although the separated specialisations in design and delivery brought benefits and led to further sub-specialisation in design, construction, and project management services, many industry reports commissioned by governments, starting with those in the UK from the mid-20th century, identified the downsides of fragmented approaches (Miller, 2000; Hale et al., 2009). This led to attempts at reintegration with a return to 'design and build' and consequently a revival of public–private partnerships (PPP) with a focus on 'value for money (VFM)' principles, considering wider stakeholders' perspectives. Evolutions in the present century include more collaborative contracts, integrating shared responsibilities where possible, for example, with partnering, alliancing, integrated project delivery incorporating various degrees of 'joint risk management' between client-consultants and contractors, relying on better relationships and relationship/stakeholder management in general and 'relational contracts' in some cases (Pietroforte and Miller, 2002; Ling et al., 2013; Kumaraswamy, 2024a).

INNOVATIONS IN PROCUREMENT AND CURRENT DRIVERS IN GENERAL

Despite the trend towards more collaborative contracts in some well-developed construction industries, some countries, including many emerging economies, may be apprehensive about the risk of collaboration degenerating into collusion in corrupt practices. Therefore, there still are many 'traditional' contracts with more rigidly demarcated responsibilities, and hence risk management and liabilities (Doloi, 2013; Kumaraswamy, 2024a). Focusing, therefore, on the stakeholder-related dimension within construction procurement that would apply in any case, this chapter looks at the interconnections between networks of stakeholders along with numerous construction and procurement issues. Can, for example, 'socially responsible' procurement practice be promoted as an alternative to the traditional practice for delivering better value in the construction of infrastructure projects? How can the UN's 2030 Agenda Sustainable Development Goals (SDGs) be best addressed in a context-specific project environment? In this context, the United Nations (UN) through the 2030 Agenda (UN, 2023) feeds well into efforts to address social procurement and innovation. Traditional value-for-money (VFM) parameters, where public sector comparison was one of the prevailing principles, considered only the comparative price advantage of the private sector delivery projects over the public sector equivalent (Cheung et al., 2009). Such a practice did not adequately address social procurement or innovation from public or societal perspectives. With the increasing sophistication of our society and the drivers for participatory practices in public decision-making, the traditional VFM trend has clearly changed to more inclusive considerations of public viewpoints and value propositions in the societal context (Doloi, 2012a, 2012b). For instance, acknowledging the Indigenous custodians of the place, the Victorian social procurement framework promotes the integration of Indigenous knowledge and culture to make the project more inclusive (Victoria Government, 2018).

Referring to the SDGs, Goal 8 requires decent work and economic growth. For instance, this could be interpreted in an Australian context to enable equitable access for Indigenous communities and aged citizens who are perceived to have been left out of participation in the economy. Goal 9 (UNDP, 2017) of the agenda lists building resilient infrastructure, promoting inclusive and sustainable industrialisation, and fostering innovation in processes. Goal 10 requires all countries to reduce inequalities through a variety of processes that include greater representation of traditionally marginalised communities. Moreover, the Procurement Act 2023 in the UK has been enacted to facilitate public procurement to be formulated in more innovative ways (UK Government, 2015; Jozepa, 2023).

As seen in the above examples, whether it is public bodies or private developers who identify needs for new-built infrastructure, then formulate projects to meet those needs, overcoming constraints and optimising ways forward requires innovative thinking across all phases of the project lifecycle. For instance, the integration of ideas, contextual knowledge, and relevant practices for realising a particular SDG or a range of SDGs through a project needs innovative approaches across all management and technological dimensions. However, innovations are not without risks or uncertainties. Indeed, risks may be intuitively considered to increase with the 'innovativeness' of the innovations as more 'things can go wrong'. For example, managerial innovations may include more collaborative procurement strategies and protocols, which would need added effort for compliance with contractual agreements and other shared responsibilities among the contracting parties (Doloi, 2013; 2024). Moreover, 'innovative procurement protocols' and 'technological innovations' can be synergistically boosted with innovative organisational and cooperative learning being introduced across supply chains and project teams. 'Joint risk management', as introduced above, can empower more innovations in planning and control protocols and technological approaches as well, to collectively address innovation-triggered risks and other unforeseen risks that could arise in such scenarios (Dulaimi and Kumaraswamy, 2000; Kumaraswamy, 2024b).

Following the above broadbrush overview, it was considered important to probe deeper in this chapter into emerging trends and dimensions of procurement delivery for supporting innovations and boosting performance in public sector infrastructure construction projects in particular. The discussion on procurement delivery is particularly viewed with a focus on an advanced economy such as Australia, where many lessons may be drawn on how strategies and ensuing practices may be adapted and applied elsewhere, including with comparisons with those in emerging economies.

Figure 12.1 highlights the eight key functional areas as envisaged for supporting innovations and boosting performance in the overall procurement–delivery cycle. As

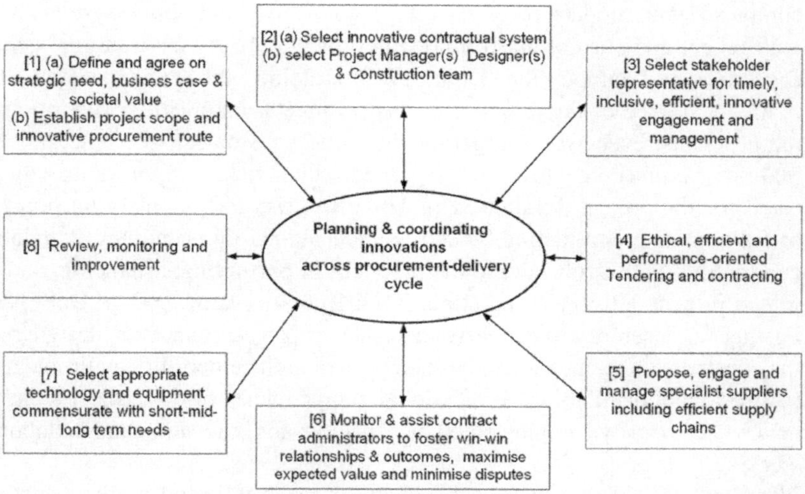

Figure 12.1 Key functional areas in the procurement–delivery cycle for supporting innovations and boosting performance

seen, upstream planning and coordinating for innovations is positioned at the core of the procurement–delivery cycle. The eight envisaged functional areas, as enumerated in Figure 12.1, are briefly discussed below.

[1] Defining strategy and establishing an innovative roadmap are seen as fundamental first steps in planning project procurement routes that align with the organisation's long-term goals and strategic objectives. Governmental requirements and priorities need to be examined, along with the project's benefits and measurable outcomes. At project inception, also depending on the context and nature of the target functionalities at the post-construction stage, a clearly defined strategy and a well-developed business case, including adequate articulation of social value outcomes, are important considerations for both the project's proponents and the key stakeholder groups. The project scope should be designed to meet the required target. This will then point to the areas where innovations may be targeted by appropriate procurement. Suitable targets may be identified based on the expected outcomes in all or a few relevant SDGs. For instance, Goal 8: Decent Work and Economic Growth can be ensured by early engagement of local and social enterprises so that they have or build the necessary capacity to be involved in the project and provide support to reach a context-specific target (Doloi, 2024).

[2] Over the past decades, while traditional contracting routes have provided a sound basis for the procurement of many construction projects, some projects have become infamous for time and cost overruns and/or quality and safety shortfalls and avoidable claims and disputes (Doloi, 2012a; 2018). Moreover, emerging

complexities in modern projects call for innovative contractual systems to deal with recent and current uncertainties in project delivery environments, market capacities, and competition (Doloi, 2013; Dulaimi and Kumaraswamy, 2000). A higher degree of flexibility is needed to meet dynamically evolving project requirements, even while targeting the same core objectives. Since all eventualities cannot be captured in black and white, risks cannot be exactly allocated for the project duration. This has given rise to the emerging popularity of collaborative contracting, which is underpinned by principles of 'relational contracting' and is reflected in the practices of partnering, alliancing, and integrated project delivery (Ling et al., 2014; Kumaraswamy, 2024a; Doloi, 2013). Furthermore, senior project personnel, such as project managers, designers, and the construction team, need to be selected to suit the emerging requirements and realities mentioned above. Professional competencies of the project team must now be necessarily complemented by capacities for innovation and collaborative working.

[3] Identification of key project stakeholders must be followed by the selection of appropriate representatives of each stakeholder, who can work together to reach a consensus on important issues, enabling smooth project progress from conceptualisation, briefing, design, and construction, through to operations and maintenance. Such stakeholders would range from the designated executives in the project governance and leadership team to regulatory authorities, local and extended communities, end-users and concerned legitimate NGOs, professional bodies, trade associations, and any affected parties such as neighbours. A 'procurement innovation logic map', including risks and benefits management plan, and underlying governance arrangements may need to be developed, carefully assessed, and implemented in relation to the project's scope and innovative targets. Stakeholders' engagement and consultation processes including feedback loops need to be established for efficient and inclusive project management (Kumaraswamy et al., 2017a).

[4] Development of ethical, efficient, and performance-oriented tendering and contracting packages is seen as important for attracting and onboarding more proactive and suitable stakeholders to be lead contractors, financiers, consultants, contractors, and suppliers, apart from social enterprises, local NGOs and community groups, marginalised community representations who get involved; so they may synergise in enhancing value-based procurement planning and innovative outcomes (Kumaraswamy, 2024c). Social value realisation protecting community heritage, culture, and environmentally and culturally sensitive natural resources, including flora and fauna, are all important considerations at this stage of project procurement planning (Doloi, 2024). A few relevant SDGs may be indirectly, if not directly, targeted in these exercises, such as Goal 10: Reduced Inequalities, Goal 11: Sustainable Cities and Communities, and Goal 16: Peace, Justice, and Strong Institutions.

[5] Identification, engagement, and management of specialist suppliers, along with the establishment of an efficient supply chain including capacity-building of local

communities and social enterprises in relation to specific project opportunities, is the fifth identified 'functional area' shown in Figure 12.1 that is expected to contribute to generating innovative solutions.

[6] Monitoring contract administrators and nurturing win–win relationships across all sections of the procurement–delivery cycle is crucial for maximising values and minimising disputes (Ling et al., 2013; 2014). Appropriate checklists and milestones need to be developed: highlighting roles, responsibilities, risk retentions, risk transfer, gain- and pain-sharing mechanisms, including management strategies for the itemised procurement packages and delivery processes. The outcome of this function is expected to contribute towards Goal 17: Partnerships for the Goals.

[7] Appropriate technology and equipment need to be sourced and mobilised to suit the project scope and specifications. These will be key enablers in facilitating innovations in the procurement–delivery cycle. Product types and availability, environmental and societal impacts, capital and operating costs, owning and leasing costs, accessibility, configuration, interfaces, operations, and expected performance and short- and long-term productivity ratings are some of the key criteria to be assessed when considering alternative options during decision-making.

[8] Review, monitoring, and continuous improvement are crucial to improving performance against targets across every front of the project procurement–delivery cycle. Accurate recording of the 'right' (carefully chosen) onsite performance data of processes, products, and equipment is a precursor to monitoring the productivity and efficiencies of suppliers, contracting parties, and management involved in various facets of procurement. Deriving appropriate key performance indicators (KPIs) for this is important in taking any needed corrective actions in the procurement–delivery cycle (Kumaraswamy et al., 2017b). Accountability, transparency, and ethical approaches to review and monitoring processes add confidence, credibility, and more opportunities for continuous improvement in procurement innovations to yield added value from built infrastructure projects.

RECENT TRIGGERS AND INNOVATIONS OF CONSTRUCTION PROJECT PROCUREMENT: EXAMPLES FROM AUSTRALIA

Having identified and described what is expected for fostering useful innovations through eight key functional areas in the procurement–delivery cycle, this section highlights some of the key triggers for supporting innovations in construction project procurement from an Australian construction industry perspective.

Public Procurement

Government spending is a major component of economic activity, and public procurement, in particular, accounts for around 12% of GDP and 29% of all government expenditures across the OECD countries (OECD, 2017). In Australia, over 40% of GDP on average is spent on public procurement, being incurred as government spending, which is one of its key functions for keeping up with the demands across multiple interrelated sectors, such as health, environmental protection, energy, transportation, information and communication technology, including research and development. Among the OECD countries, Australia's infrastructure spending is sizeable and increasing steadily. For instance, in 2022–23, there were 83,625 contracts published with a combined value of $74.8 billion. This data relates to the Commonwealth (federal government) procurement through the official channel known as 'AusTender' and does not include other forms of procurement at different state and territory levels. Of these tender-based procurements, 94.5% by volume and 88.4% by value of the contracts in 2022–23 were awarded to businesses with an Australian address (Australian Government, 2023), which shows the extent of the public goods being procured within the Australian business environment annually.

While public procurement practices in Australia have been reasonably well established, reforms of the public procurement framework are increasingly reviewed to align with the strategic goals of the rapidly changing economic, social, and environmental landscapes. With increasing public participation, there is growing awareness of the significance and necessity of social value and social procurement across public sector projects. For example, the Aboriginal Reconciliation Act 1991 (Commonwealth) is one of the key principles in the procurement of public projects, enabling the integration of the Indigenous Procurement Policy (IPP) and ensuring access and participation for First Nations people in economic activities across the country (NIAA, 2020). Aligning with this requirement, in the state of Victoria, the Victorian Aboriginal Procurement Target (VAPT) established a target of 1% of all public procurement for First Nations people and businesses.

Social Procurement Framework Redressing Social Purchasing

In order to increase public accessibility and participation, including social welfare public procurement, the Victorian Government introduced a comprehensive social procurement framework as a mandatory requirement for all public projects (Victoria Government, 2018). The core principle of the social procurement framework is leveraging the buying power involved in public projects to generate social value beyond the delivery of goods and services intended in the project's economic business case:

> Social procurement is when organisations use their buying power to generate social value above and beyond the value of the goods, services, or construction being procured. In the Victorian Government context, social value means the benefits that accrue to all Victorians when the social and sustainable outcomes in this Framework are achieved. (Victoria Government, 2018)

Over recent years, social procurement has been adopted by many organisations across governments, including the delivery of the UN's SDGs (UN, 2023). Social procurement, derived from the UN Agenda 2030 Goals, requires applying innovation towards poverty alleviation, improving jobs for marginalised groups, and fostering policies that aim to eliminate poverty. In the Australian context, social procurement mostly focuses on fostering social enterprises, Aboriginal businesses, and Australian Disability Enterprises.

Value for Money, Societal Value, and Public Perceptions

Over a long period of time, value for money (VFM) has been one of the most decisive criteria in procuring public projects in Australia. Ensuring VFM is prioritised even more in capital-intensive infrastructure projects where public–private partnerships (PPP) are used as a prime method of project procurement. The PPP modality is a favoured procurement strategy for large infrastructure projects in Victoria and across Australia. PPP is defined as 'a long-term contract between the public and private sectors where government pays the private sector to deliver infrastructure and related services on behalf, or in support, of government's broader service responsibilities' (Australian Government, 2008). PPP has served Victoria well in the past, as broader acceptance for VFM propositions in projects' business cases was reasonably high.

Conceptually, VFM entails a balancing judgement call over a range of financial and non-financial measures by careful integration of factors such as quality, lifecycle costs, fitness for purpose, risks, and opportunities in the short, medium, and long term. However, governments' focus on VFM alone has not been able to deliver all projects successfully, especially from the viewpoints of public vis-à-vis social value perspective. There are many examples, such as the East-West Link (EWL) project, where the capital investment did not make sense among a range of issues. The business case of the EWL project initially had a \$0.45 return on every dollar, which was later changed to \$0.84 and further to \$1.45 and \$1.90 or more for the purpose of lifting the project off the ground and achieving success. Even in its final iteration, the EWL would have been completed even if it had taken 56 years to pay off the debt (Aliento, 2014). However, the project was eventually cancelled due to unprecedented public backlash, including political rivalry. This type of fiasco underscores the need for not just an innovative procurement model, but more inclusive and reliable innovations, enabling the project to better integrate public support and equip the project proponents across all decisions to make it a success. It is clear that PPP is not perfect going forward due to its multi-headed management structures – changes are difficult as there are diverse parties who need to agree to the changes. In a PPP project, governments sacrifice overall control, which could be a disadvantage in managing public sentiment, as in the EWL project. In a privately funded PPP project like the EWL, the cost of debt is usually much higher as lenders will not lend at the lower rates of interest that governments can obtain. Although private partners may save costs through innovative technologies and designs, more elegant approaches to social procurement from both public and private sectors are needed.

Given the usual contrast between traditional VFM based on price point comparison (Cheung et al., 2009) versus the societal value of public built infrastructure projects (Doloi, 2012b; 2018), some of the key questions needing answers in project procurement decisions may include: Will a social procurement practice diminish or enhance overall VFM in public projects? Are the noble goals of the UN's Agenda 2030, Australia's IPP, and Victoria's VAPT mutually exclusive with the wider public's VFM expectations? Is there a way out of this dichotomy? Can innovation in procurement be the solution (Lenderink et al., 2022)? These questions clearly posit the challenges of whether innovation could be a means to address the VFM expectations while fulfilling the social procurement policies of governments. In addressing this question, an overarching question may further arise as to whether VFM and other goals should be redefined to balance the ultimately agreed social procurement objectives, taking a broader perspective, and so as to allow innovations within the relevant VFM models. Expanding on these issues, the following section delves into further discussion on construction innovations and underlying challenges from a social procurement perspective.

ADDRESSING CHALLENGES WITH INNOVATIONS IN COLLABORATION AND SOCIAL PROCUREMENT

Innovation can be defined as bringing new ideas, strategies, methodologies, and solutions that help to solve procurement-related problems (Cheung et al., 2009; Doloi, 2012b; 2018; Kumaraswamy, 2024b). We posit in this chapter that current innovations in procurement show ways of working in a collaborative partnership for good outcomes, given that collaborative procurement and delivery have been identified as the best way forward to boost project performance levels that were either lowered or had reached points of diminishing returns under traditional procurement regimes (Kumaraswamy, 2024b; Doloi, 2024). Successful innovations are underpinned by the progressive collection and convergence of building blocks of vast interconnected areas so that the resulting product of such processes adds value and contributes towards meeting the ever-evolving targets. The results of such collaborative process convergence generate synergies leading to a win–win situation that contrasts with the traditional zero-sum approach.

The world has changed, with many countries, including Australia, facing high interest rates, real estate weaknesses with a crisis in housing affordability, ongoing reductions in consumer confidence, and indications of recessionary trends. The global financial crisis was managed by many governments through Keynesian economics, which led governments to accumulate significant debts. In their 2010 study 'Growth in a Time of Debt', Reinhart and Rogoff examine the link between public debt and its impact on economic growth. This study was cited by several global institutions and governments to inform how future financial crises could be better managed (Reinhart and Rogoff, 2010).

In 'The Impact of Public Debt on Economic Growth' (2021, p503), Jack Salmon's survey analysis finds a correlation between public debt and an adverse impact on the economy. However, in that survey, two studies showed how good policy decisions could alleviate the negative effects of public debt on the economy. The caveat from these surveys is that data from low-income countries was limited, and only high-income and middle-income country data was available (Salmon, 2021). Good policies that use innovation in procurement can alleviate the impact of high levels of government debt while leading to successful outcomes.

Referring to the existing regime for social procurement targets set out in IPP and VAPT frameworks, VFM may continue to be the guiding principle in major project procurements. For instance, using the VFM lens, in November 2023, the federal government of Australia in Canberra withdrew support for 50 projects it stated were high risk in terms of costs. Additionally, based on a separate broad review, another 82 projects across Australia did not receive federal funding. In all these cases, costs were deemed to be unrealistic without balancing the intangible or indirect value outcomes. These projects would have served the public if completed but failed due to cost blowouts and because the dollars spent did not make overall sense. These examples clearly show that project-level innovations cannot be achieved with a local or project-based effort alone. Instead, a solid foundation of relevant knowledge along with innovations and capability across a hierarchy of interrelated business or organisational units are collectively responsible for driving project-level innovation leading to the target outcome.

Figure 12.2 depicts the hierarchy of organisational or business units in the form of a procurement innovation pyramid. As seen, the bottom of the pyramid is underpinned by innovations from research and development, being one of the key drivers for supporting innovations. Construction procurement is one of the leading topics among researchers across the globe, given its potential impacts, for better or worse, on built infrastructure projects. Also, given the increasing challenges in addressing issues such as high carbon footprint, construction waste, cost and time overruns, technological innovations, social procurement, etc., the contribution of the construction research community in creating innovative concepts and building capacities in the industry is potentially phenomenal. Yet, the uptake of innovative concepts and the culture of change are extremely slow in the construction industry. Lack of confidence in undertaking innovations first-hand is one of the key reasons for negativity around construction innovations among leading organisations (Saad et al., 2024).

As in the second layer up in the pyramid structure, organisational innovation can spring from ideas generated in practice. However, the innovative capacity of an organisation, which can also be boosted by its integration of research-based inputs into every facet of the project lifecycle, provides the impetus and confidence to embark on courageous disruptive changes to traditional business processes. Research-based inputs with a deep analytical understanding of multi-dimensional interrelated issues and underlying methodical approaches for deriving acceptable solutions are fundamental requirements, especially in high-value procurement of construction projects. However, confidence in the reliability of research-based innovations that may disrupt current processes could impede innovations, especially among risk-averse organisations. Thus, a cultural shift in the construction industry is important for propagating innovations across the construction industry sector.

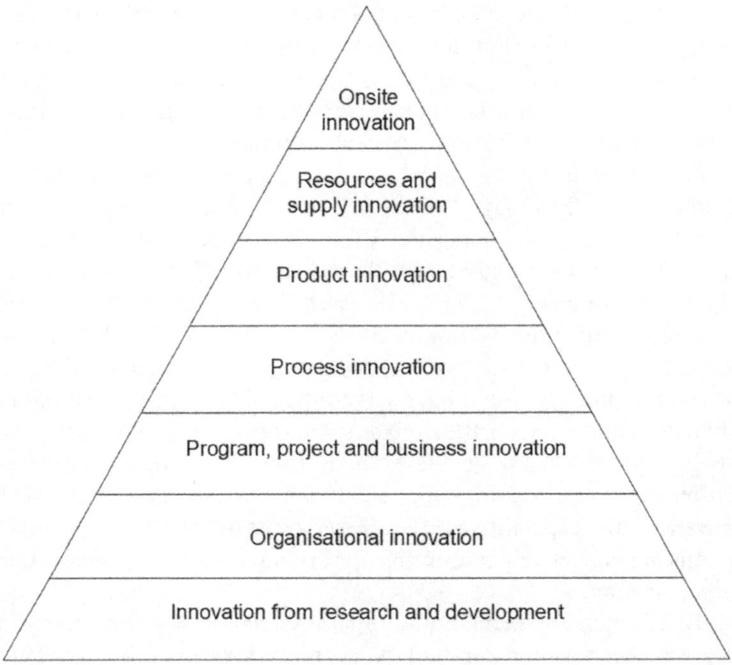

Figure 12.2 *Procurement innovation pyramid*

Innovations in programme, project, and business processes at the next broader level of the pyramid entail the adoption, integration, and optimisation of modern methods of construction through innovative strategies, techniques, and tools, including the ability to influence market forces to realign them to help meet organisational targets in win–win settings. For instance, in a particular programme of projects, a hybrid construction approach combining offsite and onsite construction module manufacture may provide a much better solution for reducing waste and lowering overall embodied energy when considering the entire business process. To facilitate such a mixed-methods approach, execution plans should reach beyond the knowledge areas of project management (PMBOK, 2021) in order to generate creative ideas for innovative processes by inspiring contracting parties in the project.

Processes and product innovations at the upper section of the pyramid are linked to construction methods, design specifications, and contract provisions, including the engagement of specialised contractors involved in downstream project delivery processes. Roles, responsibilities, stakes, and positioning of various units of the project organisation need to be carefully defined and decided, to drive the necessary structural changes for innovations in physical, cultural, and social dimensions at both process and product levels. Commitments, communication, and leadership of stakeholders should be commensurate with the responsibilities, opportunities, and challenges that need to be aligned for accessing required resources and establishing seamless supply chains to support innovative processes, leading to the overall innovations at the end of the procurement–delivery cycle. At the peak of the pyramid is the tangible measure of project-level innovations, which can be used against the target set at the outset of the procurement–delivery cycle. The following sections will discuss more on the methodical implementation of a framework as a novel contribution for fostering innovations through appropriate procurement policies and practices in construction projects.

DEVELOPING PROCUREMENT POLICY AND FOSTERING PROCUREMENT INNOVATIONS

Keeping social procurement as the core principle, Australian governments have developed national guidelines for the delivery of infrastructure projects. The Australian guidelines are not mandatory, and a problem that exists relates to the use of external consultants. In *The Big Con: How the Consulting Industry Weakens Our Businesses, Infantlizes our Governments and Warps our Economies*, Mazzucato and Collington (2023) expose problems posed by the consulting industry. Governments already have or should develop internal expertise without having to depend on external consultants. However, many projects are often arguably announced prematurely, while a business case is later developed to justify the announced decision.

Research by the Grattan Institute suggests that 'for transport infrastructure projects 89 per cent of cost overruns are due to causes such as the premature announcement – when a politician promises to build a road or rail line at a particular cost, often in the lead-up to an election, and in a marginal electorate' (Terrill et al., 2020). This is where external consultancies allegedly come in, to rubber-stamp a previous decision. Anecdotal evidence suggests that external consultancies have used 'state capture' widely in Australia and around the world to the detriment of the paying public. Australia is, at the time of writing, conducting three parliamentary inquiries into the consulting industry and its adverse influence on public institutions.

Governments need to build in-house expertise and embed innovation in the mindset of their policy analysts. Probity is an important part of procurement as bidders, who normally spend between 0.8% and 1.2% of the project costs on the preparation of their bids, would have no confidence in bidding unless probity was deeply ingrained in the public service. This requires public servants to be independent and not beholden to their political masters.

Having reviewed the above facts, it is apparent that the current status quo in large projects is untenable for the following reasons:

- Receiverships due to bankruptcy have become a common phenomenon, even among established construction organisations in recent years.
- In public–private partnership arrangements, private consortia often demand changes to risk allocation models to minimise their exposure to uncertain business risks.

- There are adversarial attitudes and a lack of collaborative approaches, as seen in many large-scale projects including the most recent West Gate Tunnel project in Melbourne, Australia (Bigbuild, 2018).
- There are deficiencies in fixed-price contracts in a dynamic business environment, with supply chain bottlenecks, global conflicts, increases in interest rates, and energy costs.

Due to the virtue of innovations being similar across many industries, innovation in procurement could also entail applying some of the lessons learned in the IT industry. In the early days of IT projects, the Project Management Institute's (PMI) project management methodology (PMBOK, 2021) and the UK Government's Prince 2 methodology (PRINCE 2) were the widely used processes to deliver projects. Over time, it became apparent that neither the clients nor the vendors achieved satisfaction from the delivered results. Clients kept seeking changes without allowing deviations in costs and time. Introducing agile methodology, where collaborative partnerships become the key, could entice parties to manage adversarial disputes more effectively.

While this chapter highlights some of the intrinsic characteristics of partnerships in public construction projects, collaborative partnerships need to be monitored through suitable value-judgement lenses that position public and societal interests at the core of the procurement process. Governmental policies need to be better aligned to context-based parameters that can transform the way contracts are awarded and projects are delivered. These policies need to take a big-picture view with proper long-term planning, including environmental shifts in a given demographic landscape. Due to the lack of a robust value-judgement framework capable of supporting apolitical decisions, some major public infrastructure projects in Victoria, such as the East West Link (EWL) project (Eastlink, 2024), Suburban Rail Loop (SRL) project (Bigbuild, 2022) and enhancements to the Geelong and Ballarat rail corridors (Bigbuild, 2009) have become political footballs. Consequently, what is proposed by one party to attract votes before elections with a competitive advantage may be discarded by the other party if they win the elections. A stable, consistent, and long-term approach is thus needed for building confidence in the contracting industry and building resilience through recruitment, training, and resourcing. These have a direct impact on employment, incomes, poverty alleviation, housing affordability, and overall social good in the community.

While the focus of this chapter is on recent and nascent innovations in construction project procurement involving relatively high capital investments, it is also important to understand how other business sectors may approach procurement to deliver success. For this, it is worthwhile to perhaps briefly compare what has happened in two specific industries, at least in the Australian context: (1) grocery (including all of the fast-moving consumer goods (FMCG, 2015) industries); and (2) residential construction. FMCG in Australia is dominated by the two largest supermarket chains – Coles (https://www.colesgroup.com.au/home/) and Woolworths (https://www.woolworths-group.com.au/). Given their market dominance and imbalance in power over their small suppliers, Coles and Woolworths can dictate many terms, even if considered unfair.

Given the market capitalisation values of Coles and Woolworths of over AU$20 billion and AU$40 billion respectively, there is a high volume of high-value procurement contracts across a range of interrelated sectors, including their partners involved in the business supply chains. Based on their reviews of business practices in 2014, the Australian Competition and Consumer Commission (ACCC) (ACCC, 2014) alleged that Coles, being one of the largest supermarket giants, had squeezed small suppliers for money without being innovative in their business and supply chain, which resulted in tarnishing Coles' brand as being arrogant and unethical (SBS, 2014). Consequently, Coles had to pay millions in compensation in addition to the reputational damage inflicted on its brand name. While this example is from a business perspective, one of the key lessons that can be drawn in the construction project context is that in good business negotiation, instead of a 'heads I win, tails you lose' approach, everyone should win, and thus 'resilience of suppliers is tremendously important' for the long-term sustainability of any business, including construction (Gang, 2023).

The second example is the high-value (over AU$86 billion) high-transaction residential construction sector in Australia. Despite being a fairly well-established industry over a long period of time, since mid-2021, over 2,000 construction companies have collapsed in Australia, according to the Australian Security and Investments Commission (ASIC) (Fitzpatrick, 2023). They failed due to a complex set of factors that include rising interest rates affecting capital, supply chain bottlenecks pushing up costs, shortages of good-quality labour, and a failure to innovate. Tens of thousands of people who had paid various amounts as deposits for homes lost their money, as they were unsecured creditors to the builder even as their houses remained unfinished. A failure to take a big-picture view of the business context is often cited as a reason for labour shortages. To provide a comparison with most information technology (IT) companies, when work is less, staff are upskilled while 'sitting on the bench' (Baranski et al., 2023). It is seldom the case that staff will be made redundant and then new staff recruited when the next project is secured. In the construction industry, however, most companies contract or hire workers for specific parcels of work. This leads to the industry being unattractive for many young people, who seek permanency and stability. Upskilling with recent technologies is also an individual pursuit rather than an organisation-driven professional development. These chronic issues eventually reflect poor procurement practices across the board, which result in the sub-standard delivery of housing projects. The above anecdotal evidence indicates that the current practices of the construction industry, especially in the residential sector, fail the litmus test of the UN's SDG 8 (Decent Work and Economic Growth) and 9 (Industry, Innovation, and Infrastructure).

In both industry examples above, there is an opportunity to innovate in managing working capital, fostering corporate social responsibility in procurement, and enabling win–win outcomes. What is required is a collaborative mindset that treats all procurement as a partnership between the client organisation and supplier organisation, while also factoring in the wider social perspectives.

Figure 12.3 depicts a framework as a way forward for developing innovative procurement–delivery strategies and practices focusing on the comprehensive pursuit of the triple bottom line (TBL), with close reference to the UN SDGs. Some of the prevailing criteria for measuring innovations within a construction project are highlighted under the three initial broad dimensions: *social innovations*, *environmental innovations*, and *economic innovations*. The resulting measures are derived with direct reference to the functional areas of the procurement–delivery cycle for promoting innovation depicted in Figure 12.1.

As seen, under social innovations, four key measures – namely *social values*, *equity and justice*, *local resources*, and *growth of social enterprises* – have the potential to collectively contribute to a total of eight SDGs, partially or entirely. These are: Goal 1: No Poverty; Goal 2: Zero Hunger; Goal 3: Good Health and Well-being; Goal 4: Quality Education; Goal 5: Gender Equity; Goal 10: Reduced Inequalities; Goal 12:

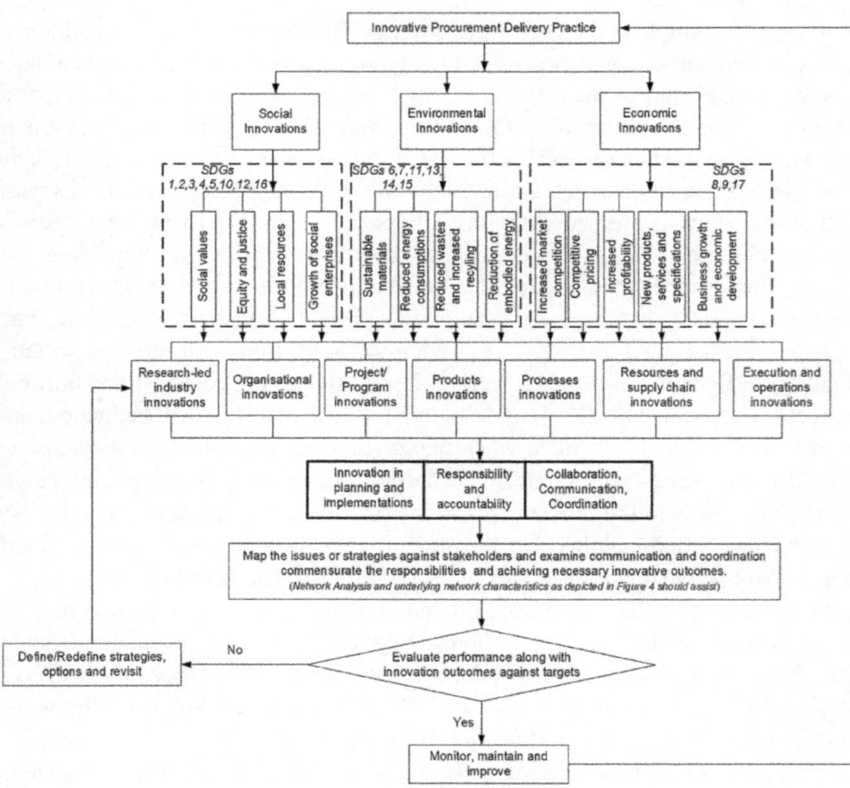

Figure 12.3 *Framework for developing innovative procurement–delivery practice*

Responsible Consumption and Production; and Goal 16: Peace, Justice, and Strong Institutions. When a project moves through development phases from concept to scoping, contracting, implementation, commissioning and handover, then to operations and eventually decommissioning, relevant measures and indicators associated with the procurement–delivery functions in each of the phases need to underpin the social innovation plans of the entire project. For establishing realistic measures and indicators in social innovations, context-specific procurement information needs to be solicited to target tangible outcomes.

Under environmental innovations, the four broad measures comprise *sustainable materials*, *reduced energy consumption*, *reduced waste and increased recycling*, and *reduction of embodied energy*. These measures collectively contribute to six more SDGs, namely: Goal 6: Clean Water and Sanitation; Goal 7: Affordable and Clean Energy; Goal 11: Sustainable Cities and Communities; Goal 13: Climate Action; Goal 14: Life Below Water; and Goal 15: Life on Land. In order to achieve these goals through environmental innovations, context-specific individual procurement–delivery planning with measurable targets will be needed and should span across the project lifecycle.

The third dimension of economic innovations incorporates five broad measures, namely: *increased market competition*, *competitive pricing*, *increased profitability*, *new products, services and specifications*, and *business growth and economic development*. These measures are again linked to three broad SDGs: Goal 8: Decent Work and Economic Growth; Goal 9: Industry Innovation and Infrastructure; and Goal 17: Partnerships for the Goals. Bottom-up procurement innovation plans need to be worked out through all project phases in the context of the project development environment in order to achieve these goals.

The following layers in the hierarchy are the interconnected innovation dimensions drawn from the procurement innovation pyramid depicted in Figure 12.2. As seen, these innovative dimensions are the direct results of effective planning, implementation, and control of the innovative procurement measures in the preceding step of the framework, meeting the targets of the respective SDGs in the project-specific context.

Having worked out and mapped the detailed innovative procurement dimensions and underlying measures leading to the contextual implementation of SDGs, one of the key remaining challenges is to develop a roadmap for innovation planning, including the identification of stakeholders and an underlying responsibility matrix for necessary onsite execution. Thus, as depicted in Figure 12.3, three key processes in the implementation strategy include 'Innovation Planning and Implementation', 'Responsibility and Accountability', and 'Collaboration, Communication and Coordination'. At this stage, one of the unanswered questions remains how to go about and implement this framework with respect to stakeholders' roles, responsibilities, collaboration, and coordination, including accountability, so that innovative procurement is achieved in a particular project context and it is possible to evaluate, compare, and monitor against the targets.

STAKEHOLDERS' COMMUNICATION AND COORDINATION IN THE PROCUREMENT DELIVERY PROCESS

As demonstrated throughout the chapter, procurement delivery in projects depends on the planning and execution of the procurement functions across each phase of the project lifecycle. Stakeholders' roles, responsibilities, and abilities in performing the functions effectively in the context-specific environment are crucial for driving innovations in procurement–delivery practices at the project level. As different stakeholders are involved in performing different functions in each phase, efficiencies in interactions, communication, and coordination among the stakeholders in relation to performing the functions are important to assess so that the practice is continuously improved in performing procurement functions and meeting targets for ensuring innovations in projects.

Social network analysis (SNA) is one of the widely used mapping tools for assessing the interactions between stakeholders in performing procurement functions in projects. SNA can be used for both one-mode or two-mode network studies for measuring interactions in terms of information flow between the respective stakeholders within the network structures (Doloi, 2012a; 2018). One-mode networks entail measuring interactions and information flow between the respective stakeholders while performing a single function, and this is often termed a 'stakeholder–stakeholder' network. Two-mode networks entail measuring interactions and information

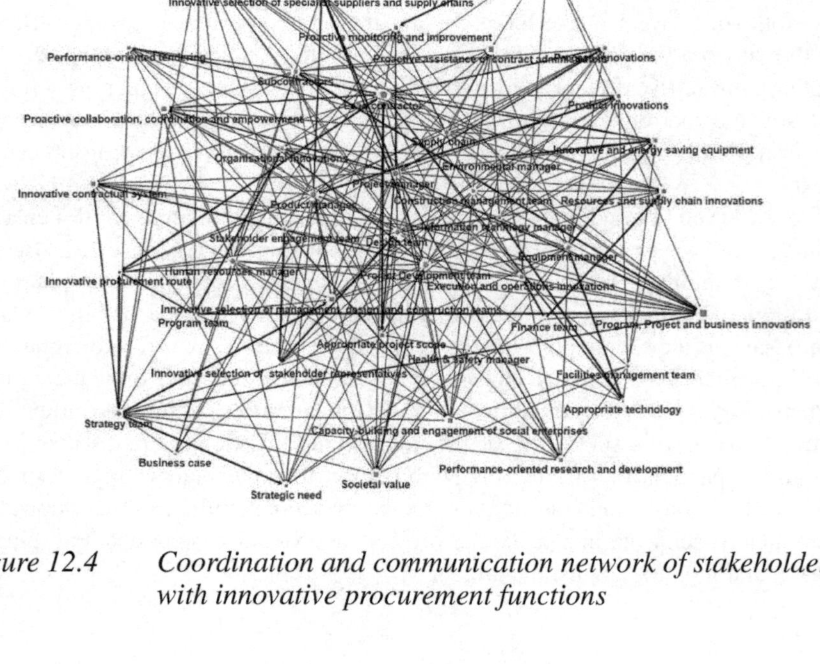

Figure 12.4 Coordination and communication network of stakeholders with innovative procurement functions

flow between the respective stakeholders and multiple functions, and this is termed a 'stakeholder–functions' network.

Figure 12.4 depicts a two-mode network highlighting the interactions between the typical stakeholders and the procurement functions, portraying the information flow and responsibility-sharing in a procurement implementation process. There are 19 stakeholders (shown as circular nodes) and 24 procurement-related functions (shown as square nodes) in the network. The network is analysed using a dataset hypothetically created using a Likert scale of 0–5 depicting the responsibilities (0: no responsibility; 5: highest responsibility; 1–4 are intermediate values) of a stakeholder in performing particular procurement functions. The sizes of the nodes are determined by the 'Eigenvector' network measures, which depict the 'power of influence' of a particular node in relative terms to the other similar nodes in the network (Doloi, 2018). The thickness of the connecting lines highlights the amount of information being shared (e.g., thinner line: less information flow; thicker line: more information flow) among nodes in the network while performing the functions. Arrows depict the direction of the information flow in a directed network structure.

Table 12.1 shows the numerical values of the Eigenvector centralities as measures of 'power of influence' in the stakeholders–procurement-related functions network. As seen, higher numbers, which are highlighted for both stakeholders and procurement functions, demonstrate the higher power of influence for achieving the target outcomes. In other words, the responsibilities and ownership of the stakeholders in effectively performing the procurement functions can be improved by increasing the communication, coordination, interaction, and information flow among the stakeholders and the procurement functions.

The study of SNA-based network characteristics of the stakeholders and procurement function is certainly a way forward for any organisation to take into account for planning and implementation in achieving desired outcomes of procurement innovation. While the above is a hypothetical example, such analysis can go much deeper and across multiple levels and dimensions. Overall, this chapter demonstrates value and ways forward for procurement innovations by integrating broad areas, dimensions, and functions, and including a bottom-up approach that also targets relevant SDGs. This also closely aligns with idealised procurement models available in OECD (1991; 2017) and UN literature (2023).

CONCLUSION

It seems fitting to conclude this chapter with an apt recap of the UN's 2030 Agenda:

> The 2030 Agenda for Sustainable Development, adopted by all United Nations Member States in 2015 … provides a shared blueprint for peace and prosperity for people and the planet, now and into the future. At its heart are the 17 Sustainable Development Goals (SDGs), which are an urgent call for action by all countries – developed and developing – in a global partnership. (UN, 2023)

Table 12.1 Eigenvector centralities as measures of power of influence in stakeholders–procurement-related functions network

Stakeholders	**Eigenvector centrality (EC)**	**Procurement-related functions**	**Eigenvector centrality (EC)**
Programme team	0.217	Strategic need	0.180
Strategy team	**0.301**	Business case	0.103
Project development team	**0.322**	Societal value	0.167
Design team	**0.245**	Appropriate project scope	**0.240**
Finance team	0.070	Innovative procurement route	**0.240**
Product manager	0.281	Innovative contractual system	0.184
Project manager	**0.392**	Innovative selection of management, design, and construction teams	0.208
Information technology manager	0.120	Innovative selection of stakeholder representatives	0.193
Lead contractor	**0.487**	Performance-oriented tendering	0.192
Sub-contractors	0.218	Innovative selection of specialist suppliers and supply chains	**0.250**
Supply chain	0.164	Proactive collaboration, coordination, and empowerment	**0.246**
Environmental manager	0.116	Performance-oriented and ethical conduct	0.235
Stakeholder engagement team	0.146	Proactive assistance of contract administrators	0.201
Construction management team	0.155	Appropriate technology	0.141
Audit and review team	0.108	Innovative and energy-saving equipment	0.234
Facilities management team	0.077	Capacity-building and engagement of social enterprises	0.140
Health and safety manager	0.040	Proactive monitoring and improvement	0.167
Human resources manager	0.158	Performance-oriented research and development	0.208
Equipment manager	0.167	Organisational innovations	0.234
-	-	Programme, project, and business innovations	**0.300**
-	-	Product innovations	0.193

Stakeholders	Eigenvector centrality (EC)	Procurement-related functions	Eigenvector centrality (EC)
-	-	Process innovations	0.212
-	-	Resources and supply chain innovations	0.172
-	-	Execution and operations innovations	0.133

Recent research indicates that current procurement processes of public infrastructure in Australia encapsulate the spirit and intentions of the SDGs to some extent, but there is a lot more to be done. Drawing on these observations, this chapter aims to provide a framework and core concepts for developing a clear roadmap towards the much-needed innovations in procurement strategies, protocols, and practices that could provide the long-awaited launch pad for delivering successful outcomes in projects.

In defining and describing the targeted innovative procurement delivery in the project development lifecycle, the discussion in the chapter started by providing a broad overview of the identified key functional areas and their significance in the procurement–delivery cycle, which eventually supports innovations as a project proceeds, leading to enhanced overall performance. The discussion further demonstrated how project-based innovations are generated, starting from broad bases of research and development, organisation-level input, programme- and business-level initiatives, process- and product-level innovations, to innovative ways of engaging and controlling resourcing and supply chains.

Having established the broad bases of innovative procurement in general, the discussion next moved on to an ambitious framework for facilitating innovative procurement–delivery practices in the context of an individual project. The starting point of the innovative delivery framework has been derived from the traditional triple bottom line (TBL) consideration, which was then further refined by breaking it down into a range of measurable outcomes linked to 17 SDGs. Collective outcomes of the SDGs were then linked to the seven broad innovative step-wise progressions, as illustrated in Figure 12.2. The focus of the final part of the framework was on responsibilities and accountabilities among the relevant stakeholders in delivering the respective procurement-related functions. The underlying requirements of efficient coordination and communication among the stakeholders and collaboration in an innovative procurement drive were examined using a demonstrable SNA methodology. The processes of comparing progress with the target were also highlighted, and these included monitoring and continuous improvement of the innovations in the overall procurement–delivery cycle.

REFERENCES

ACCC (2014). https://www.accc.gov.au/media-release/accc-takes-further-action-against-coles-for-alleged-unconscionable-conduct-towards-its-suppliers (accessed in April 2024).

Aliento, W. (2014). The business case publicly falls down for Melbourne's East-West, *Infrastructure and Transport*, The Fifth Estate Newsletter. https://thefifthestate.com.au/urbanism/infrastructure/the-business-case-publicly-falls-down-for-melbournes-east-west-link/

Australian Government (2008). Infrastructure investment & project delivery, Infrastructure Australia, Canberra. https://www.infrastructure.gov.au/infrastructure-transport-vehicles/infrastructure-investment-project-delivery

Australian Government (2023). Statistics on Australian Government Procurement Contracts, Canberra. https://www.finance.gov.au/government/procurement/statistics-australian-government-procurement-contracts-

Baranski, E., Lindberg, C., Gilligan, B., Fisher, J.M., Canada, K., Heerwagen, J., Kampschroer, K., Sternberg, E. and Mehl, M.R. (2023). Personality, workstation type, task focus, and happiness in the workplace, *Journal of Research in Personality*, 103. doi.org/10.1016/j.jrp.2022.104337.

Battersby, L. (2003). Master Builder Project Delivery System and Designer Construction Knowledge, *Journal of Construction Engineering and Management*, 129(6). doi:10.1061/(ASCE)0733-9364(2003)129:6(635).

Bigbuild (2009). Victoria Government. https://bigbuild.vic.gov.au/projects/ballarat-line-upgrade# (accessed in April 2024).

Bigbuild (2018). Victoria Government. https://bigbuild.vic.gov.au/projects/west-gate-tunnel-project (accessed in April 2024).

Bigbuild (2022). Victoria Government. https://bigbuild.vic.gov.au/projects/regional-rail-revival (accessed in April 2024).

Cheung, E., Chan, A.P.C. and Kajewski, S. (2009). Enhancing value for money in public private partnership projects: Findings from a survey conducted in Hong Kong and Australia compared to findings from previous research in the United Kingdom, *Journal of Financial Management of Property and Construction*, 14(1), pp7–20.

Doloi, H. (2012a). Assessing stakeholders' influence on Social Performance of Infrastructure Projects, *Facilities*, 30(11), pp531–550.

Doloi. H. (2012b). Understanding impacts of time and cost related construction risks on operational performance of PPP projects, *International Journal of Strategic Property Management*, 16(3), pp316–337.

Doloi, H. (2013). Empirical analysis of traditional contracting and relationship agreements for procuring partners in construction projects, *Journal of Management in Engineering*, ASCE, 29(3), pp224–235.

Doloi, H. (2018). A community centric model for evaluating social value in projects, *Journal of Construction Engineering and Management*, ASCE, May, 144(5).

Doloi, H. (2024). Social value in procurement – Incorporating inclusivity and equity in a societal context. In Kumaraswamy, M.M. (ed.), *Routledge Handbook of Construction Project Procurement & Delivery – Fundamentals, Trends and Imperatives*. Taylor & Francis, pp314–330.

Dulaimi, M. and Kumaraswamy, M.M. (2000). Procuring for Innovation: The Integrating Role of Innovation in Construction Procurement, ARCOM 2000 Conference, Glasgow, Sept., Proceedings, 1, pp303–312.

EastLink (Melbourne) (2024, April 26). https://en.wikipedia.org/wiki/EastLink_(Melbourne)

Fitzpatrick, L. (2023). ASIC data shows 1,200 construction companies collapsed. https://original.newsbreak.com/@luke-fitzpatrick-560905/2971895940044-asic-data-shows-1-200-construction-companies-collapsed

FMCG (2015). https://corporatefinanceinstitute.com/resources/valuation/fast-moving-consumer-goods-fmcg/

Gang, P. (2023). 11th Annual Global Supply Chain Excellence Summit USC Marshall, Global Supply Chain Institute, USA, 9–10 August.

Hale, D., Shrestha, P., Gibson Jr, G. and Migliaccio, G. (2009). Empirical comparison of design/build and design/bid/build project delivery methods, *Journal of Construction Engineering and Management*, 135(7), pp579–587. doi:10.1061/(ASCE)CO.1943-7862.0000017

Harp, D.W. (1991). Historical background: Low bid concept, *Transportation Research Circular*, 386, 43–47.

Jefferies, M. and Rowlinson, S. (2016). Public-Private Partnerships and relationship-based procurement approaches: An introduction. In Jefferies, M. and Rowlinson, S. (eds), *New Forms of Procurement: PPP and Relational Contracting for the 21st Century*, Routledge, pp1–12.

Jozepa, I. (2023). Procurement Bill 2022–2023. House of Commons Library, UK, Research Briefing, No. 0042. https://researchbriefings.files.parliament.uk/documents/CBP-9402/CBP-9402.pdf (accessed 24 December 2023)

Kumaraswamy, M.M. (2024a). Collaborative contracting and relationship-based teamworking. In Kumaraswamy, M.M. (ed.), *Routledge Handbook of Construction Project Procurement & Delivery – Fundamentals, Trends and Imperatives,* Taylor & Francis, pp184–199.

Kumaraswamy, M.M. (2024b). Project opportunities, innovations and risk management. In Kumaraswamy, M.M. (ed.), *Routledge Handbook of Construction Project Procurement & Delivery – Fundamentals, Trends and Imperatives*, Taylor & Francis, pp37–59.

Kumaraswamy, M.M. (2024c). Ethics, professionalism and the construction industry. In Kumaraswamy, M.M. (ed.), *Routledge Handbook of Construction Project Procurement & Delivery – Fundamentals, Trends and Imperatives,* Taylor & Francis, pp60–73.

Kumaraswamy, M.M., Wong, K.K.W. and Chung, J.K.H. (2017a). Focusing megaproject strategies on sustainable best value of stakeholders, *Built Environment Project and Asset Management Journal*, Special Issue on 'Emerging Issues in the Built Environment Sustainability Agenda', edited by Thanuja Ramachandra and Gayani Karunasena, 7(4), pp441–455.

Kumaraswamy, M., Mahesh, G., Mahalingam, A., Loganathan, S. and Kalidindi, S.N. (2017b). Developing a clients' charter and construction project KPIs to direct and drive industry improvements, *Built Environment Project and Asset Management Journal*, Special Issue on 'Securing Clients' Objectives throughout Construction Project Life Cycles', edited by Chamila Ramanayaka and Monty Sutrisna, 7(3), pp253–270.

Lenderink, B., Halman, J.I. and Voordijk, H. (2022). Innovation and public procurement: From fragmentation to synthesis on concepts, rationales and approaches, *Innovation: The European Journal of Social Science Research*, 35(4), pp650–674.

Ling, F.Y.Y., Ning, Y., Ke, Y.J. and Kumaraswamy, M.M. (2013). Modeling relational transaction and relationship quality among team members in public projects in Hong Kong, *Automation in Construction*, 36, pp16–24.

Ling, F.Y.Y., Ke, Y., Kumaraswamy, M.M. and Wang, S.Q. (2014). Key relational contracting practices affecting the performance of public construction projects in China, *Journal of Construction Engineering and Management*, ASCE, 140(1).

Mazzucato, M. and Collington, R. (2023). *The Big Con: How the Consulting Industry Weakens our Businesses, Infantilizes our Governments and Warps our Economies*, Allen Lane.

Miller, J.B. (2000). *Principles of Public and Private Infrastructure Delivery*, Kluwer Academic.

NIAA (2020). *Indigenous Procurement Policy*, Canberra. https://www.niaa.gov.au/indigenous-affairs/economic-development/indigenous-procurement-policy-ipp
OECD (1991). *The Nature of Innovation and the Evolution of the Productive System. Technology and Productivity – the Challenge for Economic Policy*, Paris, pp302–314.
OECD (2017). *Public Procurement for Innovation Good Practices and Strategies*, OECD Public Governance Reviews. https://doi.org/10.1787/9789264265820-en
Pietroforte, R. and Miller, J.B. (2002). Procurement methods for US infrastructure: Historical perspectives and recent trends. *Building Research & Information*, 30(6), pp425–434.
PMBOK (2021). *A Guide to Project Management Body of Knowledge*, Project Management Institute (PMI), USA.
PRINCE 2. https://www.prince2.com/
Reinhart, C.M. and Rogoff, K.S. (2010). Growth in a time of debt, *American Economic Review: Papers & Proceedings*, 100(May), pp573–578.
Saad, A.M., Dulaimi, M. and Arogundade, S. (2024). Modern methods of construction (MMC) and innovation negativism in the UK public sector, *Built Environment Project and Asset Management*, 14(2), 147–163. doi:10.1108/BEPAM-06-2023-0108.
Salmon, J. (2021). The impact of public debt on economic growth, *Cato Journal*, 41(3).
SBS (2014). Coles fined for squeezing suppliers. https://www.sbs.com.au/news/article/coles-fined-for-squeezing-suppliers/tsuofnk61
Terrill, M., Emslie, O. and Moran, G. (2020). *The Rise of Megaprojects: Counting the Costs*, The Grattan Institute Report. https://iccpm.com/grattan-institute-reviews-major-project-overruns/
UK Government (2015). *Public Procurement Policy – Directives, Regulations, Policies and Guidance Relating to the Procurement of Supplies, Services and Works for the Public Sector*, OGC, UK Government. https://www.scribd.com/document/72138907/Introduction-to-Public-Procurement
UK Government (2023). *Transforming Public Procurement – Our Innovation Ambition (HTML) – Guidance*. https://www.gov.uk/government/publications/transforming-public-procurement-our-innovation-ambition/transforming-public-procurement-our-innovation-ambition-html
UN (2023). *Global Sustainable Development Report (GSDR)*, OECD Public Governance Reviews, OECE Publishing, Paris. https://sdgs.un.org/#goal_section
UNDP (2017). Goal 9: Industry, Innovation and Infrastructure, New York. https://www.undp.org/sustainable-development-goals/industry-innovation-and-infrastructure
Victoria Government (2018). *Social Procurement – Victorian Government Approach*. https://www.buyingfor.vic.gov.au/social-procurement-victorian-government-approach

PART IV

INNOVATIVE TECHNOLOGICAL SYSTEMS AND ORGANISATIONS

13. Developing a tool for measuring innovation for construction organisations

Hossein Sadeghzadeh Fasaghandis, Suzanne Wilkinson and Mostafa Babaeian Jelodar

INTRODUCTION

Having a successful and innovative building and construction industry creates wealth for countries, especially as construction is a main contributor to employment and economic growth. Internationally, the performance of the building industry in terms of productivity, efficiency, and quality achievement is considered low when compared with other industries, and this low performance is often attributed to a lack of innovation, including a slow uptake of new technologies (Schoenborn, 2012; Fernando et al., 2021). Innovation has improved the performance of an organisation by enhancing productivity, financial outcomes, and competitiveness (Jansen et al., 2006; DIISR, 2011; Daft et al., 2010). Within the construction industry, researchers have shown that innovation leads to efficiency and better business performance (Wandahl et al., 2011: Panuwatwanich et al., 2008). Investing in innovation is thought to bring economic growth and higher international competitiveness (Smith & Estibals, 2011).

The construction industry is recognised as a slow innovator and low-technology sector (Dansoh et al., 2017). Many countries have focused on developing a culture of innovation in construction, aiming to become technologically advanced and efficient. However, they face challenges in introducing innovative technologies and processes (Shibeika & Harty, 2015).

The main aim of this chapter is to improve innovation within construction organisations through the development of a tool for companies to measure innovation. By using an innovation measurement tool, companies can understand how well they are doing in providing innovative solutions and where they can make changes.

THE PROBLEM OF INNOVATION IN THE CONSTRUCTION INDUSTRY

McMeel and Sweet (2016) highlighted three critical challenges within the construction sector that underscore the importance of fostering innovation:

- *Creating a nurturing market environment*: There is a need for a market and operating environment that supports the nurturing, establishment, and growth of innovation within the construction industry.
- *Providing leadership and guidance*: The industry requires strong leadership and guidance to navigate emerging technologies, anticipate future industry trends, and prepare for transformative changes.
- *Upskilling and knowledge exchange*: Developing a skilled workforce and facilitating knowledge exchange among industry stakeholders are essential for embracing emerging technologies and driving innovation forward.

Innovation has become increasingly vital due to its significant impact on economic development and productivity enhancement (Smith & Estibals, 2011). Challenges in construction have put pressure on companies to implement innovation to survive. However, reports suggest that the construction industry has been slow to embrace change (Gambatese & Hallowell, 2011a; Gambatese & Hallowell, 2011b; Hemström et al., 2017; Dansoh et al., 2017).

Several factors contribute to the difficulty of fostering innovation within the construction sector. These include conservative project management approaches, apprehensions about potential losses, concerns regarding the costs associated with innovation, and time and resource constraints (Hampson, 2005). Additionally, limited financial resources allocated for innovation within construction organisations contribute to the industry's reputation as a low innovator and underperforming sector (Dulaimi et al., 2002; Hampson & Manley, 2001; Miozzo & Dewick, 2004). Furthermore, many construction companies lack formal systems to recognise, evaluate, and develop innovative ideas effectively, leading to missed opportunities and the loss of talented individuals within the industry (Loosemore, 2015).

Defining innovation has been a subject of considerable debate among scholars, but many of these scholars agree on elements of innovation, including creating ideas and new knowledge and income generation. Drucker (1985) originally referred to innovation as how an entrepreneur either generates new resources that produce wealth or improves the ability of existing resources to generate greater value. Similarly, Becheikh et al. (2006) and McAdam et al. (1998) emphasised the transformation of inventive ideas into marketable propositions, commenting on innovation's practical impact to generate income. Aouad et al. (2010) viewed innovation as the creation and adoption of new knowledge to enhance the value of products, processes, and services. Loosemore (2014) also emphasised innovation as a process by which organisations transform ideas into improved products, services, or processes to enhance productivity and efficiency. The more recent definition, from the OECD Oslo Manual (2018), defined innovation as 'a new or improved product or process (or combination thereof) that differs significantly from the unit's previous products or processes' (p. 20).

Innovation can manifest in different domains, including product, process, and organisational. Product innovation focuses on developing new offerings that deliver value to customers, while process innovation enhances the efficiency and effectiveness of delivery systems. Organisational innovation aims to optimise internal structures and systems to drive performance (Hampson & Manley, 2001; Shahbazpour, 2010). Innovation can also be classified from incremental to radical, depending on the magnitude of change required within an organisation to facilitate the innovation (Cooper, 1998). Additionally, innovation can be strategic or emergent, based on its alignment with organisational strategy (Koch & Leitner, 2008).

Measuring and monitoring innovation is crucial because it gives the organisation an idea of how well new innovations are doing, and whether what is being created is adding value to the organisation. However, construction organisations' ability to measure and track innovations is poor, and there has been a need for guidance on measuring innovation within the construction industry (Loosemore, 2014; Dansoh et al., 2017).

A FRAMEWORK FOR MEASURING INNOVATION IN ORGANISATIONS

Much of the literature has focused on how construction companies manage innovation (e.g., Dikmen et al., 2005; Rutten et al., 2009; Cohen, 2010; Song & Chen, 2014). These studies describe the technological innovation process, innovation management, drivers/barriers, and organisational attributes from the point of view of individual organisations.

Innovation has been found to be related to different organisational attributes, such as leadership, collaborative partnering, organisational structures, knowledge management, communication, research and development, work environment, and the ways in which external influences impact the organisation (Park et al., 2004; Aouad et al., 2010; Ozorhon et al., 2014, Song & Chen, 2014; Loosemore, 2015; Dansoh et al., 2017). More in-depth analysis of the literature showed that, overall, five key categories emerged which are key to the impact of organisations on innovation development for construction companies:

- workforce skills and human resources management
- organisational structure
- business strategy
- technology
- external influences.

The five categories were seen to be critical for organisations to focus on if they want to become innovative, that is, they were designed to measure the innovation potential of the organisations. Breaking these categories down led to a series of indicators for each category. The indicators emphasise the ability of the organisation to deliver innovations from various views (e.g., developing the ways in which a business strategy could indicate a strong focus on innovation). Each category and its indicators were cross-checked against the literature to ensure that the tool had captured the full range of category impacts on construction organisation innovation.

For instance:

> The 'workforce skills and human resources management' category had 16 indicators, which included: a measure for adaptability and flexibility; entrepreneurialism; collaboration; and partnerships. For each indicator, a review of relevance was undertaken. For example, collaboration among departments and employees was seen as important and was emphasised in the literature. Lack of collaboration was reported as a barrier for innovation in the construction industry (Chen, 2014).
>
> The 'organisational structure' category had nine indicators, including: innovation champion; lessons learned; and incentives for innovation. Incentives, for instance, were chosen because incentives such as government support and tax incentives were believed to encourage innovation (Oslo Manual, 2018).
>
> The 'business strategy' category had nine indicators, including: strategic planning; innovation strategy; and communication strategy. Crawford (2021) believed that it is necessary to develop an innovation strategy and Green et al. (2010) emphasised the importance of innovation strategy to understand the contribution of innovation towards the organisation's economic growth.
>
> The 'technology' category had seven indicators, including technical expertise and digital technologies. Many authors (Project Management Institute, 2014; Blayse & Manley, 2004; Ofori, 2008) expressed the need for technical expertise in an organisation for innovation to be effectively adopted. This need for technical expertise includes the organisation's leadership implementing digital technologies which are required if organisations are to operate effectively in an innovative environment.
>
> The 'external influences' category had 27 indicators, including: subsidies; market demand; competition; and contracts. The Oslo Manual (2018) stressed that it is necessary to understand the characteristics of suppliers, the structure of demand in the existing and future markets that the organisation is aiming at, the markets for finance and markets for labour, and information on the level of competition and standards that exist.

The five categories and indicators enabled tool development for construction organisations wanting to innovate. The developed innovation measurement tool measures the potential of the construction organisation to innovate. A sixth category was added to the tool (business innovation score) – a way of capturing the business innovation score was developed which measures the current situation of the organisation with respect to innovation. The business innovation score represents how many innovations the company has achieved, for instance, in the products, process changes, and changes to activities. Capturing the business innovation score in the tool shows how organisations are currently performing. The business innovation score is captured through a developed set of targeted questions, for instance: how the organisation organises for innovation; the process for developing innovations; the numbers of innovations per year, etc. Overall, the innovation measurement tool presents 36 questions that collectively measure the 'actual' innovation (through applying the business innovation score) of organisations.

Figure 13.1 shows the Business Innovation Measurement Framework used to test innovation potential and actual innovations in construction companies. An 'overall score' can be produced, which is an average of 'potential for innovation' and 'actual innovation'.

The developed tool gives organisations the capability to instantly generate results once their scores are entered into the tool for each indicator (potential for innovation categories) and questions answered (actual innovation). The results can be analysed and innovation gaps across the categories identified. The results for each company can also be compared with other companies. Ultimately, an industry benchmark can be developed and used as a basis for ongoing support and improvement of innovation uptake within the construction industry.

TESTING THE TOOL AND FINDINGS

The business innovation tool was developed using an Excel spreadsheet to allow each indicator to be scored on a scale of 1–5 (where '1' is strongly disagree and '5' is strongly agree) – for instance, using simple questions such as 'Does your organisation have technical expertise in emerging technologies?' (score 1 (low) to score 5 (high)) (Category: technology; Indicator: technical expertise), or 'Does your organisation have the latest industry digital technologies?' (Category: technology; Indicator: digital technology). Each indicator is given a score (1–5) by a selection of members of the organisation. A focus group could be used to provide the scores as an alternative to individual responses. In the test case, amalgamated scores from each member of the organisation made up an overall mean score for the indicator and an overall category score. This is the 'innovation potential score' for each category. Combining the 'innovation potential scores' across the five categories allows an 'overall potential innovation score' to be calculated. The business innovation score provides evidence of what is currently happening in the organisation. Combining the business innovation scores from members of the organisation creates an overall business innovation score. The combined scores from potential innovation score and the business innovation score give an overall score (overall innovation score: potential + actual score in Figure 13.1). Results for each organisation show how they are progressing with implementing innovation within their organisations. Scores show what they need to do to improve their innovation potential and what their current scores are, reflecting how they are currently innovating. Ultimately, scores can be compared with other organisations to develop an industry norm and to track improvements in the industry.

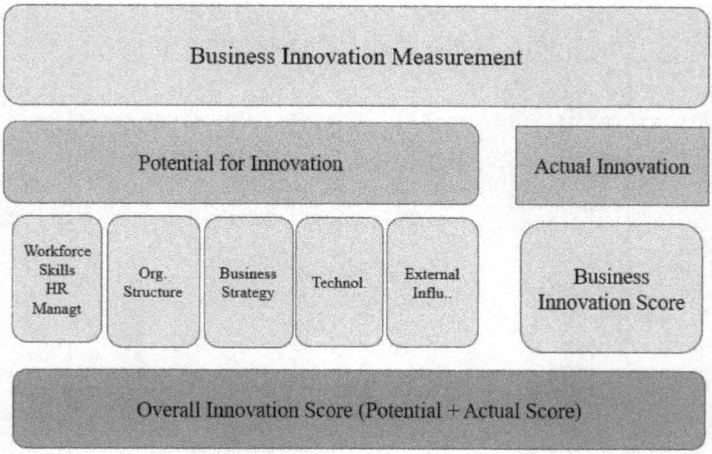

Figure 13.1 *Business Innovation Measurement Framework*

INNOVATION POTENTIAL: TECHNOLOGY

Company demographics for each participant were collected. Table 13.1 shows typical demographics information of the targeted respondents in one of the case study organisations. Demographics are important as they allow for clustering of individuals into discipline-specific groups to see if they answer differently from other groups (for instance, project managers). In this case, there were not enough participants to allow for clustering, but with wider adoption of the tool, the demographic information would be useful for data analytics and future training and decision-making.

The interviewees were given questions asking to what extent the following indicators were present in their organisations. Table 13.2 represents the results obtained for this set of questions from the five interviewed employees at Case Study 1 (CS1) for the 'technology' category.

The table illustrates the respondents' thoughts about innovation with respect to technology and generally shows there is room for improvement in technological uptake. In particular, the results show that the organisation is relatively low on technical innovation awareness. One of the key indicators for an organisation is to make sure that the employees are aware of how to innovate, and what their contribution is to developing innovations within the organisation. The technical side of the organisation showed some room for improvement in their expertise and technology use and in the general design capabilities. Again, by focusing on ways to encourage and develop technological expertise, innovation capability within the organisation would improve.

Table 13.1 *Results for Case Study 1 – company demographics*

Question	**CS1 – Int1**	**CS1 – Int2**	**CS1 – Int3**	**CS1 – Int4**	**CS1 – Int5**
Role	H&S After-sales care systems	Project management	Management	Administration + Accounts	Contracting
Sector history	Private sector	Private	Private	Private	Public + Private
Profession	Administration + H&S	Project manager	Project management + Builder	Accounts	Project manager/ builder
Designation	Administration	Project manager	Project manager	Accountant	CEO + Board member
Highest academic qualification	Human resources management	Master of Science Project Management	Level 4 Carpentry + Level 4 Site Supervisor	School certificate	Carpentry (NZQA level 4)
Years of experience	0–5	16–20	11–15	0–5	21
Years of management	0–5	6–10	6–10	0–5	11–15
Age range	21–29	40–49	50–59	50–59	30–39

Note: Table 13.1 shows typical demographics information of the targeted respondents in Case Study 1.

Table 13.2 *Results of Case Study 1 for technology indicators*

Indicator Code	Indicator Title	CS1 – Int1	CS1 – Int2	CS1 – Int3	CS1 – Int4	CS1 – Int5	Indicator average
Tech 1	**Innovation awareness**	2	3	1	4	4	2.8
Tech 2	**Problem-solving techniques**	3	4	3	4	3	3.4
Tech 3	**Innovative designers/designer understanding**	3	4	3	3	4	3.4
Tech 4	**Design capabilities**	2	3	3	3	4	3
Tech 5	**Presence of technology**	3	3	4	3	4	3.4
Tech 6	**Technical expertise**	3	3	3	3	3	3
Tech 7	**Digital technologies**	3	2	2	4	4	3
Respondents' average	–	2.71	3.14	2.71	3.42	3.71	–
Overall score	–	–	–	–	–	–	3.14

Note: Table 13.2 shows the results of category technology indicators for Case Study 1.

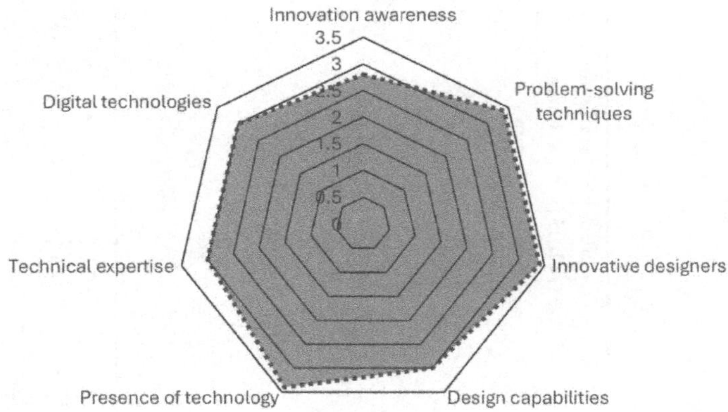

Figure 13.2 Illustration of 'technology' category scores for Case Study 1

Immediate feedback to the participants could be accessed via a spider diagram view, which shows their results and the overall results of the organisation. Figure 13.2 shows the outputs for the organisation for overall results for the 'technology' category.

The tool was then used to create an opportunity to delve deeper into the results of the analysis through discussion with the individuals about their scores, which provided a richer picture of organisational innovating behaviour across the key categories (e.g., technology generally) and within each category (e.g., at each indicator level).

The results of the discussions showed that the case study organisation believes they have technically well-qualified people able to meet the needs of the organisation. The organisation believes that they are technologically innovative and create a culture where the use of innovative solutions and different tools for problem-solving are encouraged. However, there does not appear to be any reward system for improving technological solutions within the organisation.

Expanding on the use of technological innovations, caution is exercised in the uptake of new technology. When new technology becomes available, there is a belief that there is no point in purchasing it, including software, if the organisation does not have the systems, capabilities, and forward work to make use of this new technology. On the other hand, there is a belief that innovative technology is fundamental to achieving increased profits, so the choice of investment needs to be carefully considered. Generally, the organisation feels that the industry is low in technology uptake and there is an awareness of the need to improve. Overall, when considering the innovation potential with respect to technology, it is clear that the case study organisation has some ways to improve, but their current decision-making takes into account the limited budget and the costs of investment. These technology scores are indicative of the scores received for the other categories. There was some acknowledgement that a tool for developing innovations is useful in helping the organisation understand what they need to do and in providing a platform for dialogue within the organisation on how to best innovate.

In terms of innovation activities, the respondents were given a list of activities believed to be useful for improving innovation generation and adoption within the organisation. The respondents were asked to what extent, during the past year, their organisation had undertaken activities such as research and experimental development, developing intellectual property, engaging in employee training, or software development. The organisation was also asked about their planned annual innovation expenditure and what percentage of this expenditure is directed towards innovation development. These questions, alongside other inquiries about the business environment of the organisation, created a more detailed picture of how their business innovates.

The research presented in this chapter has shown the testing of the innovation tool for the 'technology' category. The whole testing for the case study organisations comprised all the indicators across the five categories: workforce skills and human resources management, organisational structure, business strategy, technology, and external influences.

When examining the overall case study organisation, the results showed that there are positive attributes of the organisation that can enhance their innovation potential. For instance, a focus on employing highly qualified people is a particular strength. Aligning employees with the organisation's vision enhances the organisation's ability to innovate by instilling a common goal. The organisation is also focused on cost reduction, productivity, and developing more effective processes.

One aspect for improvement is the investment in research and development. Currently, the organisation outsources for expertise and appears to have a low appetite for in-house research and development. Discussions within the organisation on the findings could allow it to determine if this strategy needs modification, as it may improve innovation uptake.

When examining the organisational structure and the ability to innovate, the findings show that there could be improvements made in, for instance, enhancing their knowledge management system and being more systematic about the lessons learned and how these are communicated within the organisation. One of the risks to the company is the impact of external factors and the effect these have on the organisation's performance and their ability to innovate.

FUTURE TOOL ITERATIONS AND FURTHER DEVELOPMENT

This chapter highlights the development of a tool for companies to diagnose where they are good at innovation and where they could improve. For each of the categories, there is room for improvement. The tool was initially tested on two companies, and in addition to the actual questions about innovations, they were asked about the usability of the tool and any refinements required. The tool proved easy to understand, and by creating a quick output graph (spider diagram), the tool outputs could be readily used as a means of having conversations about whether any of the indicators need to be improved, and how. Given that not all the questions were relevant to one position in the organisation, there needs to be a wide number of participants from different backgrounds to give a realistic picture of how innovation behaviour operates. The organisation could re-run the tool through annual workshops to check on changes and progress. The case study participants felt that the tool could help the leadership make some decisions, such as improved training, which could further improve the capabilities of the organisation to innovate. An annual review would be an option proposed by the organisation so that regular information could be used by the organisation to develop a future innovation culture. On the specifics of the tool, there was a belief that the categories could be simplified and more focused, including changing the distribution via software on a website or app. Hence, the organisation could tailor the tool to their particular needs.

CONCLUSION

This chapter shows the development of a tool to aid organisational innovation within the construction industry. The chapter outlines the key categories that can be used to assess how well an organisation is doing with respect to creating innovations. The organisations that tested the tool, and the case study reported in this chapter, showed that a simple visual tool could provide them with evidence to communicate and make decisions on what to focus on and how to invest in innovations in the future. The five categories were believed to be those that would provide the outlook for the organisations to enact change. The methodology used for developing the tool could be easily followed and could provide wider benefits of innovation potential for any construction organisation. The tool provides enough information to allow an organisation to undertake a diagnostic test or to update annually to see whether any implemented decision changes the indicator scores and affects the innovation culture of the organisation. Feedback about this tool shows it could be further developed or altered so that it is tailored for a company, is more user-friendly, and can operate on different platforms. One of the key findings from the case study was the need to gain different perspectives on how innovation is working within an organisation. Again, the tool provided options to assess how individuals might view different aspects of innovation potential, and how different groups might view organisational innovation. With the knowledge gained from the diagnostics, organisations can change their decisions, invest more in innovation, and understand how each team or individual could contribute to developing innovation within the organisation. By measuring the construction organisation's innovation potential, future strategies can be developed to create better outcomes for construction.

REFERENCES

Aouad, G., Ozorhon, B., & Abbott, C. (2010). Facilitating innovation in construction: Directions and implications for research and policy. Construction Innovation: Information, Process, Management, 10(4), 374–394. https://doi.org/10.1108/14714171011083551

Becheikh, N., Landry, R., & Amara, N. (2006). Lessons from innovation empirical studies in the manufacturing sector: A systematic review of the literature from 1993–2003. *Technovation*, *26*(5–6), 644–664.

Blayse, A. M., & Manley, K. (2004). Key influences on construction innovation. *Construction Innovation*, *4*(3), 143–154.

Chen, H. L. (2014). Innovation stimulants, innovation capacity, and the performance of capital projects. *Journal of Business Economics and Management*, *15*(2), 212–231.

Cohen, W. M. (2010). Fifty years of empirical studies of innovative activity and performance. *Handbook of the Economics of Innovation*, *1*, 129–213.

Cooper, J. R. (1998). A multidimensional approach to the adoption of innovation. *Management Decision*, *36*(8), 493–502.

Crawford, R. (2021). Focused innovation policy: Lessons from international experience [NZPC Working paper No. 2021/03]. NZPC. Available from www.productivity.govt.nz

Daft, R. L., Murphy, J., & Willmott, H. (2010). Organization theory and design (10th ed.). South-Western Cengage Learning.

Dansoh, A., Oteng, D., & Frimpong, S. (2017). Innovation development and adoption in small construction firms in Ghana. *Construction Innovation*, *17*(4), 511–535.

DIISR – Department of Innovation, Industry, Science and Research. (2011). *Australian innovation system report & NBSP.* Canberra, Australia: Department of Innovation, Industry, Science and Research.

Dikmen, I., Birgonul, M. T., & Artuk, S. U. (2005). Integrated framework to investigate value innovations. *Journal of Management in Engineering, 21*(2), 81–90.

Drucker, P. (1985). *Entrepreneurship and innovation: Practices and principles.* New York: Harper & Row.

Dulaimi, M. F., Ling, F. Y., Ofori, G., & Silva, N. D. (2002). Enhancing integration and innovation in construction. *Building Research & Information*, *30*(4), 237–247.

Fernando, S., Panuwatwanich, K., & Thorpe, D. (2021). Introducing an innovation promotion model for construction projects. *Engineering, Construction and Architectural Management*, *28*(3), 728–746.

Gambatese, J. A., & Hallowell, M. (2011a). Enabling and measuring innovation in the construction industry. *Construction Management and Economics*, *29*(6), 553–567.

Gambatese, J. A., & Hallowell, M. (2011b). Factors that influence the development and diffusion of technical innovations in the construction industry. *Construction Management and Economics*, *29*(5), 507–517.

Green, R., Liyanage, S., Pitsis, T., Scott-Kemis, D., & Agarwal, R. (2010). *The OECD Innovation Strategy: Getting a head start on tomorrow.* Paris: OECD Publishing.

Hampson, K. D. (2005). The unholy alliance: Collaboration and innovation in property and construction. *AIQS Building Economist Magazine*, 4–13.

Hampson, K. D., & Manley, K. (2001). Construction innovation and public policy in Australia. In *Innovation in construction: An international review of public policies* (pp. 31–57). London: Spon Press.

Hemström, K., Mahapatra, K., & Gustavsson, L. (2017). Architects' perception of the innovativeness of the Swedish construction industry. *Construction Innovation*, *17*(2), 244–260. https://doi.org/10.1108/CI-06-2015-0038

Jansen, J. J., Van Den Bosch, F. A., & Volberda, H. W. (2006). Exploratory innovation, exploitative innovation, and performance: Effects of organizational antecedents and environmental moderators. *Management Science*, *52*(11), 1661–1674.

Koch, R., & Leitner, K. (2008). The dynamics and functions of self-organization in the fuzzy front end: Empirical evidence from the Austrian semiconductor industry. *Creativity and Innovation Management*, *17*(3), 216–226.

Loosemore, M. (2014). Innovate or perish? Exploring some of the myths of construction innovation. Paper presented at the *Australasian Journal of Construction Economics and Building-Conference Series*, *2*(2), 44–55.

Loosemore, M. (2015). Construction innovation: Fifth generation perspective. *Journal of Management in Engineering*, *31*(6), 04015012.

McAdam, R., Armstrong, G., & Kelly, B. (1998). Investigation of the relationship between total quality and innovation: A research study involving small organisations. *European Journal of Innovation Management*, *1*(3), 139–147.

McMeel, D., & Sweet, K. (2016). *ER11 Roadmap for supporting innovation in the New Zealand construction sector.* BRANZ Research Report. https://www.branz.co.nz/pubs/research-reports/er11/

Miozzo, M., & Dewick, P. (2004). *Innovation in construction: A European analysis.* Cheltenham, UK and Northampton, MA, USA: Edward Elgar Publishing.

Ofori, G. (2008). Leadership for future construction industry: Agenda for authentic leadership. *International Journal of Project Management*, *26*(6), 620–630.

Oslo Manual. (2018). Guidelines for collecting, reporting and using data on innovation. *The Measurement of Scientific, Technological and Innovation Activities.* 4th edition. OECD.

Ozorhon, B., Abbott, C., & Aouad, G. (2014). Integration and leadership as enablers of innovation in construction: Case study. *Journal of Management in Engineering*, *30*(2), 256–263.

Panuwatwanich, K., Stewart, R. A., & Mohamed, S. (2008). The role of climate for innovation in enhancing business performance: The case of design firms. *Engineering, Construction and Architectural Management*, *15*(5), 407–422.

Park, M., Nepal, M. P., & Dulaimi, M. F. (2004). Dynamic modeling for construction innovation. *Journal of Management in Engineering*, *20*(4), 170–177.

Project Management Institute. (2014). *A guide to the Project Management Body of Knowledge: PMBOK® Guide*. Project Management Institute.

Rutten, M. E., Dorée, A. G., & Halman, J. I. (2009). Innovation and interorganizational cooperation: A synthesis of literature. *Construction Innovation*, *9*(3), 286–297. doi:10.1108/14714170910973501

Schoenborn, J. (2012). *A case study approach to identifying the constraints and barriers to design innovation for modular construction.* Master's thesis, Virginia Tech. http://hdl.handle.net/10919/32397

Shahbazpour, M. (2010). *Strategic manufacturing system and process innovation. A framework for small and medium sized enterprises*, PhD thesis. University of Auckland.

Shibeika, A., & Harty, C. (2015). Diffusion of digital innovation in construction: A case study of a UK engineering firm. *Construction Management and Economics*, *33*(5–6), 453–466.

Smith, K., & Estibals, A. (2011). Innovation and research strategy for growth. https://api.semanticscholar.org/CorpusID:153043081

Song, M., & Chen, Y. (2014). Organizational attributes, market growth, and product innovation. *Journal of Product Innovation Management*, *31*(6), 1312–1329.

Wandahl, S., Jacobsen, A., Heidemann Lassen, A., Bolvig Poulsen, S., & Sørensen, H. (2011). User-driven innovation in a construction material supply network. *Construction Innovation*, *11*(4), 399–415.

14. Readiness of the construction industry to adopt the modern methods of construction: a scoping review

Ali M. Saad, Mohammed Fadhil Dulaimi, Chris Gorse and Suhaib Arogundade

INTRODUCTION

The shift towards modern methods of construction (MMC) could be described as a transformational stance in which on-site activities are significantly minimised, with cost and time-consuming activities being diverted off-site. Such a transformation is expected to have a considerable influence on the quality, productivity, and safety of construction projects (Abdul Nabi & El-adaway, 2021). Recognising MMC values, the government devoted comprehensive support to promote and encourage the use of MMC in construction projects (MHCLG, 2017). Such unconditional government support can be linked to the belief that MMC is capable of meeting deadlines, reducing environmental impacts, ensuring more cost certainty, minimising waste, and achieving better quality and safety (MHCLG, 2019). Hence, from a bird's eye view, MMC seems a reasonable construction methodology in an industry well known for its regular navigation through uncertainty.

When comparing MMC with traditional construction methods, advantages like construction waste reductions (Akinade et al., 2015; Monahan & Powell, 2011; Pacultová et al., 2019; Tan et al., 2020), construction safety enhancements (Ahn et al., 2020; Sutrisna & Goulding, 2019, p. 278), and sustainability improvements (Alwan et al., 2017; Goulding et al., 2015; Švajlenka & Kozlovská, 2018) are effectively prevailing. Moreover, such a transformation is believed to offer SMEs better scope to have a more significant role in the industry (Dowsett et al., 2019; Gbadamosi et al., 2020; Pablo & London, 2020) and is pinpointing critical cost reductions (Akmam Syed Zakaria & Amtered El-Abidi, 2021; Goulding et al., 2015; Hasan & Lu, 2021; O'Connor et al., 2014; Pan & Sidwell, 2011; Švajlenka & Kozlovská, 2018).

Several advantages of using MMC have also been recognised to align with contemporary technological adoptions like digitalisation (An et al., 2020; Iuorio et al., 2019; Martinez et al., 2021; Moon et al., 2020), which has a significant influence on construction's productivity (Akmam Syed Zakaria & Amtered El-Abidi, 2021; Goh & Goh, 2019). Therefore, it can be argued that MMC also has the ability to address the key challenges that have undermined the performance and standing of the construction industry for decades. However, despite the demonstrated advantages offered by the literature, MMC uptake is still slow (Taylor, 2022). More worrying is the collapse of MMC organisations, which could further discourage the adoption of this technology (Saad et al., 2023b). Consequently, it can be argued that the wider promotion of MMC would benefit industry stakeholders by capturing and extracting value from such a change in the construction methodology.

However, improving MMC uptake in the sector, especially among clients, is not a straightforward process. Construction clients are seen to discourage new initiatives, especially those requiring a radical shift from best practice, and prefer traditionally embedded norms (Eriksson et al., 2014). This is reflected in the poor uptake of MMC, believed not to exceed 8% in all new developments (Branson, 2020), with a very low adoption among public clients (Charlson & Dimka, 2021). It has been inferred that clients tend to lock their design and construction requirements before considering a construction innovation, an event that blocks any gateway for future use of MMC (Saad et al., 2023a; Viking & Lidelöw, 2015). Nevertheless, Chen and Perez (2018) discuss that contractors can similarly play a role in better communicating confidence; for example, their capability to provide an impressive portfolio could yield more attraction. However, the same study reports that such clients would strictly require their exact needs and demands to be met, especially since they are accompanied by large budgets. Hence, the initial exploration of the literature implies the need to understand both clients' decision-making and MMC businesses' capabilities to communicate confidence.

Subsequent to screening and analysing over 500 published articles relevant to the modern methods of construction (MMC) from a variety of academic databases, a rationale supporting the understudied nature of the social and decision-making aspects emerges across the existing body of knowledge. The rest of this chapter describes the development of such a rationale on the basis of multiple phases in (a) MMC characteristics and their influence on clients' decision-making process, (b) studying the characteristics of clients and the influence on their own decision-making process, and (c) studying the competency factors forming MMC firms' business model to communicate confidence. Data collected systematically and empirically will feed into the key considerations involved in adequately developing a public procurement agreement that fosters construction innovations across the sector.

NEED TO EXPLORE ADOPTION OF INNOVATION

MMC has acquired stakeholders' interest due to evidence suggesting its effectiveness compared with traditional construction methods. The uptake has been identified to be particularly low in the public sector (Charlson & Dimka, 2021). Research efforts relate the low adoption rates to attributes ranging between technicalities, perceptions, and attitudes (El-Abidi & Ghazali, 2015; Oti-Sarpong, 2020). Farmer (2016) describes MMC as a saviour due to the substantial ability to prevent an 'inexorable decline' in the construction industry. The report encourages the sector to enhance their uptake of MMC as 'the only way forward' to address contemporary construction challenges. Charlson and Dimka (2021), similarly, infer the need for more MMC adoption due to the emergence of housing crises, with the inability of traditional methods to meet the increasing needs. There is a need, therefore, to equally look at the decision-making process from the same nuanced view as the majority of scholars investigating the associated technicalities.

The adoption of an innovation in the sector and the decision-making processes of clients are two important aspects that require further investigation. This means that in order to shed light on the reasons that can explain the reluctance of the sector towards MMC, it is key to critically examine the innovation decision of the decision-makers. To simplify the complexities associated with studying this process, Rogers' (2003) theory offers constructs that can measure and highlight the perceptions and characteristics of public clients, a critical approach that would unravel the influence on public clients' favourable decisions. The same constructs have been used in other construction innovation research such as Besklubova et al. (2021), Xu et al. (2020), and Mead et al. (2020). Such an approach has proven to be effective when studying the influence of an innovation in construction. The theory explains how an innovation's complexity, compatibility, trialability, observability, and relative advantage compared with conventional methods would impact its adoptability and prospects across a specific decision-making unit.

EXPLORING THE DECISION-MAKING PROCESSES

Evidence suggests that despite the potential of MMC to improve the industry's performance, pressure from housing shortages, and governmental support, it is not yet widely accepted among clients generally (Eriksson et al., 2014; Larsson et al., 2014; Viking & Lidelöw, 2015), resulting in its low uptake. It is therefore important to critically examine the influence of MMC's perceived attributes on building clients' perception and influencing their decision-making relevant to theory, which in return would enhance our understanding of the critical factors undermining better MMC uptake in the sector. This approach can be argued to pinpoint a gap within the literature by aligning our enhanced understanding of public clients' decision-making to assess the competency of MMC businesses to communicate confidence for the construction market.

It is argued that the sector's growth is influenced by the ability of clients to drive innovation (Lindblad & Karrbom Gustavsson, 2021), and public clients are often viewed as the largest construction clients in the industry (Langston & Zhang, 2021). Studying clients, therefore, would enable research and practice to understand how the client's characteristics influence their decision-making towards MMC, supporting businesses to develop business models with better alignment – a stance that has been indicated to differ from the private sector (Sutrisna & Goulding, 2019). The capabilities of firms to meet the sector's strict requirements mean that they can better communicate confidence and change current perceptions (Hyun et al., 2021). Clients are aware of their influence on innovation in the construction context and expect a mature market with proven capabilities to deliver projects that reflect their values (Agapiou, 2021). Hence, an overarching understanding of the influence of the characteristics of construction clients on their decision-making process towards MMC adoption would help in explaining the low MMC uptake across the public sector.

NEED TO EXPLORE THE BUSINESS SIDE

The analysis of business models of MMC organisations can be argued to accommodate limitations that raise questions regarding their suitability as they enter the market. The extant body of knowledge encourages changing existing business models to better align with the nature of MMC projects compared with conventional practices (Wuni & Shen, 2019). These business models need to critically communicate confidence to ensure MMC is increasingly becoming a necessity (Ho et al., 2017), and sustain a dynamic capability to align with changes in demand (Goulding et al., 2013). Goulding et al. (2015) infer the requirement for alternative business models that would learn from other sectors and influence wider MMC adoption. MMC organisations are called upon to create business models that can promote confidence in the market (Brege et al., 2014). This argument bridges the previous sections of this chapter as it highlights the need to develop business models with an ability to aggregate demand and build upon the knowledge gained from the investigation on public clients' decision-making processes. The existing body of knowledge hereby supports the argument of the importance of suitable business models to align with clients' characteristics, and by that, drive wider uptake.

An exploration of academic research reveals criticism regarding the incongruity of the business strategies employed by MMC organisations. This section provides a review of the literature in search of a gap that may aid in building a solid research rationale.

The Issue of Value Offering

One of the strategies employed by organisations is the offering of values as an attempt to encourage adoption (Nielsen & Kyhnau, 2015). However, values associated with innovations are often not sufficient justifications for adoption to occur (Rogers, 2003). Such a paradigm, hence, demands strategies to locate and offer the values customers want (Payne et al., 2017), which is achieved by embracing MMC as a transition to fit with the needs of an external environment rather than proposing values by default (Hairstans & Smith, 2018). Moreover, O'Connor et al. (2014) report the necessity of clearly defining values at an early stage as a critical factor influencing clients' decisions to adopt MMC. The inability of clients to adequately comprehend the values offered by MMC providers nurtures a perception that the innovation is complex and unsuitable (Ezzeddine & García de Soto, 2021), limiting adoption due to the lack of trust (Hedgren & Stehn, 2014). Trust, in this context, is found to be crucial by Rahman (2014), who reports that addressing clients' suspicion of MMC, through adequately offering the values that matter, can drive favourable decisions. Therefore, this chapter assumes that a divergence exists to distinguish the business competency of organisations communicating MMC values compared with those lagging.

The Issue of Outreach

Literature depicts that the utilisation of effective means of exposure is a crucial prerequisite to reaching customers. In the case of MMC, the particular focus on clients motivates the use of public procurement agreements as channels linking both supply and demand. The literature describes that these frameworks reduce the lengthy procurement process, boost confidence, and promote innovation (Charlson & Dimka, 2021). Similarly, Oti-Sarpong et al. (2022) describe leveraging MMC projects through such agreements as a conduit that would facilitate broader adoption. In contrast, other scholars suggest that marketing construction innovations is encouraged through increased use of social media (Butt et al., 2021), relevant webinars (Tinazzi, 2021), and construction events (Razkenari et al., 2020). Hence, a supposed knowledge gap exists within the literature as the effective outreach of MMC organisations remains empirically undetermined.

The Issue of Relationships

Sustaining healthy customer relationships is crucial to any business endeavour, a matter that is contemplated through communication (Metallo et al., 2018). Reasonably, focusing on maximising customers' positive experiences enhances overall profits (Li et al., 2020), which is explained by the extended ability of a firm to substantiate value (Ibarra et al., 2018). However, limited research focuses on MMC businesses, as Diaz Lopez et al. (2019) argue that any change in the conventional process would require a change in how business relationships are built. An effective relationship with clients, hence, is one that extends an organisation's knowledge of their customers' needs and wants (Gbadamosi et al., 2020). To achieve this, MMC businesses are encouraged to leverage these relationships in pursuit of awareness of their clients' changing requirements (Brege et al., 2014). Therefore, better understanding how MMC organisations build client relationships may underscore wider coherence on their business competency in the UK.

The Issue of Cost

The matter of cost is a long-standing issue associated with any innovation, dictating the direct affiliation between organisations' pricing and their survival (Iheanachor et al., 2021). Such importance is determined by demand, as cost remains one of the major influences on innovation adoption (Charlson & Dimka, 2021). Saiyed (2019), moreover, reports that this element is influenced and affected by the organisation's business strategies to justify the same. In the construction context, De Schutter et al. (2018) establish that the cost structure of a construction innovation varies to comprise expertise, labour, and material costs in addition to the innovation's cost, which reasonably implies that cost is innovation-specific, and it is illogical to state that a standard and optimal cost structure exists among all organisations. Revealing whether the business competency of MMC organisations is influenced by how cost is being offered, therefore, is essential.

The Issue of Scale

It is rational that a robust revenue stream from one or more sources is a condition for any firm's survival in a business environment (DaSilva & Trkman, 2014). However, literature is discreet about how MMC businesses are seeking revenue, and whether any particular route is more effective in driving more public client projects. Despite revenue being understudied in relevant research, MMC businesses are believed to be enhancing their revenue streams by exerting input where profitability is maximised (Goh & Goh, 2019). In this context, Gbadamosi et al. (2020, p. 17) link both demand's 'visibility' with supply's 'profitability' as critical associates. Because current widespread adoption is not enough to support core revenues, a change is required in business models to better achieve diversification (Hyun et al., 2021). This could, however, be considered a changing reality amidst the government's progression to mandate MMC in their budgets (Agapiou, 2021; Oti-Sarpong et al., 2022). Hence, it is imperative to understand whether MMC organisations are seeking any specific strategies in pursuit of revenue and whether these strategies are enabling better penetration of the public construction market.

NEED FOR A THEORETICAL UNDERPINNING

Since the issues seem interrelated between the three pillars of the innovation itself, the decision-makers and their stance on the innovation, and the business side and the ability of providers to communicate confidence, a framework could be seen as useful to establish the much-needed theoretical underpinning. The conceptualisation presented in this chapter is derived from a proliferation of MMC studies, particularly from the past decade, mostly denoting that these are associated with a tendency to focus on the technical aspects rather than on how these aspects influence the decision-making and adoption process responsible for promoting MMC in the construction sector. It is logical to hypothesise that the adopted research trends, albeit increasing, are limited in the sense of investigating the relational structures in how supply can influence demand, and vice versa, to the extent that makes this chapter among the very first to critically conceptualise the increasing attention on MMC. This chapter therefore is inclined to suggest that previous works are context-specific and disjointed, requiring an overarching need for more generalisation and empirical investigation in accordance with a robust theoretical underpinning.

Clients play a vital role in driving industry change towards accepting and embracing innovation. MMC can offer solutions beyond merely meeting public demand, but can also benefit the overall construction sector. Therefore, a better understanding of the decision-making process can highlight the pressure points that influence clients' adoption. By focusing on the client-oriented stance rather than the regular innovation-oriented approach reflected in vast literature, this section conceptualises that the application of the Diffusion of Innovation theory would illuminate and explain the decision-making process and the formation of an innovation-adoption decision. Additionally, subsequent to extracting and critically reviewing previous research efforts, this chapter encourages uncovering the equally important potential role of MMC businesses, contemplated by the business environment, enabling the study to extend knowledge beyond clients themselves, to also understand the divergences in how confidence is being reinforced and communicated in the sector. Additionally, the conceptualisation includes questioning mechanisms that would unlock reliable assessments of the business competency of MMC organisations in the UK. Such understanding enables this study to benefit from the application of various theories, in an attempt to rationalise the role of the interactions between the business environment and MMC organisations that are outperforming others in the sector. Therefore, there is an unrealised need to conceptualise the flow of innovation adoption in construction through the lens of decision-making, business competency, and contingency theory.

The emerging paradigm identifies broad research efforts that have MMC technical attributes at the forefront of their interests. Alternatively, this chapter calls on scholars to focus on clients' decision-making and MMC organisations' business competencies as understudied areas that may carry the needed rationale to explain unsatisfactory MMC uptake. This conceptualisation aims to act as a foundation for subsequent empirical investigations. As shown in Figure 14.1, the developed framework depicts the key arguments made in relation to their theoretical constructs.

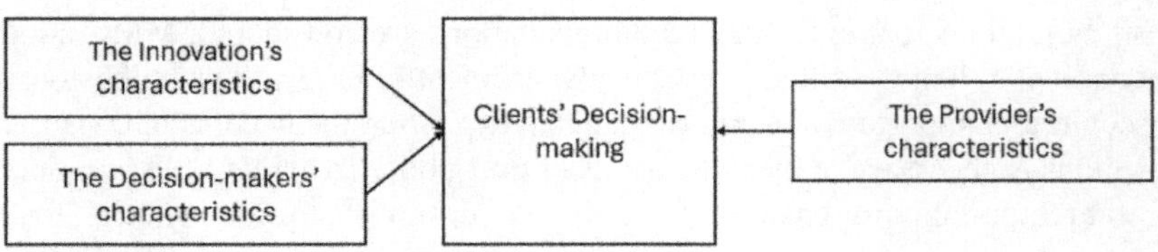

Figure 14.1 Theoretical framework

NEED FOR A RESEARCH AGENDA

The chapter calls for a research agenda that would aid in revealing and addressing the main factors believed to be damaging to the adoption of innovations in the construction sector and highlight the problematic and complex decision-making process. Instead of realising the inherent advantages and potential of MMC, the general perception seems to be yet residing upon its historical failures rather than its growing and invaluable potential. However, such resistance to change can be minimised through key changes to the current practices in procurement. This chapter therefore depicts the key recommendations believed to be necessary for an effective procurement that would enhance the prospects of MMC adoption in the sector. The chapter recommends:

- the development of a longitudinal comparison criteria that is established beyond cross-sectional comparisons with traditional construction, inferring that comparing MMC with traditional construction is generic and unfair with the lack of clear benchmarks that consider the long-term benefits and the innovation's characteristics;
- the establishment of a commonly accessible platform of historical evidence based on information-sharing and continuous communication between clients. The platform comprises the success stories and the challenges encountered by clients;
- that client organisations rethink existing requirements that are designed to procure traditional construction but are misaligned with MMC. This includes replacing short-term piloting with long-term commitments, redefining local and social values, and changing how these are measured and reported;
- that communication between supply and demand is enhanced and effectively pioneered to align MMC with the sector needs as key attributes to the diffusion of MMC in the UK construction sector;
- unblocking a relationship history between clients and MMC organisations, and paving a route with fewer restrictions to entering the construction market for those substantiating effective innovations pioneering different, and new, methods to delivering places.

The study predicts that until such recommendations are considered, MMC adoption will continue to be low, influencing the advancement of innovations in the construction sector. It is important to acknowledge that exploiting the benefits of construction innovations is necessary to meet the needs of the public. To achieve this exploitation, the research community has a responsibility to ensure that circumstances for construction innovations are appropriate for their diffusion. Hence, construction scholars, construction businesses, decision-makers, and policymakers are encouraged to consider the recommendations of this study to accelerate changing the ways places are traditionally delivered.

CONCLUSION

This chapter presents a critical examination of the existing gaps in literature towards proposing the key approaches that may explain the low uptake of a supposedly effective construction methodology. Scholars need to rely on theories to establish the key constructs that can then inform the competency of MMC organisations, allowing them to communicate confidence to the construction market. Based on analysing the existing body of knowledge, clients' perceptions and characteristics towards construction innovations are topics that have not received much attention from the construction management community, and doing so would support efforts to identify the critical influences on the decision-making process and MMC organisations' business competency, as well as studying the core competencies necessary for the market to embrace and effectively adopt a fundamental change in how places are delivered.

REFERENCES

Abdul Nabi, M., and I. H. El-adaway. 2021. "Risk-based approach to predict the cost performance of modularization in construction projects". J. Constr. Eng. Manag., 147 (10): 04021133. https://doi.org/10.1061/(ASCE)CO.1943-7862.0002159

Agapiou, A. 2021. "An exploration of the best value perceptions of small housebuilding developers towards offsite construction." Sustainability, 13 (7): 4054. https://doi.org/10.3390/su13074054

Ahn, S., L. Crouch, T. W. Kim, and R. Rameezdeen. 2020. "Comparison of Worker Safety Risks between Onsite and Offsite Construction Methods: A Site Management Perspective." J. Constr. Eng. Manag., 146 (9): 05020010. https://doi.org/10.1061/(asce)co.1943-7862.0001890

Akinade, O. O., L. O. Oyedele, M. Bilal, S. O. Ajayi, H. A. Owolabi, H. A. Alaka, and S. A. Bello. 2015. "Waste minimisation through deconstruction: A BIM based Deconstructability Assessment Score (BIM-DAS)." Resour. Conserv. Recycl., 105: 167–176. https://doi.org/10.1016/j.resconrec.2015.10.018

Akmam Syed Zakaria, S., and K. M. Amtered El-Abidi. 2021. "Economic effects of migrant labor on industrialized building system (IBS) adoption in the Malaysian construction industry." Archit. Eng. Des. Manag., 17 (1–2): 50–66. https://doi.org/10.1080/17452007.2020.1781587

Alwan, Z., P. Jones, and P. Holgate. 2017. "Strategic sustainable development in the UK construction industry, through the framework for strategic sustainable development, using Building Information Modelling." J. Clean. Prod., 140: 349–358. https://doi.org/10.1016/j.jclepro.2015.12.085

An, S., P. Martinez, M. Al-Hussein, and R. Ahmad. 2020. "BIM-based decision support system for automated manufacturability check of wood frame assemblies." Autom. Constr., 111 (June 2019): 103065. https://doi.org/10.1016/j.autcon.2019.103065

Besklubova, S., M. J. Skibniewski, and X. Zhang. 2021. "Factors affecting 3D printing technology adaptation in construction." J. Constr. Eng. Manag., 147 (5): 04021026. https://doi.org/10.1061/(asce)co.1943-7862.0002034

Branson, A. 2020. "Why 2020 is the year of modular (RICS)." R. Inst. Chart. Surv. https://ww3.rics.org/uk/en/modus/built-environment/construction/why-2020-is-the-year-of-modular.html

Brege, S., L. Stehn, and T. Nord. 2014. "Business models in industrialized building of multi-storey houses." Constr. Manag. Econ., 32 (1–2): 208–226. https://doi.org/10.1080/01446193.2013.840734

Butt, B., R. V. Jones, and A. Fuertes. 2021. "Opportunities and barriers to business engagement in the UK domestic retrofit sector: An industry perspective." Build. Serv. Eng. Res. Technol., 42 (3): 293–305. https://doi.org/10.1177/0143624420975640

Charlson, J., and N. Dimka. 2021. "Design, manufacture and construct procurement model for volumetric offsite manufacturing in the UK housing sector." Constr. Innov., 21 (4): 800–817. https://doi.org/10.1108/CI-10-2019-0108

Chen, Y., and Y. Perez. 2018. "Business model design: Lessons learned from Tesla Motors." In da Costa, P., Attias, D. (eds) *Towards a Sustainable Economy. Sustainability and Innovation.* Cham: Springer, 53–69. https://doi.org/10.1007/978-3-319-79060-2_4

DaSilva, C. M., and P. Trkman. 2014. "Business model: What it is and what it is not." Long Range Plann., 47 (6): 379–389. https://doi.org/10.1016/j.lrp.2013.08.004

De Schutter, G., K. Lesage, V. Mechtcherine, V. N. Nerella, G. Habert, and I. Agusti-Juan. 2018. "Vision of 3D printing with concrete: Technical, economic and environmental potentials." Cem. Concr. Res., 112 (August): 25–36. https://doi.org/10.1016/j.cemconres.2018.06.001

Diaz Lopez, F. J., T. Bastein, and A. Tukker. 2019. "Business model innovation for resource-efficiency, circularity and cleaner production: What 143 cases tell us." Ecol. Econ., 155 (March 2017): 20–35. https://doi.org/10.1016/j.ecolecon.2018.03.009

Dowsett, R., M. Green, M. Sexton, and C. Harty. 2019. "Projecting at the project level: MMC supply chain integration roadmap for small housebuilders." Constr. Innov., 19 (2): 193–211. https://doi.org/10.1108/CI-07-2017-0059

El-Abidi, K. M. A. and F. E. M. Ghazali. 2015. "Motivations and limitations of prefabricated building: An overview" Applied Mechanics and Materials, 802 (October): 668–675.

Eriksson, P. E., S. Olander, H. Szentes, and K. Widén. 2014. "Managing short-term efficiency and long-term development through industrialized construction." Constr. Manag. Econ., 32 (1–2): 97–108. https://doi.org/10.1080/01446193.2013.814920

Ezzeddine, A., and B. García de Soto. 2021. "Connecting teams in modular construction projects using game engine technology." Autom. Constr., 132: 103887. https://doi.org/10.1016/j.autcon.2021.103887

Farmer, M. (2016). Modernise or die: The Farmer Review of the UK construction labour market.

Gbadamosi, A. Q., L. Oyedele, A. M. Mahamadu, H. Kusimo, M. Bilal, J. M. Davila Delgado, and N. Muhammed-Yakubu. 2020. "Big data for Design Options Repository: Towards a DFMA approach for offsite construction." Autom. Constr., 120 (August): 103388. https://doi.org/10.1016/j.autcon.2020.103388

Goh, M., and Y. M. Goh. 2019. "Lean production theory-based simulation of modular construction processes." Autom. Constr., 101 (November 2018): 227–244. https://doi.org/10.1016/j.autcon.2018.12.017

Goulding, J., M. Arif, F. Pour-Rahimian, and M. Sharpe. 2013. "Offsite construction: Sustainable innovation futures using SMART approaches." CIB World Build. Conf. Brisbane, Australia.

Goulding, J. S., F. Pour Rahimian, M. Arif, and M. D. Sharp. 2015. "New offsite production and business models in construction: Priorities for the future research agenda." Archit. Eng. Des. Manag., 11 (3): 163–184. https://doi.org/10.1080/17452007.2014.891501

Hairstans, R., and R. E. Smith. 2018. "Offsite HUB (Scotland): Establishing a collaborative regional framework for knowledge exchange in the UK." Archit. Eng. Des. Manag., 14 (1–2): 60–77. https://doi.org/10.1080/17452007.2017.1314858

Hasan, M., and M. Lu. 2021. "Error propagation model for analyzing project labor cost budget risks in industrial construction." J. Constr. Eng. Manag., 147 (4): 04021007. https://doi.org/10.1061/(asce)co.1943-7862.0002010

Hedgren, E., and L. Stehn. 2014. "The impact of clients' decision-making on their adoption of industrialized building." Constr. Manag. Econ., 32 (1–2): 126–145. https://doi.org/10.1080/01446193.2013.833340

Ho, W., K. Lam, T. Ma, T. Chan, and H. Tsang. 2017. "Age composition and survival of public housing stock in Hong Kong." Int. Res. Conf. 2017 Shap. Tomorrow's Built Environ. Conf. Proc., 435–446.

Hyun, H., Y.-M. Lee, H.-G. Kim, and J.-S. Kim. 2021. "Framework for Long-Term Public Housing Supply Plan Focusing on Small-Scale Offsite Construction in Seoul." Sustainability, 13 (10): 5361. https://doi.org/10.3390/su13105361

Ibarra, D., J. Ganzarain, and J. I. Igartua. 2018. "Business model innovation through Industry 4.0: A review." Procedia Manuf., 22: 4–10. https://doi.org/10.1016/j.promfg.2018.03.002

Iheanachor, N., Y. David-West, and I. O. Umukoro. 2021. "Business model innovation at the bottom of the pyramid – A case of mobile money agents." J. Bus. Res., 127 (January): 96–107. https://doi.org/10.1016/j.jbusres.2021.01.029

Iuorio, O., A. Wallace, and K. Simpson. 2019. "Prefabs in the North of England: Technological, environmental and social innovations." Sustainability, 11 (14): 3884. https://doi.org/10.3390/su11143884

Langston, C., and W. Zhang. 2021. "DfMA: Towards an integrated strategy for a more productive and sustainable construction industry in Australia." Sustainability, 13 (16): 9219. https://doi.org/10.3390/su13169219

Larsson, J., P. E. Eriksson, T. Olofsson, and P. Simonsson. 2014. "Industrialized construction in the Swedish infrastructure sector: Core elements and barriers." Constr. Manag. Econ., 32 (1–2): 83–96. https://doi.org/10.1080/01446193.2013.833666

Li, Y., J. Dai, and L. Cui. 2020. "The impact of digital technologies on economic and environmental performance in the context of industry 4.0: A moderated mediation model." Int. J. Prod. Econ., 229 (May 2019): 107777. https://doi.org/10.1016/j.ijpe.2020.107777

Lindblad, H., and T. Karrbom Gustavsson. 2021. "Public clients ability to drive industry change: The case of implementing BIM." Constr. Manag. Econ., 39 (1): 21–35. https://doi.org/10.1080/01446193.2020.1807032

Martinez, P., B. Barkokebas, F. Hamzeh, M. Al-Hussein, and R. Ahmad. 2021. "A vision-based approach for automatic progress tracking of floor paneling in offsite construction facilities." Autom. Constr., 125 (December 2020): 103620. https://doi.org/10.1016/j.autcon.2021.103620

Mead, T., S. Jeanrenaud, and J. Bessant. 2020. "Factors influencing the application of nature as inspiration for sustainability-oriented innovation in multinational corporations." Bus. Strateg. Environ., 29 (8): 3162–3173. https://doi.org/10.1002/bse.2564

Metallo, C., R. Agrifoglio, F. Schiavone, and J. Mueller. 2018. "Understanding business model in the Internet of Things industry." Technol. Forecast. Soc. Change, 136 (June 2017): 298–306. https://doi.org/10.1016/j.techfore.2018.01.020

MHCLG. 2017. The role of international evidence in housing policy: A report from the Ministry of Housing, Communities and Local Government to the Public Accounts Committee.

MHCLG. 2019. Modern methods of construction: Introducing the MMC Definition Framework. Ministry of Housing, Communities & Local Government.

Monahan, J., and J. C. Powell. 2011. "An embodied carbon and energy analysis of modern methods of construction in housing: A case study using a lifecycle assessment framework." Energy Build., 43 (1): 179–188. https://doi.org/10.1016/j.enbuild.2010.09.005

Moon, S., N. Ham, S. Kim, L. Hou, J. H. Kim, and J. J. Kim. 2020. "Fourth industrialization-oriented offsite construction: Case study of an application to an irregular commercial building." Eng. Constr. Archit. Manag., 27 (9): 2271–2286. https://doi.org/10.1108/ECAM-07-2018-0312

Nielsen, C., and J. Kyhnau. 2015. "How to create products and services customers want." J. Bus. Model., 3 (1): 81–92.

O'Connor, J. T., W. J. O'Brien, and J. O. Choi. 2014. "Critical success factors and enablers for optimum and maximum industrial modularization." J. Constr. Eng. Manag., 140 (6): 04014012. https://doi.org/10.1061/(ASCE)CO.1943-7862.0000842

Oti-Sarpong, K. and G. Burgess. 2020. *Offsite manufacturing and construction industry transformation: A multi-level sociotechnical transitions perspective.* University of Cambridge Repository. https://doi.org/10.17863/CAM.70388

Oti-Sarpong, K., R. S. Shojaei, Z. Dakhli, G. Burgess, and M. Zaki. 2022. "How countries achieve greater use of offsite manufacturing to build new housing: Identifying typologies through institutional theory." Sustain. Cities Soc., 76 (September 2021): 103403. https://doi.org/10.1016/j.scs.2021.103403

Pablo, Z., and K. A. London. 2020. "Stable relationality and dynamic innovation: Two models of collaboration in SME-driven offsite manufacturing supply chains in housing construction." Eng. Constr. Archit. Manag., 27 (7): 1553–1577. https://doi.org/10.1108/ECAM-07-2019-0346

Pacultová, K., A. Klegová, K. Karásková, L. Obalová, A. Martaus, and J. Vlček. 2019. "Theoretical and methodological aspects of transition to a circular economy." Waste Forum, (4): 295–414.

Pan, W., and R. Sidwell. 2011. "Demystifying the cost barriers to offsite construction in the UK." Constr. Manag. Econ., 29 (11): 1081–1099. https://doi.org/10.1080/01446193.2011.637938

Payne, A., P. Frow, and A. Eggert. 2017. "The customer value proposition: evolution, development, and application in marketing." J. Acad. Mark. Sci., 45 (4): 467–489. https://doi.org/10.1007/s11747-017-0523-z

Rahman, M. M. 2014. "Barriers of implementing modern methods of construction." J. Manag. Eng., 30 (1): 69–77. https://doi.org/10.1061/(asce)me.1943-5479.0000173

Razkenari, M., A. Fenner, A. Shojaei, H. Hakim, and C. Kibert. 2020. "Perceptions of offsite construction in the United States: An investigation of current practices." J. Build. Eng., 29 (April 2019): 101138. https://doi.org/10.1016/j.jobe.2019.101138

Rogers, E. M. 2003. *Diffusion of Innovations.* New York: Free Press.

Saad, A. M., M. Dulaimi, S. Arogundade, S. L. Zulu, and C. Gorse. 2023a. "Modern methods of construction (MMC) and innovation negativism in the UK public sector." Built Environ. Proj. Asset Manag. https://doi.org/10.1108/BEPAM-06-2023-0108

Saad, A. M., M. Dulaimi, and S. L. Zulu. 2023b. "A systematic review of the business contingencies influencing broader adoption: Modern Methods of Construction (MMC)." Buildings, 13 (4): 878. https://doi.org/10.3390/buildings13040878

Saiyed, A. A. M. 2019. "The role of leadership in business model innovation: A case of an entrepreneurial firm from India." New Engl. J. Entrep., 22 (2): 70–88. https://doi.org/10.1108/NEJE-08-2019-0040

Sutrisna, M., and J. Goulding. 2019. "Managing information flow and design processes to reduce design risks in offsite construction projects." Eng. Constr. Archit. Manag., 26 (2): 267–284. https://doi.org/10.1108/ECAM-11-2017-0250

Švajlenka, J., and M. Kozlovská. 2018. "Houses based on wood as an ecological and sustainable housing alternative: Case study." Sustainability, 10 (5): 1502. https://doi.org/10.3390/su10051502

Tan, T., W. Lu, G. Tan, F. Xue, K. Chen, J. Xu, J. Wang, and S. Gao. 2020. "Construction-oriented design for manufacture and assembly guidelines." J. Constr. Eng. Manag., 146 (8): 04020085. https://doi.org/10.1061/(asce)co.1943-7862.0001877

Taylor, M. D. 2022. "A definition and valuation of the UK offsite construction sector: Ten years on." Int. J. Constr. Manag., 22 (15): 2877–2885. https://doi.org/10.1080/15623599.2020.1829783

Tinazzi, U. 2021. "Dissemination and cultural change as an alternative business value proposition in the construction industry: Manni Group case study." J. Creat. Value, 7 (2): 183–188. https://doi.org/10.1177/23949643211049772

Viking, A., and S. Lidelöw. 2015. "Exploring industrialized housebuilders' interpretations of local requirements using institutional logics." Constr. Manag. Econ., 33 (5–6): 484–494. https://doi.org/10.1080/01446193.2015.1050966

Wuni, I. Y., and G. Q. P. Shen. 2019. "Holistic review and conceptual framework for the drivers of offsite construction: A total interpretive structural modelling approach." Buildings, 9 (5): 117. https://doi.org/10.3390/buildings9050117

Xu, X., G. Wang, D. Cao, and Z. Zhang. 2020. "BIM Adoption for Facility Management in Urban Rail Transit: An Innovation Diffusion Theory Perspective." Adv. Civ. Eng., (T. Dede, ed.), 2020: 1–12. https://doi.org/10.1155/2020/8864221

15. Emerging technologies in health and safety management: current perspectives and future prospects

C. Mahammedi, T.E. Butt and M.K.S. Al-Mhdawi

1. INTRODUCTION

The construction industry, renowned for its high-risk nature, is characterised by the utilisation of heavy machinery, exposure to hazardous materials, and work conducted at significant heights, among other challenges (Ahn et al., 2020). Given these inherent risks, prioritising robust health and safety measures is imperative to safeguard the well-being of workers (Sorensen et al., 2021). The significance of health and safety in the construction sector cannot be overstated. Beyond being a moral and legal obligation, a comprehensive approach to safety not only protects the lives of workers but also plays a pivotal role in enhancing the overall efficiency and productivity of a project (Maqsoom et al., 2023; Shabani, Jerie and Shabani, 2023).

The statistics provided by the Health and Safety Executive (HSE) in the UK shed light on the significant and far-reaching impact of workplace injuries in the construction sector (Health and Safety Executive, 2023b). The financial burden alone, with an estimated cost of £16.2 billion in 2019/20, underscores the economic consequences of these incidents on society. Despite ongoing efforts to enhance safety protocols, the construction industry continues to face a daunting challenge regarding fatalities. The reported increase in fatal injuries from 29 to 45 in the span of a year (2022/23) raises concerns about the effectiveness of existing safety measures and the need for more stringent regulations or enforcement. The non-fatal injury rate of 2,880 per 100,000 workers in the construction industry is alarming, emphasising the prevalence of accidents on construction sites. Comparing this rate with the combined rate of all other industries at 1,650 per 100,000 workers highlights the disproportionate risk faced by those employed in construction. This stark difference underscores the unique hazards and challenges inherent in the construction sector, demanding targeted and industry-specific safety interventions. Given these figures, the fact that approximately 2.9% of workers in the construction sector suffered injuries is a cause for concern (Daniel, 2022). This rate is notably higher than the average injury rate across all industries, which stands at 1.6%.

This discrepancy points to the need for tailored safety measures and interventions within the construction industry to bring these injury rates more in line with national averages. However, the HSE report paints a sobering picture of the state of health and safety in the construction sector. The financial and human costs of workplace injuries and fatalities underscore the urgency of addressing safety issues in this industry. The incorporation of Construction 4.0 principles, utilising digital technologies and automation, is vital for improving safety in the construction sector. This approach stresses collaboration among industry stakeholders, regulatory bodies, and workers to utilise technologies such as building information modeling (BIM), the Internet of Things (IoT), robotics, and artificial intelligence (AI) for proactive risk management and accident prevention. It's essential to empower workers with the necessary skills for successful implementation.

Maintaining health and safety within the construction sector involves a comprehensive and systematic approach, meticulously designed to proactively identify, assess, and manage risks at every stage of project development and execution. This encompasses a range of strategies, protocols, and measures aimed at safeguarding the well-being of workers, stakeholders, and the surrounding environment (Hoeft and Trask, 2022). This lifecycle spans planning, design, construction, operation, and dismantling or demolition. Each phase involves unique challenges and activities tailored to ensure the well-being of workers and the public. During planning, there is an emphasis on anticipating risks and implementing preventive measures. The design phase integrates safety features, considering structural integrity and ergonomic layouts (Oloke, 2020; Hoeft and Trask, 2022). The construction phase emphasises on-site safety protocols, worker training, and protective measures. The operation phase maintains ongoing safety standards, while the dismantling or demolition phase requires careful planning to manage potential hazards associated with structure removal (Nenonen, Kivistö-Rahnasto and Vasara, 2015).

Maintaining a safe working environment in the construction industry is paramount and requires robust health and safety regulations that address specific risks associated with construction activities, including fall protection, hazard communication, scaffolding, and electrical safety (Sanni-Anibire et al., 2020). These regulations form a comprehensive framework for creating a secure workplace, guided by fundamental principles such as thorough risk assessments and compliance with regulatory requirements, emphasising the duty of care toward employees (Mzyece, 2015). Emergency preparedness, hierarchy of controls, ongoing inspections, and robust enforcement mechanisms further contribute to continued compliance with safety measures (Dekker, 2014). Record-keeping serves as a crucial element for accountability and evaluating the effectiveness of implemented measures (Aksorn and Hadikusumo, 2008). Additionally, the integration of cutting-edge technologies such as BIM, virtual reality (VR), and AI brings unprecedented tools for hazard identification and risk management, enabling construction executives to usher in a new era of safety practices and enhance the overall well-being of industry workers (Elghaish et al., 2021).

With the advent of Industry 4.0, which is the integration of digital technologies and automation into various sectors, the construction industry has also witnessed significant advancements (Javaid et al., 2021; Newman et al., 2021). This has led to the emergence of the concept of Health and Safety in Construction Industry 4.0, which focuses on utilising technology to enhance safety measures and mitigate risks

(Smallwood and Allen, 2023). Health and Safety in Construction Industry 4.0 encompasses a range of technologies and practices that aim to improve safety standards on construction sites. One of the key aspects is the use of IoT devices, which enable the collection and analysis of real-time data (Ghosh, Edwards and Hosseini, 2021). For example, wearable devices can monitor the vital signs of workers, detect any signs of fatigue or stress, and alert supervisors in case of any emergencies. This allows for proactive measures to be taken to prevent accidents and ensure the well-being of workers (He, He and Wen, 2023). Another important aspect of Health and Safety in Construction Industry 4.0 is the use of drones for site inspections and monitoring (Nwaogu et al., 2023). Drones equipped with cameras can capture high-resolution images and videos of construction sites, which can be analysed by AI algorithms (Javaid et al., 2022). This can help identify potential hazards, such as unstable structures or unsafe work practices, and prompt necessary actions to mitigate risks. Drones can also be used to conduct surveys in hard-to-reach areas, reducing the need for workers to be exposed to dangerous environments (Elghaish et al., 2021; Namian and Turkan, 2023). Additionally, BIM technology plays a significant role in enhancing safety in the construction industry (Getuli et al., 2017; Tender et al., 2022). BIM enables the creation of digital models that simulate the construction process, allowing for the identification of potential safety issues before the actual construction begins (Muzafar, 2021). This helps in designing safer structures and planning efficient workflows. BIM also facilitates coordination between different stakeholders, such as architects, engineers, and contractors, ensuring that safety considerations are integrated into every stage of the construction project (Musarat et al., 2022).

Previous studies have identified the potential of emerging technologies like BIM, augmented reality (AR), VR, AI, robotics, and drones to enhance safety through real-time communication and monitoring (Zhou et al., 2012; Li et al., 2018). These innovations can proactively address safety concerns, potentially reducing accidents, boosting productivity, and enhancing overall well-being (Fang et al., 2016; Guo et al., 2021). However, existing knowledge lacks a comprehensive overview of current and evolving technologies in Health and Safety in Construction 4.0 (Dou et al., 2023). This chapter aims to fill this gap by presenting a wide-ranging account of digital technologies, including digital twin applications. It introduces a novel concept of incorporating health and safety considerations throughout the construction lifecycle, facilitating a comparative analysis between traditional and modern approaches. Covering stages from design to deconstruction, the chapter also identifies knowledge gaps, providing a foundation for future research and development in the field.

Consequently, this chapter aims to comprehensively examine existing health and safety approaches in the context of Construction 4.0, emphasising the understanding and documentation of diverse strategies employed in this evolving framework. An innovative perspective is introduced by integrating health and safety considerations throughout the entire lifecycle of construction projects, covering both traditional and Construction 4.0 lifecycles. The chapter also explores the impact of advanced technologies like BIM, virtual reality, artificial intelligence, and drones on health and safety practices in Construction 4.0. The chapter conducts a thorough review of the current state-of-the-art, identifying knowledge gaps and defining their scopes. The goal is to pave a clear path for future research and development by articulating these knowledge gaps and their boundaries, offering a roadmap for advancing health and safety practices in Construction 4.0.

2. HEALTH AND SAFETY LEGISLATION AND REGULATIONS IN THE UK

2.1 The Health and Safety at Work etc. Act 1974

The Health and Safety at Work etc. Act 1974, commonly known as HSWA, is the primary legislation governing occupational health and safety in Great Britain (Great Britain, 2023). It originated from the report of Robens (1972), which highlighted issues in existing legislation and advocated for a streamlined, self-regulatory approach. The core principle of the Robens report aimed to simplify laws and promote individual accountability (Mzyece, 2015). The HSWA requires employers to ensure the health, safety, and welfare of employees by providing a safe working environment, training, information, and supervision (Hughes and Ferrett, 2011; Griffith and Howarth, 2014). It also grants employees the right to refuse work if they believe it poses a serious health and safety risk (Lèfstedt, 2011). The Act establishes a framework for risk assessment and mitigation, particularly in industries like construction, emphasising the importance of proper training, equipment use, and maintaining a safe work environment to prevent accidents (Sirrs, 2016b; Guo et al., 2021).

2.2 Construction (Design and Management) Regulations 2015

The Construction (Design and Management) Regulations 2015 (CDM) is significant UK legislation stemming from a 1992 European Union directive, with its introduction in the UK in 1995 aimed at improving health and safety in construction by assigning responsibilities to stakeholders (Health and Safety Executive, 2023a). While contributing to overall advancements, concerns persist about the implementation of CDM regulations, an area with limited empirical research. CDM 2015 differs from CDM 2007 by not explicitly requiring clients to check the adequacy of resources for the principal designer (PD) or the principal contractor (PC), referring instead to the standardised prequalification questionnaire, PAS 91 (Patterson, Davies and Trebes, 2015).

A study by Mzyece (2015) investigates the practical implementation of the CDM regulations, revealing a significant lack of collaboration among duty holders, a key requirement in the regulations. The study identifies weak correlations between the discharge of duties and their perceived importance, with 50% of CDM Coordinator (CDM-C) duties and 25% of PC duties exhibiting misalignment between discharge and perceived importance. This suggests a lack of understanding regarding duty importance for improved health and safety management. Surprisingly, duties of the

PC and CDM-C show statistically significant relations to perceived difficulty, indicating that difficulty does not hinder their discharge. The study also raises concerns about the suitability of CDM-Cs, with designers expressing dissatisfaction with their input during planning and construction. In summary, the study highlights ongoing challenges in the collaboration, accountability, and compliance of duty holders within the CDM framework.

2.3 Reporting of Injuries, Diseases and Dangerous Occurrences Regulations 2013

RIDDOR, or the Reporting of Injuries, Diseases and Dangerous Occurrences Regulations 2013 in the UK, mandates the reporting and recording of work-related accidents, diseases, and dangerous events (Health and Safety Executive, 2013). It applies to employers, self-employed individuals, and those in control of work premises. In relation to the Health and Safety (England) Regulations 2013, reporting is required for incidents leading to death, serious injuries, work-related diseases, and dangerous occurrences like explosions or gas leaks. RIDDOR aims to enhance workplace safety by identifying risks and patterns through incident reporting, enabling the development of strategies to prevent future incidents and promote overall well-being for workers. However, limitations exist. Underreporting may occur due to fear, lack of awareness, or perceptions of insignificance, leading to inaccurate incident data. The complexity of regulations poses challenges, with employers struggling to understand reporting criteria, resulting in inconsistencies across organisations. Subjectivity in reporting decisions introduces variations among employers, impacting reporting consistency. RIDDOR focuses on post-incident reporting, lacking emphasis on preventive measures. Mental health issues are inadequately addressed. While a valuable tool, ongoing efforts are needed to enhance awareness, simplify reporting criteria, and foster a holistic approach to health and safety beyond mere reporting.

2.4 Control of Substances Hazardous to Health (COSHH) Regulations 2002

The Control of Substances Hazardous to Health (COSHH) Regulations 2002 is significant legislation in the UK aimed at safeguarding workers from the harmful effects of hazardous substances encountered in their work. Employers must assess risks and implement control measures to minimise exposure, covering storage, handling, and disposal. Despite being a crucial framework, COSHH faces challenges due to its broad scope, and potentially complex compliance for employers dealing with multiple substances and industries. Additionally, there may be delays in updating regulations to address emerging substances, and small and medium-sized enterprises (SMEs) may struggle with limited resources. Nevertheless, COSHH remains essential for managing and minimising workplace risks, with ongoing efforts to address limitations, enhance awareness, and update regulations for improved worker health and safety (Health and Safety Executive, 2001).

2.5 Other Regulations

In addition, several other regulations play a crucial role in ensuring health and safety in the construction industry. The Management of Health and Safety at Work Regulations 1999 offer detailed guidance on risk assessment and management, particularly essential in a sector where diverse hazards are prevalent. The Work at Height Regulations 2005 focus on the safety of workers in elevated positions, outlining specific requirements for equipment use, risk assessments, and training. The Construction (Head Protection) Regulations 1989 mandate the use of suitable and well-maintained head protection in situations posing a risk of injury. The Construction (Health, Safety and Welfare) Regulations 1996 concentrate on welfare aspects, addressing sanitary facilities, accommodation, and general welfare arrangements at construction sites. The Lifting Operations and Lifting Equipment Regulations 1998 (LOLER) govern the safe use of lifting equipment, emphasising proper planning, supervision, and execution of lifting operations. The Provision and Use of Work Equipment Regulations 1998 (PUWER) set standards for the safe use, maintenance, and suitability of work equipment, including machinery and tools. Lastly, electrical safety on construction sites is addressed by regulations such as the Electricity at Work Regulations 1989, ensuring comprehensive guidelines for maintaining a secure working environment.

Overall, legislation and regulations in the UK are essential for maintaining health and safety standards in various settings. These laws place legal obligations on employers to ensure the well-being of their employees and the general public. By complying with these regulations, employers can create a safe and healthy working environment, thereby reducing the risk of accidents, injuries, and illnesses. It is important for future professionals to be aware of this legislation and the regulations to ensure their own safety and that of those around them.

3. EXPLORING EMERGING TECHNOLOGIES FOR ENHANCING HEALTH AND SAFETY ACROSS THE CONSTRUCTION LIFECYCLE

3.1 Health and Safety in the Planning Stage

The planning phase is appropriately highlighted as a critical stage in which a thorough risk assessment is advocated (Willumsen et al., 2019; Lyu et al., 2020; Okudan, Budayan and Dikmen, 2021). For example, in the site selection phase, a study conducted by Kumar and Bansal (2016) explored the application of geographic information systems (GIS) for identifying a safe site, considering factors like proximity to emergency services and access to clean water. This may be perceived as an attempt to oversimplify intricate decision-making processes. As interest broadens to encompass various aspects, several studies (Rikalovic, Cosic and Lazarevic, 2014; Kumar and Bansal, 2016; Karakuş et al., 2020; Zambrano-Asanza, Quiros-Tortos and Franco, 2021; Balew et al., 2022) have introduced tools for site selection that comprehensively consider diverse factors, including environment, safety, security, sustainability, accessibility, and energy consumption. These tools aim to address the complexity of decision-making by assessing the impact on the local ecosystem and optimising the use or repurposing of existing infrastructure.

Light detection and ranging (LiDAR) technology plays a pivotal role in improving efficiency and safety in construction. By automating inspections, LiDAR reduces the need for manual, labor-intensive processes, as highlighted by Puri and Turkan (2020). This not only enhances overall efficiency but also mitigates the exposure of workers to potentially hazardous environments. Additionally, LiDAR enables remote monitoring of construction sites, reducing the necessity for physical presence in high-risk areas, as emphasised by Rao et al. (2022). This remote monitoring capability enhances safety by minimizing personnel exposure to dangerous conditions. In the context of health and safety, a research study led by Jeelani and Gheisari (2021) delves into the challenges associated with the use of unmanned aerial vehicles (UAVs) equipped with LiDAR sensors in construction. The study categorises the risks posed by UAVs into three groups. First, physical risks involve the potential for UAVs or their components to physically collide with workers or construction entities due to malfunctions, operational errors, or adverse environmental conditions. Second, attentional costs encompass distractions caused by visual or auditory stimuli from UAVs, along with cognitive distraction arising from the mere awareness of their presence. Lastly, the study explores the psychological impact of UAVs on workers, including effects such as acute stress, cognitive overload, and sensory overload. The integration of LiDAR into UAVs adds a technological dimension to these safety considerations, emphasising the importance of balancing technological advancements with the well-being of construction workers.

The use of AI in the planning stage of construction projects offers significant potential benefits in terms of analysing historical data to predict potential safety hazards and risks on construction sites (Pan and Zhang, 2021; Baduge et al., 2022; Regona et al., 2022). This proactive approach allows for better safety planning and the implementation of preventive measures. The use of AI in the planning stage offers significant potential benefits for health and safety, but it comes with challenges that need careful consideration. For example, the effectiveness of AI depends on the quality and unbiased nature of the data it is trained on. If historical data contains biases or inaccuracies, the AI system may produce unreliable safety predictions (Yaseen et al., 2020).

Studies might argue that while digital twins and BIM offer advancements in visualisation and data integration, their effectiveness in hazard identification during the planning phase could be overstated (Getuli et al., 2020; Alizadehsalehi et al., 2020; Swallow and Zulu, 2020). The complexity of real-world construction projects and the myriad of potential hazards may not always be fully captured or accurately represented in the digital realm. Furthermore, critics might express concerns about the learning curve associated with adopting and implementing these technologies. They may argue that the proficiency required to harness the full capabilities of digital twins and BIM might lead to challenges for professionals in the industry, particularly those who are not well versed in the intricacies of these advanced tools. While digital twins and BIM undeniably offer promising opportunities for innovation in the construction sector, it is essential to critically assess their practical implications and address any potential limitations to ensure that their adoption aligns with the industry's evolving needs and expectations. Using BIM in the planning stage of construction projects can contribute to health and safety in various ways, but it also presents certain considerations and challenges. BIM allows for the identification of safety risks and hazards during the planning stage by providing a detailed 3D model of the construction project. This helps in proactively addressing potential safety issues before they arise during the construction phase (Getuli et al., 2020; Alizadehsalehi et al., 2020; Swallow and Zulu, 2020). Also, the visual representation provided by BIM facilitates better safety planning. Stakeholders can assess the safety implications of different design options and make informed decisions to enhance on-site safety. The implementation of BIM in the planning stage may require training for construction professionals, introducing a learning curve. Some construction firms may face challenges in adopting and integrating BIM into their existing workflows.

VR technology offers significant benefits for enhancing health and safety in construction projects. It enables risk assessment and simulation, providing opportunities to identify hazards before they occur. VR facilitates immersive training experiences for workers, improving safety knowledge and readiness. Collaborative planning sessions in VR enhance communication and coordination among stakeholders. The technology also aids in site planning and design review, ensuring early identification of safety issues. VR simulations for emergency scenarios enhance preparedness, while remote inspections and accessibility assessments improve overall safety and inclusivity. Additionally, VR enables behavioral analysis, informing safety interventions.

3.2 Health and Safety in the Design Stage

The concept of "health and safety through design" emphasises the significance of incorporating safety considerations right from the outset when designing products, systems, or environments (Manuele, 2008). The focus is on proactively recognising and addressing potential safety risks during the design phase, with the goal of avoiding the necessity to deal with safety concerns as an afterthought or incorporating additional safety features later in the development process (Stangel and Witeczek, 2015).

Ensuring health and safety during the design phase in Construction 4.0 involves leveraging advanced technologies, fostering collaborative processes, and integrating data-driven decision-making (Smallwood and Allen, 2023). In this era of digital transformation, it's crucial to proactively identify risks, monitor situations in real time, and comply with ever-evolving regulatory standards (Newman et al., 2021). To achieve this, the industry is leveraging advanced tools such as BIM and VR to simulate and address potential hazards before construction even begins (Puri and Turkan, 2020). Moreover, collaboration among architects, engineers, and health and safety professionals through digital platforms enhances communication and ensures a comprehensive approach to safety measures (Stangel and Witeczek, 2015). This commitment extends to integrating user-friendly safety protocols directly into the design, empowering on-site workers and promoting adherence to the highest safety standards. The industry is dedicated to continuous monitoring, dynamically adapting safety protocols, and prioritizing the upskilling of the workforce. All of these efforts underline the industry's dedication to creating a secure construction environment in the context of Construction 4.0 (Newman et al., 2021).

In design, utilising AI algorithms to optimise efficiency, cost-effectiveness, and structural integrity shows promise in reducing accident risks (Pan and Zhang, 2021). Striking a balance is necessary, ensuring safety remains a key focus amidst various design goals. AI streamlines the design process through virtual prototyping, allowing proactive identification and correction of safety issues prior to construction (Yaseen et al., 2020). Yet, simulation accuracy relies on precise input data, emphasising the need for real-world conditions. AI also aids compliance with building codes, lowering non-compliance risks. Despite these benefits, potential interpretation errors or regulatory changes pose vulnerabilities, highlighting the need for human oversight to validate AI-driven code compliance. This integrated approach mitigates risks associated with interpretation errors and regulatory adjustments in the complex landscape of building codes and safety regulations (Baduge et al., 2022).

BIM fundamentally transforms construction by prioritising safety from the outset. Through detailed virtual models, hazards are identified early, minimising accidents and rework (Azhar, 2011). BIM not only aids in safety planning but also fosters collaborative decision-making, optimising safety measures and ensuring compliance (Kassem et al., 2017). This proactive approach extends to safety training, where virtual reality simulations better prepare workers for on-site execution (Li et al., 2015). Moreover, continuous monitoring with BIM enhances safety practices, contributing to the creation of safer construction sites (Hossain et al., 2023).

These advancements seamlessly integrate with smart heating, ventilation, and air conditioning systems, which leverage sensor technology to dynamically adjust ventilation rates based on occupancy patterns. Research has shown the effectiveness of such systems in reducing airborne contaminants while enhancing energy efficiency (Li et al., 2018). As buildings become smarter, occupancy monitoring systems play a pivotal role in optimising space utilisation and ensuring safety compliance (Liu et al., 2023). Real-time feedback systems further empower occupants and facility managers to proactively maintain indoor environmental conditions, utilising predictive analytics for preemptive interventions (Roskams and Haynes, 2019).

In parallel, AR technology emerges as a game-changer in safety training, offering immersive simulations for practicing emergency procedures (Yazdi, 2024). Meanwhile, the IoT revolutionises on-site safety with wearable sensors and location-tracking technologies, ensuring real-time monitoring of worker health and enabling swift response during emergencies (Chen, Lai and Lin 2020).

3.3 Health and Safety in the Construction Stage

The integration of Industry 4.0 technologies into the construction stage undoubtedly brings promising advancements in health and safety practices (Elghaish et al., 2021). The incorporation of digital twins, BIM, sensors, wearable devices, drones, and various other technologies presents an innovative approach to minimising risks on construction sites (Boje et al., 2020; Banerjee and Nayaka, 2022). However, a critical examination reveals both commendable aspects and potential challenges associated with these technological interventions. BIM, a pivotal element of Industry 4.0, holds the promise of enhancing health and safety during the construction phase through the visualisation of the entire construction project, helping identify potential safety risks and hazards (Sattineni, 2014). This proactive approach enables early mitigation strategies to be implemented, reducing the likelihood of accidents (Manuele, 2008). Also, BIM's clash detection capabilities can be applied to ensure that safety systems, such as emergency exits and fire suppression systems, do not conflict with other elements in the building. This helps in designing and constructing safer buildings (Ma and Wu, 2020; Chen, Lai and Lin, 2020; Banerjee and Nayaka, 2022; Hasannejad, Sardrud and Shirzadi Javid, 2023). Moreover, BIM promotes collaboration among different stakeholders, including health and safety professionals. Improved communication through a shared model allows for better coordination of safety measures and ensures that everyone involved is on the same platform (Ghaffarianhoseini et al., 2017).

The use of sensors in construction presents notable health and safety advantages, but it necessitates a thorough assessment of both effectiveness and potential challenges, as highlighted by Awolusi, Marks and Hallowell (2018). As per Rao et al. (2022), these sensors facilitate real-time monitoring of construction sites, identifying hazards and monitoring environmental conditions, equipment status, and worker activities. The integration of real-time monitoring data empowers informed decision-making in construction projects, enabling project managers to discern trends and identify potential risks. This capability facilitates proactive measures to address emerging issues promptly. When coupled with alarm systems, sensors play a pivotal role in providing swift responses during emergencies, such as increases in hazardous gas levels or equipment malfunctions (Rane, 2023). Moreover, these sensors extend their monitoring capabilities to construction equipment, mitigating the risk of failures and ensuring the safety of workers on site. As an illustrative example, Zhou and Ding (2017) have proposed an IoT-based safety barrier warning system for creating a safer underground construction site in China. However, it is crucial to note that the system's limitations include the reliance on small-scale trials with a short application period. This aspect raises concerns about the realistic verification of the system's capacity and accuracy, particularly in terms of optimising the positioning algorithm to enhance accuracy to the decimeter level on site. Furthermore, addressing the compression and storage of the substantial monitoring data is essential to ensure the stability of the system over an extended period. Additionally, there is a need for further research to explore the deep utilization and mining of the vast amount of big data generated by the proposed system during prolonged monitoring periods (He, He and Wen, 2023). This comprehensive approach will contribute to refining the system's effectiveness, addressing its limitations, and fostering continuous improvement in safety measures within the construction industry.

Drones are pivotal for enhancing construction safety through aerial inspections and safety assessments, reducing reliance on manual processes (Zhou and Ding, 2017). However, safety considerations such as collision risks and technical malfunctions highlight the need for strict regulations and proper training. A review conducted by Clarke and Moses (2014) identifies a considerable range of gaps and uncertainties that need to be addressed, such as limited battery life, necessitating contingency plans. In emergency response, drones equipped with sensors swiftly assess incidents, but limited payload capacity poses constraints. Seamless integration into emergency procedures requires close coordination with services, emphasising effective communication. Adherence to evolving drone regulations is paramount for legal and safe operations, necessitating construction companies to stay informed and adapt proactively.

Moreover, the incorporation of robotics, AR, and VR technologies for safety training is a commendable strategy (Li et al., 2018). Simulating real-life scenarios prepares workers for potential risks. However, the effectiveness of these technologies depends on the quality of simulation and the adaptability of workers to virtual training environments. As technology continues to shape safety practices in the construction industry, it is imperative to strike a balance between innovation and addressing concerns related to accuracy, privacy, and regulatory compliance (Clarke and Moses, 2014). This holistic approach ensures that technological advancements contribute positively to on-site well-being and safety standards, creating a synergistic and comprehensive safety framework in the construction sector.

3.4 Health and Safety in the Operation Stage

The integration of Construction 4.0 technologies into the operation stage undoubtedly brings promising advancements in health and safety practices (Newman et al., 2021; Smallwood and Allen, 2023). AI is instrumental in advancing health and safety measures during building operations (Pan and Zhang, 2021). AI-powered sensors enable real-time monitoring of parameters, while drones and robots equipped with AI automate safety inspections, minimising human exposure to risks (Baduge et al., 2022). Smart sensors integrated into traditional personal protective equipment (PPE) enhance worker safety by monitoring vital signs and environmental conditions (Elghaish et al., 2021). AI-powered cognitive computing systems analyse data to provide recommendations for optimizing safety protocols and responding to emergencies. Overall, AI contributes to heightened situational awareness, efficient inspections, improved PPE, and informed decision-making, fostering a safer operational environment in buildings (Pan and Zhang, 2021).

BIM is traditionally associated with the design and construction phases of a project. The focus on health and safety is often stronger during these stages, while its application and integration during the operation and maintenance phase may receive less attention (Hoeft and Trask, 2022). BIM has evolved beyond its traditional role by incorporating real-time insights from IoT devices and sensor technologies. These sensors monitor crucial aspects such as occupancy levels, air quality, temperature, and structural conditions in real time. The wealth of data collected by these sensors serves as a foundation for informed decision-making within the BIM model (Rao et al., 2022).

VR training and AR technology stand out as transformative tools reshaping health and safety practices during the operational stage of buildings. VR simulations, as evidenced by studies conducted by Li et al. (2018), Getuli et al. (2020) and Yazid (2024), excel in providing immersive emergency scenario training, refining situational awareness, and honing decision-making skills crucial for effective response strategies. Similarly, AR technology, showcased by research from Li et al. (2018) and Chen, Lai and Lin (2020), empowers technicians with hands-free access to real-time data and repair instructions, amplifying competency while reducing errors during maintenance tasks. Moreover, the integrated deployment of IoT sensors into building infrastructure offers a proactive approach to safety management. He, He and Wen (2023) illustrate how IoT sensors enable real-time monitoring of environmental conditions, facilitating early intervention to avert potential hazards such as gas or water leaks. Leveraging historical data for preemptive detection of safety issues, as championed by Ghosh, Edwards, and Hosseini (2021), the predictive maintenance capabilities of IoT ensure timely interventions to mitigate risks and maintain a safe operational environment.

Smart lighting systems, as illuminated by studies conducted by Chew et al. (2017) and Xu et al. (2022), offer multifaceted benefits ranging from enhanced visibility to dynamic control based on occupancy and natural light levels. These systems not only enhance occupant comfort but also ensure safe egress during emergencies or power outages, significantly contributing to health and safety measures during the operational phase of buildings. Furthermore, the deployment of robotic technologies equipped with sensors and cameras, as demonstrated by research from Javaid et al. (2022), significantly reduces human exposure to hazardous conditions in construction, maintenance, and industrial settings during the operational lifespan of buildings. These advancements not only improve safety, but also enhance efficiency in executing tasks, ensuring a secure working environment for building occupants and maintenance personnel alike.

In parallel, wearable health monitoring devices emerge as pivotal tools for continuous tracking of vital signs and biometric data, supporting health and safety initiatives during the operational stage of buildings. Studies by Abuwarda et al. (2022) and Rane, Choudhary, and Rane (2023) underscore the potential of wearables in enhancing occupant safety by enabling early detection of medical emergencies and facilitating swift assistance during crises, thereby contributing to the overall well-being of building occupants.

3.5 Health and Safety in the Deconstruction Stage

Implementing Industry 4.0 technology for health and safety in the demolition phase involves deploying advanced technologies to monitor, manage, and improve safety conditions (Winston, 2010). One of the key advancements is providing workers with wearable devices embedded with biometric sensors to monitor their vital signs and overall health (Abuwarda et al., 2022). These wearables can also include GPS tracking to monitor the location of workers in real time, ensuring their safety in hazardous zones (Awolusi, Marks and Hallowell, 2018).

Another important aspect of Industry 4.0 in deconstruction or demolition is the use of smart helmets or glasses with AR capabilities to enhance workers' situational awareness (Rane, Choudhary and Rane, 2023). AR overlays can provide real-time information, safety alerts, and instructions, improving communication and reducing the likelihood of accidents (Li et al., 2018).

Furthermore, drones equipped with cameras and sensors can be used to conduct aerial surveillance of the demolition site (Javaid et al., 2022). Drones can identify potential safety hazards, monitor the progress of demolition, and provide real-time data for decision-making (Elghaish et al., 2021). In addition, robotic systems can be employed for precision-controlled demolition tasks, reducing the need for human exposure to dangerous areas (Nwaogu et al., 2023). These robots can be equipped with advanced sensors and cameras to navigate and perform tasks more safely and efficiently (Namian and Turkan, 2023).

The creation of detailed 3D models by BIM is acknowledged for aiding comprehensive deconstruction and hazard identification (Hoeft and Trask, 2022). However, the effectiveness of this visual roadmap in enabling proactive risk mitigation may depend on the accuracy of the input data and the proficiency of users (Rane, 2023). The claim that BIM enhances risk assessments through data incorporation is valid, but a more nuanced exploration of the reliability and precision of hazard identification is needed. The accuracy of risk assessments may vary based on the completeness and accuracy of the data, raising questions about the potential for oversight in certain scenarios (Getuli et al., 2016). Also, efficient waste management resulting from the demolition process with BIM is highlighted as a strength (Banerjee and Nayaka, 2022).

4. CHALLENGES AND OPPORTUNITIES

4.1 Barriers to Implementing Health and Safety Lifecycle in Construction Industry 4.0

The implementation of 4.0 technology in the construction industry holds significant potential for enhancing health and safety lifecycle practices, but a critical examination reveals several challenges and barriers that impede its effective integration (Marefat, Toosi and Mahmoudi Hasankhanlo, 2019). The foremost obstacle is the substantial cost associated with implementation (Rane, 2023). Construction 4.0 technologies demand a considerable upfront investment for acquiring and installing equipment, coupled with the necessity of training workers to proficiently operate and maintain these systems. This financial burden poses a challenge for construction companies, as justifying such an investment becomes complicated.

Additionally, the complexity of integrating Construction 4.0 technologies into existing construction processes exacerbates the implementation challenges (Newman et al., 2021). Compatibility issues with current systems and workflows contribute to the complexity of this task, further hindering a seamless adoption of automated solutions (Smallwood and Allen, 2023). A key difficulty lies in the demand for a higher level of technical expertise among workers. Adequate training is imperative for the operation and maintenance of these systems, yet a shortage of skilled personnel amplifies the hurdles faced during implementation.

Integrating new health and safety protocols into existing construction processes and systems presents another challenge (Musarat et al., 2022). Legacy systems may not easily accommodate the integration of digital technologies, leading to resistance and slow adoption. The construction industry is subject to various regulations and standards related to health and safety. Integrating new protocols may require ensuring compliance with these regulations, which can be a complex process (Tender et al., 2022). Navigating the regulatory landscape and adapting existing processes to meet new standards can pose a significant challenge (Patterson, Davies and Trebes, 2015).

The construction industry faces cultural challenges rooted in its historical practices, hindering the adoption of new health and safety measures associated with Industry 4.0 (Tender et al., 2022). Resistance to change from both workers and management poses a significant obstacle. Additionally, the industry's intricate regulatory environment complicates the adaptation of existing regulations to accommodate innovative technologies, potentially slowing down implementation (Swallow and Zulu, 2020).

Successful integration of health and safety practices in Industry 4.0 requires a crucial cultural shift (Mittal et al., 2018). Leaders must commit to prioritizing health and safety as integral components of the construction process (Elghaish et al., 2021). Collaboration among employers, workers, regulators, and technology providers is essential for fostering this change. Leadership commitment plays a pivotal role in driving cultural change, and technology providers must align solutions with industry practices. This collaborative approach ensures that technology becomes an enabler, not a disruptor, promoting a safer and more efficient construction environment. Moreover, reliability and technical challenges, including system failures and connectivity issues, can compromise worker safety (Gao, Gonzalez and Yiu, 2019). Skill shortages are another significant barrier, as Industry 4.0 demands a workforce with advanced technical skills. This shortage may limit the successful implementation of health and safety protocols (Delgado et al., 2019).

Addressing security and privacy concerns emerges as a significant hurdle in adopting Construction 4.0 technologies (Atkinson et al., 2022). For example, the use of sensors, cameras, and data-collecting devices in automation raises the stakes for safeguarding sensitive information (Gao, Gonzalez and Yiu, 2019). The potential cybersecurity risks associated with automation systems create a reluctance to fully embrace these technologies. Furthermore, the dynamic nature of construction sites, characterised by changes in weather, uneven terrain, and unexpected obstacles, complicates the adaptation of automation systems to handle such variations (Mittal et al., 2018; Delgado et al., 2019).

The integration of Industry 4.0 technologies into the construction industry poses various challenges related to regulatory compliance, particularly in the realm of health and safety (Soltani, Maxwell and Rashidi, 2023). It is essential to adhere to established occupational safety and health regulations, ensuring that technologies such as wearables, drones, and IoT devices meet safety standards (Newman et al., 2021). These standards encompass factors like permissible exposure limits, fire protection, safety protocols, and mandatory training for workers (Abuwarda et al., 2022).

Given that health and safety technologies in Construction 4.0 involve handling sensitive data, it becomes imperative to comply with data privacy regulations, such as Europe's General Data Protection Regulation (GDPR) (Mastrolembo Ventura et al., 2023). Robust cybersecurity measures are also necessary to protect against potential data breaches (Banerjee and Nayaka, 2022). Moreover, compliance must extend to respecting worker rights and ethical considerations, emphasising transparency in data collection practices and obtaining informed consent from workers (Li et al., 2022).

Meeting relevant certifications and standards is crucial for Industry 4.0 technologies in construction, requiring adherence to criteria set by occupational safety organisations and industry-specific standards (Teisserenc and Sepasgozar, 2021). Environmental considerations tied to health and safety are subject to regulation, emphasising adherence to environment, health, and safety regulations to ensure sustainable construction practices (Tender et al., 2022; Musarat et al., 2022; Hoeft and Trask, 2022).

Specific training and certification requirements for workers interacting with Construction 4.0 technologies are often mandated by regulatory compliance (Gao, Gonzalez and Yiu, 2019). Timely and accurate reporting of safety incidents is essential, and Industry 4.0 technologies must facilitate compliance with regulatory requirements (Health and Safety Executive, 2013). Understanding legal liability in the context of Construction 4.0 adoption necessitates a review of insurance policies to ensure coverage for associated risks (Ivanov, Dolgui and Sokolov, 2019).

As Construction 4.0 continues to evolve, proactive engagement with regulatory bodies is crucial for influencing the development of effective and forward-looking regulatory frameworks (Guo et al., 2021). A collaborative approach involving construction companies, technology providers, regulators, and industry associations is essential for navigating regulatory considerations and ensuring a safe and legally compliant construction environment (Sirrs, 2016a). It is also important to stay informed about evolving regulations and integrate compliance measures into the design and implementation of Industry 4.0 technologies (Clarke and Moses, 2014).

4.2 Opportunities for Improvement and Innovation

Implementing health and safety in the construction industry, especially within the framework of Industry 4.0, presents numerous opportunities for improving safety, efficiency, and overall project outcomes (Badri, Boudreau-Trudel and Souissi, 2018).

One of the opportunities of implementing health and safety in Industry 4.0 is with risk assessment and prediction (Moktadir et al., 2018). With the use of data analytics and AI to analyse historical safety data, identify patterns, and predict potential risks, proactive measures can be taken to prevent accidents (Pan and Zhang, 2021). Furthermore, the implementation of predictive modeling can foresee potential safety hazards based on real-time data from construction sites, helping with timely risk mitigation (Baduge et al., 2022). The implementation of automated systems to monitor and ensure compliance with health and safety regulations reduces the risk of regulatory violations (Puri and Turkan, 2020; Pan and Zhang, 2021).

Another advantage of integrating health and safety measures in Construction 4.0 involves leveraging big data for decision-making (Elghaish et al., 2021). Through the application of big data analytics to derive insights from large datasets related to safety incidents, near-misses, and other relevant information, strategic decision-making for enhanced safety measures can be implemented. Additionally, comparing safety performance across projects and organisations helps identify best practices and areas for improvement (Li et al., 2018; Newman et al., 2021).

The automation and robotics aspect of implementing health and safety in Industry 4.0 offers further opportunities (Javaid et al., 2021) – for example, implementing automated systems for safety inspections, or utilising drones and robotic devices to assess and report on safety compliance (Javaid et al., 2022). In terms of VR and AR, developing simulations for safety training allows workers to experience and practice handling potential hazards in a controlled virtual environment (Chen, Lai and Lin, 2020). Also, using AR for remote inspections enables experts to assess construction sites without physically being present, reducing exposure to hazardous conditions (Li et al., 2018).

Real-time monitoring and communication present yet another opportunity (Rao et al., 2022). Utilising IoT devices like wearable sensors and smart helmets to oversee workers is one way to implement this approach (Abuwarda et al., 2022). This enables immediate response in case of emergencies. Moreover, utilizing advanced communication platforms ensures quick dissemination of safety information, alerts, and emergency notifications (Awolusi, Marks and Hallowell, 2018).

By embracing these opportunities within the health and safety lifecycle in Construction 4.0, organisations can significantly enhance safety outcomes, reduce accidents, and improve overall project efficiency (Newman et al., 2021; Elghaish et al., 2021).

5. CONCLUSION, FUTURE TRENDS, AND RECOMMENDATIONS

In conclusion, the construction industry grapples with persistent challenges in health and safety, resulting in frequent injuries and loss of life despite regulatory efforts. Emerging technologies such as BIM, AR, VR, AI, robotics, and drones offer promising solutions by enabling real-time communication, monitoring, and visualisation. These technologies facilitate proactive safety measures, potentially reducing accidents, increasing productivity, and enhancing the overall well-being of construction industry workers.

Despite the potential benefits, the existing literature reveals a significant gap in understanding the state-of-the-art of current technologies and techniques related to health and safety in construction, especially within the framework of Construction 4.0. The chapter emphasises the need for a comprehensive overview, encompassing both currently employed digital technologies and those in rapid development, such as the digital twin for health and safety.

An innovative approach introduced in the chapter involves integrating health and safety considerations throughout the entire construction lifecycle, including conventional/traditional phases and Construction 4.0 lifecycle stages. This holistic perspective spans from the design phase through construction, operation, maintenance, and even the end-of-life stage, incorporating health and safety measures intrinsically.

The chapter further explores various technologies and techniques, including BIM, VR, AI, and drones, among others. By conducting a thorough state-of-the-art analysis, identifying knowledge gaps, and defining their respective scopes, the study aims to pave a clear path for future research and development in the realm of health and safety in construction. This comprehensive understanding is crucial for advancing the industry toward a safer, more efficient, and technologically integrated future.

REFERENCES

Abuwarda, Z., Mostafa, K., Oetomo, A., Hegazy, T. and Morita, P. (2022) 'Wearable devices: Cross benefits from healthcare to construction'. *Automation in Construction*, 142, p. 104501.

Ahn, S., Crouch, L., Kim, T.W. and Rameezdeen, R. (2020) 'Comparison of worker safety risks between onsite and offsite construction methods: A site management perspective'. *Journal of construction engineering and management*, 146(9), p. 05020010.

Aksorn, T. and Hadikusumo, B.H. (2008) 'Critical success factors influencing safety program performance in Thai construction projects', *Safety Science*, 46(4), pp. 709–727.

Alizadehsalehi, S., Yitmen, I., Celik, T. and Arditi, D. (2020) 'The effectiveness of an integrated BIM/UAV model in managing safety on construction sites'. *International Journal of Occupational Safety and Ergonomics*, 26(4), pp. 829–844.

Atkinson, E., Spillane, J., Bradley, J. and Brooks, T. (2022) 'Challenges in the adoption of mobile information communication technology (M-ICT) in the construction phase of infrastructure projects in the UK'. *International Journal of Building Pathology and Adaptation*, 40(3), pp. 327–344.

Awolusi, I., Marks, E. and Hallowell, M. (2018) 'Wearable technology for personalized construction safety monitoring and trending: Review of applicable devices', *Automation in Construction*, 85, pp. 96–106.

Azhar, S. (2011) 'Building information modeling (BIM): Trends, benefits, risks, and challenges for the AEC industry'. *Leadership and Management in Engineering*, 11(3), pp. 241–252.

Badri, A., Boudreau-Trudel, B. and Souissi, A.S. (2018) 'Occupational health and safety in the industry 4.0 era: A cause for major concern?', *Safety Science*, 109, pp. 403–411.

Baduge, S.K., Thilakarathna, S., Perera, J.S., Arashpour, M., Sharafi, P., Teodosio, B., Shringi, A. and Mendis, P. (2022) 'Artificial intelligence and smart vision for building and construction 4.0: Machine and deep learning methods and applications'. *Automation in Construction*, 141, p. 104440.

Balew, A., Alemu, M., Leul, Y. and Feye, T. (2022) 'Suitable landfill site selection using GIS-based multi-criteria decision analysis and evaluation in Robe town, Ethiopia'. *GeoJournal*, 87(2), pp. 895–920.

Banerjee, A. and Nayaka, R.R. (2022) 'A comprehensive overview on BIM-integrated cyber physical system architectures and practices in the architecture, engineering and construction industry', *Construction Innovation*, 22(4), pp. 727–748.

Boje, C., Guerriero, A., Kubicki, S. and Rezgui, Y. (2020) 'Towards a semantic Construction Digital Twin: Directions for future research'. *Automation in Construction*, 114, p. 103179.

Chen, Y., Lai, Y. and Lin, Y. (2020) 'BIM-based augmented reality inspection and maintenance of fire safety equipment', *Automation in Construction*, 110, p. 103041.

Chew, I., Karunatilaka, D., Tan, C.P. and Kalavally, V. (2017) 'Smart lighting: The way forward? Reviewing the past to shape the future', *Energy and Buildings*, 149, pp. 180–191.

Clarke, R. and Moses, L.B. (2014) 'The regulation of civilian drones' impacts on public safety', *Computer Law & Security Review*, 30(3), pp. 263–285.

Daniel, G. (2022) *More than 2 million construction working days lost a year to ill health, says HSE*. Available at: https://www.building.co.uk/news/more-than-2-million-construction-working-days-lost-a-year-to-ill-health-says-hse/5120644.article#:~:text=all%2Dindustry%20rate.-,There%20were%2059%2C000%20cases%20of%20non%2Dfatal%20work%2Drelated%20injury,across%20all%20industries%20(1.6%25) (accessed: 7 December 2023).

Dekker, S.W. (2014) 'The bureaucratization of safety', *Safety Science*, 70, pp. 348–357.

Delgado, J.M.D., Oyedele, L., Ajayi, A., Akanbi, L., Akinade, O., Bilal, M. and Owolabi, H. (2019) 'Robotics and automated systems in construction: Understanding industry-specific challenges for adoption', *Journal of Building Engineering*, 26, p. 100868.

Dou, Y. et al. (2023) 'Tracking the research on ten emerging digital technologies in the AECO industry', *Journal of Construction Engineering and Management*, 149(3), 03123003.

Elghaish, F., Matarneh, S., Talebi, S., Kagioglou, M., Hosseini, M.R. and Abrishami, S. (2021) 'Toward digitalization in the construction industry with immersive and drones technologies: A critical literature review', *Smart and Sustainable Built Environment*, 10(3), pp. 345–363.

Fang, Y., Cho, Y.K. and Chen, J. (2016) 'A framework for real-time pro-active safety assistance for mobile crane lifting operations', *Automation in Construction*, 72, pp. 367–379.

Gao, Y., Gonzalez, V.A. and Yiu, T.W. (2019) 'The effectiveness of traditional tools and computer-aided technologies for health and safety training in the construction sector: A systematic review', *Computers & Education*, 138, pp. 101–115.

Getuli, V., Ventura, S.M., Capone, P. and Ciribini, A.L. (2016) 'A BIM-based construction supply chain framework for monitoring progress and coordination of site activities', *Procedia Engineering*, 164, pp. 542–549.

Getuli, V., Ventura, S.M., Capone, P. and Ciribini, A.L. (2017) 'BIM-based code checking for construction health and safety', *Procedia Engineering*, 196, pp. 454–461.

Getuli, V., Capone, P., Bruttini, A. and Isaac, S. (2020) 'BIM-based immersive Virtual Reality for construction workspace planning: A safety-oriented approach', *Automation in Construction*, 114, p. 103160.

Ghaffarianhoseini, A., Tookey, J., Ghaffarianhoseini, A., Naismith, N., Azhar, S., Efimova, O. and Raahemifar, K. (2017) 'Building Information Modelling (BIM) uptake: Clear benefits, understanding its implementation, risks and challenges', *Renewable and Sustainable Energy Reviews*, 75, pp. 1046–1053.

Ghosh, A., Edwards, D.J. and Hosseini, M.R. (2021) 'Patterns and trends in Internet of Things (IoT) research: Future applications in the construction industry', *Engineering, Construction and Architectural Management*, 28(2), pp. 457–481.

Great Britain (2023) *Health and Safety at Work etc. Act 1974*. Available at: https://www.legislation.gov.uk/ukpga/1974/37/contents (accessed: 18 November 2023).

Griffith, A. and Howarth, T. (2014) *Construction health and safety management*. Routledge.

Guo, B.H., Weston, R., Jianphinitnan, P., Liu, W., Scheepbouwer, E., van der Walt, D. and Goh, Y.M. (2021) 'A regulatory perspective on safety in design practices in New Zealand', *Safety Science*, 141, p. 105352.

Hasannejad, A., Sardrud, J.M. and Shirzadi Javid, A.A. (2023) 'BIM-based clash detection improvement automatically', *International Journal of Construction Management*, 23(14), pp. 2431–2437.

He, Y., He, J. and Wen, N. (2023) 'The challenges of IoT-based applications in high-risk environments, health and safety industries in the Industry 4.0 era using decision-making approach', *Journal of Innovation & Knowledge*, 8(2), 100347.

Health and Safety Executive (2001) 'Control of Substances Hazardous to Health (COSHH) Regulations'.

Health and Safety Executive (2013) *Reporting accidents and incidents at work*. Health and Safety Executive.

Health and Safety Executive (2023a) *Construction (Design and Management) Regulations – A timeline*. Available at: https://www.hse.gov.uk/construction/cdm/2015/timeline.htm (accessed: 9 December 2023).

Health and Safety Executive (2023b) *Costs to Great Britain of workplace injuries and new cases of work-related Ill Health – 2021/22*. Available at: https://www.hse.gov.uk/statistics/cost.htm (accessed: 7 December 2023).

Hoeft, M. and Trask, C. (2022) 'Safety built right in: Exploring the occupational health and safety potential of BIM-based platforms throughout the building lifecycle', *Sustainability*, 14(10), 6104.

Hughes, P. and Ferrett, E. (2011) *Introduction to health and safety at work*. Routledge.

Ivanov, D., Dolgui, A. and Sokolov, B. (2019) 'The impact of digital technology and Industry 4.0 on the ripple effect and supply chain risk analytics', *International Journal of Production Research*, 57(3), pp. 829–846.

Javaid, M., Haleem, A., Singh, R.P. and Suman, R. (2021) 'Substantial capabilities of robotics in enhancing industry 4.0 implementation', *Cognitive Robotics*, 1, pp. 58–75.

Javaid, M., Khan, I.H., Singh, R.P., Rab, S. and Suman, R. (2022) 'Exploring contributions of drones towards Industry 4.0', *Industrial Robot: The International Journal of Robotics Research and Application*, 49(3), pp. 476–490.

Jeelani, I. and Gheisari, M. (2021) 'Safety challenges of UAV integration in construction: Conceptual analysis and future research roadmap', *Safety Science*, 144, 105473.

Karakuş, C.B., Demiroğlu, D., Çoban, A. and Ulutaş, A. (2020) 'Evaluation of GIS-based multi-criteria decision-making methods for sanitary landfill site selection: the case of Sivas city, Turkey', *Journal of Material Cycles and Waste Management*, pp. 1–19.

Kumar, S. and Bansal, V.K. (2016) 'A GIS-based methodology for safe site selection of a building in a hilly region', *Frontiers of Architectural Research*, 5(1), pp. 39–51. Available at: https://doi.org/10.1016/j.foar.2016.01.001

Lèfstedt, R. (2011) *Reclaiming health and safety for all: An independent review of health and safety legislation.* https://assets.publishing.service.gov.uk/government/uploads/system/uploads/attachment_data/file/66790/lofstedt-report.pdf

Li, H., Lu, M., Chan, G. and Skitmore, M. (2015) 'Proactive training system for safe and efficient precast installation', *Automation in Construction*, 49, pp. 163–174.

Li, X., Yi, W., Chi, H.L., Wang, X. and Chan, A.P. (2018) 'A critical review of virtual and augmented reality (VR/AR) applications in construction safety', *Automation in Construction*, 86, pp. 150–162.

Li, Y., Soomro, M.A., Khan, A.N., Han, Y. and Xue, R. (2022) 'Impact of ethical leadership on employee turnover intentions in the construction industry', *Journal of Construction Engineering and Management*, 148(7), 04022054.

Liu, Z., Zhang, X., Sun, Y. and Zhou, Y. (2023) 'Advanced controls on energy reliability, flexibility and occupant-centric control for smart and energy-efficient buildings', *Energy and Buildings*, 297, 113436.

Lyu, H.M., Sun, W.J., Shen, S.L. and Zhou, A.N. (2020) 'Risk assessment using a new consulting process in fuzzy AHP', *Journal of Construction Engineering and Management*, 146(3), 04019112.

Ma, G. and Wu, Z. (2020) 'BIM-based building fire emergency management: Combining building users' behavior decisions', *Automation in Construction*, 109, 102975.

Manuele, F.A. (2008) 'Prevention through design (PtD): History and future', *Journal of Safety Research*, 39(2), pp. 127–130.

Maqsoom, A., Musarat, M.A., Mubbasit, H., Alaloul, W.S., Ashraf, H., Rabbani, M.B.A. and Shaheen, I. (2023) 'Extrinsic workforce diversity factors: An impact of employee characteristics on productivity', *Ain Shams Engineering Journal*, 14(10), 102170.

Marefat, A., Toosi, H. and Mahmoudi Hasankhanlo, R. (2019) 'A BIM approach for construction safety: Applications, barriers and solutions', *Engineering, Construction and Architectural Management*, 26(9), pp. 1855–1877.

Mastrolembo Ventura, S. et al. (2023) 'Enhancing safety on construction sites: A UWB-based proximity warning system ensuring gdpr compliance to prevent collision hazards', *Sensors*, 23(24), 9770.

Mittal, S., Khan, M.A., Romero, D. and Wuest, T. (2018) 'A critical review of smart manufacturing & Industry 4.0 maturity models: Implications for small and medium-sized enterprises (SMEs)', *Journal of Manufacturing Systems*, 49, pp. 194–214.

Moktadir, M.A., Ali, S.M., Kusi-Sarpong, S. and Shaikh, M.A.A. (2018) 'Assessing challenges for implementing Industry 4.0: Implications for process safety and environmental protection', *Process Safety and Environmental Protection*, 117, pp. 730–741.

Musarat, M.A., Alaloul, W.S., Irfan, M., Sreenivasan, P. and Rabbani, M.B.A. (2022) 'Health and safety improvement through Industrial Revolution 4.0: Malaysian construction industry case', *Sustainability*, 15(1), p. 201.

Muzafar, M. (2021) 'Building information modelling to mitigate the health and safety risks associated with the construction industry: A review', *International Journal of Occupational Safety and Ergonomics*, 27(4), pp. 1087–1095.

Mzyece, D. (2015) 'An investigation into the implementation of the Construction (Design and Management) regulations in the construction industry'.

Namian, M. and Turkan, Y. (2023) 'Safety management of drones in construction'. In *Handbook of Construction Safety, Health and Well-being in the Industry 4.0 Era* (pp. 52–64). Routledge.

Nenonen, S., Kivistö-Rahnasto, J. and Vasara, J. (2015) 'Safety considerations during different stages of a project life cycle in the manufacturing industry', *Human Factors and Ergonomics in Manufacturing & Service Industries*, 25(1), pp. 12–27.

Newman, C. et al. (2021) 'Industry 4.0 deployment in the construction industry: A bibliometric literature review and UK-based case study', *Smart and Sustainable Built Environment*, 10(4), pp. 557–580.

Nwaogu, J.M. et al. (2023) 'Application of drones in the architecture, engineering, and construction (AEC) industry', *Automation in Construction*, 150, 104827.

Okudan, O., Budayan, C. and Dikmen, I. (2021) 'A knowledge-based risk management tool for construction projects using case-based reasoning', *Expert Systems with Applications*, 173, 114776.

Oloke, D. (2020) *Refurbishment projects: Health and safety management.* ICE Publishing.

Pan, Y. and Zhang, L. (2021) 'Roles of artificial intelligence in construction engineering and management: A critical review and future trends', *Automation in Construction*, 122, 103517.

Patterson, R., Davies, M. and Trebes, B. (2015) 'Briefing: Construction (Design and Management) Regulations 2015 and NEC contracts', *Proceedings of the Institution of Civil Engineers-Management, Procurement and Law*, 168(3), pp. 115–120.

Puri, N. and Turkan, Y. (2020) 'Bridge construction progress monitoring using LiDAR and 4D design models', *Automation in Construction*, 109, 102961.

Rane, N. (2023) 'Integrating Building Information Modelling (BIM) and Artificial Intelligence (AI) for smart construction schedule, cost, quality, and safety management: challenges and opportunities', *Cost, Quality, and Safety Management: Challenges and Opportunities*, September 16.

Rane, N., Choudhary, S. and Rane, J. (2023) 'Leading-edge wearable technologies in enhancing personalized safety on construction sites: A review'. https://papers.ssrn.com/sol3/papers.cfm?abstract_id=4641480

Rao, A.S., Radanovic, M., Liu, Y., Hu, S., Fang, Y., Khoshelham, K., Palaniswami, M. and Ngo, T. (2022) 'Real-time monitoring of construction sites: Sensors, methods, and applications', *Automation in Construction*, 136, 104099.

Regona, M., Yigitcanlar, T., Xia, B. and Li, R.Y.M. (2022) 'Opportunities and adoption challenges of AI in the construction industry: A PRISMA review', *Journal of Open Innovation: Technology, Market, and Complexity*, 8(1), p. 45.

Rikalovic, A., Cosic, I. and Lazarevic, D. (2014) 'GIS based multi-criteria analysis for industrial site selection', *Procedia Engineering*, 69, pp. 1054–1063.

Robens, A. (1972) *Safety and health at work: Report of the Committee, 1970–72*. HM Stationery Office.

Roskams, M. and Haynes, B. (2019) 'Predictive analytics in facilities management: A pilot study for predicting environmental comfort using wireless sensors', *Journal of Facilities Management*, 17(4), pp. 356–370.

Sanni-Anibire, M.O., Mahmoud, A.S., Hassanain, M.A. and Salami, B.A. (2020) 'A risk assessment approach for enhancing construction safety performance', *Safety Science*, 121, pp. 15–29.

Sattineni, A. (2014) *A decision support framework for site safety monitoring using RFID and BIM*. University of Salford.

Shabani, T., Jerie, S. and Shabani, T. (2023) 'The impact of occupational safety and health programs on employee productivity and organisational performance in Zimbabwe', *Safety in Extreme Environments*, pp. 1–12.

Sirrs, C. (2016a) 'Accidents and apathy: The construction of the "Robens Philosophy" of occupational safety and health regulation in Britain, 1961–1974', *Social History of Medicine*, 29(1), pp. 66–88.

Sirrs, C. (2016b) *Health and Safety in the British Regulatory State, 1961–2001: The HSC, HSE and the Management of Occupational Risk.*

Smallwood, J. and Allen, C. (2023) 'Practitioners' perceptions of the potential impact of Industry 4.0 on construction health and safety', *Journal of Engineering, Design and Technology*, 21(2), pp. 486–501.

Soltani, S., Maxwell, D. and Rashidi, A. (2023) 'The state of Industry 4.0 in the Australian construction industry: An examination of industry and academic point of view', *Buildings*, 13(9), pp. 23–24.

Sorensen, G., Dennerlein, J.T., Peters, S.E., Sabbath, E.L., Kelly, E.L. and Wagner, G.R. (2021) 'The future of research on work, safety, health and wellbeing: A guiding conceptual framework', *Social Science & Medicine*, 269, 113593.

Stangel, M. and Witeczek, A. (2015) 'Design thinking and role-playing in education on brownfields regeneration. Experiences from Polish-Czech cooperation', *Architecture, Civil Engineering, Environment*, (4), pp. 19–28.

Swallow, M. and Zulu, S. (2020) 'Students' awareness and perception of the value of BIM and 4D for site health and safety management', *Journal of Engineering, Design and Technology*, 18(2), pp. 414–430.

Teisserenc, B. and Sepasgozar, S. (2021) 'Adoption of blockchain technology through digital twins in the construction industry 4.0: A PESTELS approach', *Buildings*, 11(12), 670.

Tender, M. et al. (2022) 'Emerging technologies for health, safety and well-being in construction industry', *Industry 4.0 for the Built Environment: Methodologies, Technologies and Skills*, pp. 369–390.

Willumsen, P., Oehmen, J., Stingl, V. and Geraldi, J. (2019) 'Value creation through project risk management', *International Journal of Project Management*, 37(5), pp. 731–749.

Winston, N. (2010) 'Regeneration for sustainable communities? Barriers to implementing sustainable housing in urban areas', *Sustainable Development*, 18(6), pp. 319–330. Available at: https://doi.org/10.1002/sd.399

Xu, M., Nie, X., Li, H., Cheng, J.C. and Mei, Z. (2022) Smart construction sites: A promising approach to improving on-site HSE management performance. *Journal of Building Engineering*, 49, 104007.

Yaseen, Z.M. et al. (2020) 'Prediction of risk delay in construction projects using a hybrid artificial intelligence model', *Sustainability*, 12(4), 1514.

Yazdi, M. (2024) 'Augmented reality (AR) and virtual reality (VR) in maintenance training'. In *Advances in Computational Mathematics for Industrial System Reliability and Maintainability* (pp. 169–183). Cham: Springer Nature Switzerland.

Zambrano-Asanza, S., Quiros-Tortos, J. and Franco, J.F. (2021) 'Optimal site selection for photovoltaic power plants using a GIS-based multi-criteria decision making and spatial overlay with electric load', *Renewable and Sustainable Energy Reviews*, 143, 110853.

Zhou, C. and Ding, L.Y. (2017) 'Safety barrier warning system for underground construction sites using Internet-of-Things technologies', *Automation in Construction*, 83, pp. 372–389.

Zhou, W., Whyte, J. and Sacks, R. (2012) 'Construction safety and digital design: A review', *Automation in Construction*, 22, pp. 102–111.

16. Stakeholder integration and open innovation in sustainable construction projects

Shaima AlHarmoodi and Mohammed Fadhil Dulaimi

1. INTRODUCTION

With the increasing global focus on sustainability and competitiveness, the construction industry faces growing pressure to innovate. Large, multidisciplinary construction projects require the effective integration of diverse stakeholders to ensure successful innovation outcomes. However, there is a lack of empirical research exploring the relationship between stakeholder integration and open innovation in the context of sustainable construction projects.

This chapter aims to examine how stakeholder integration influences open innovation in sustainable construction projects. The objectives are to develop and test a conceptual framework that highlights the key factors affecting innovation effectiveness, particularly the role of leadership, team identity, and stakeholder integration. The research seeks to offer practical guidance for construction firms to diagnose and improve their innovation strategies.

2. INNOVATION

Innovation is crucial in today's fast-paced world, driving progress, economic growth, and solutions to complex challenges. The diversity of innovation areas has resulted in various definitions and perspectives on its nature. Joseph Schumpeter is recognized as a pioneer in emphasizing innovation's role in economics, defining it in *The Theory of Economic Development* as "new combinations" that generate dynamic changes in economic growth. These combinations can include new or improved products, methods, markets, sources of supply, or organizational structures within industries (Hagedoorn 1996). However, early definitions have faced criticism for being too vague.

The definition of innovation has become more complex in the past decade as a result of rapid developments in the social, political, and technological domains. The term's complexity is seen from multiple perspectives. Researchers have characterized innovation as a transformation that arises from a multitude of alterations (Manuylenko et al. 2015). Innovation, as described by McKinley et al. (2014), refers to any change in the existing structures of products, services, or processes.

Chistov et al. (2021) emphasize the advantages of industrial symbiosis in promoting innovation on a large scale (Chistov et al. 2021). Others argue that innovation should be examined alongside the circular economy concept, which encompasses activities such as recycling, remanufacturing, refurbishing, and reuse. This approach aims to tackle global challenges, such as minimizing excessive waste resulting from the consumption of both renewable and non-renewable resources (Sanni & Verdolini 2022). In addition, the concept of "collaborative circular-oriented innovation" has been introduced, which integrates circular economy ideas into technological and commercial innovation (Brenner & Drdla 2023).

Effective collaboration with external stakeholders, including suppliers and enterprises, regardless of whether they operate in the same industry as competitors, is highlighted as crucial at the meso level (Chistov et al. 2021). The importance of collaborating with clients or end-users and the role crowdsourcing plays in innovation are also emphasized. Other essential variables are the magnitude of research and development (R&D), the administration of knowledge in external collaborations, and the intermediary work that connects the organization with important external partners (Rauter et al. 2017). Chistov et al. (2021) adopt a comprehensive perspective on them at the micro level, emphasizing their importance in shaping the creative outcomes of businesses that prioritize innovation.

Integrating innovation is a complex and challenging process that presents numerous challenges, hindering firms from fully benefiting from the gains of innovation. The hurdles can be classified into two distinct categories: external forces versus internal forces (Maldonado-Guzmán et al. 2017). Internally, the organization faces obstacles related to leadership, inadequate management experience, and inferior personnel abilities (Dubouloz et al. 2021). External impediments arise from factors within the broader business environment, such as governmental regulations, market forces, financial constraints, macroeconomic challenges, limited access to technology providers, and regulatory complexities (Deschryvere et al. 2020; Modranský et al. 2020). Additionally, lack of competent staff, budgetary limits, inadequate IT knowledge, and regulatory constraints are discussed as well.

3. INNOVATIVE PRACTICES AND SUSTAINABILITY

Over the past few decades, nations have increasingly recognized innovation as a key driver of economic growth and competitiveness. This shift has been amplified by global challenges, including population growth, the depletion of fossil fuels, climate change, biodiversity loss, and the 2008 global financial crisis, all of which have highlighted the need for sustainable innovation (Toole et al. 2013; Yigitcanlar & Suharto 2015; Herazo & Lizarralde 2015). Innovation, when aligned with sustainability goals, not only enhances productivity but also promotes environmental and social well-being, ensuring a balanced future.

Sustainability, once seen as conflicting with technological advancement, is now interwoven with innovation across environmental, social, and economic dimensions

(Fatima et al. 2023). This integration has led to the concept of sustainability-oriented innovation, or sustainable innovation, which emphasizes embedding sustainability considerations into the entire innovation process, from ideation to commercialization (Miranda et al. 2023). Companies increasingly adopt open innovation models, collaborating with external partners to access the knowledge and resources required for sustainable innovation (Mishra et al. 2023).

Sustainable innovations can take various forms, including technological, organizational, and social innovations (Bigliardi et al. 2023). Technological innovations focus on new processes and products, organizational innovations modify business practices to reduce environmental impact, and social innovations foster new behaviors that encourage sustainable lifestyles. These innovations are often labeled as sustainable, eco, green, or environmental innovations (Bigliardi et al. 2023), highlighting their role in advancing both corporate sustainability and broader sustainable development goals (Akanbi et al. 2020).

4. INNOVATION AND SUSTAINABILITY IN THE CONSTRUCTION INDUSTRY

Whereas the rate of change in the construction industry has been slow and the sector is considered relatively stable, innovation has become a more vital source of competitive advantage and a major incentive for construction organizations to accommodate changes in their complex products and processes (Eaton et al. 2006; Aouad et al. 2010; Ozorhon 2013).

The construction industry is often regarded as a mature sector, known for its slow pace of change. However, innovation is essential for gaining a competitive advantage within this field (Eaton et al. 2006). According to Slaughter (2000, p. 2), innovation in construction is defined as "a nontrivial improvement in a product, process, or system that is actually used and which is novel to the company developing or using it." This definition highlights the importance of successfully implementing new ideas, processes, or products to enhance an organization's efficiency and performance, a concept that has been widely recognized in the construction industry (Egbu et al. 1998; Sexton & Barrett 2005).

Emerging technologies like artificial intelligence (AI) are paving the way for cost-effective innovation in the construction industry. By adopting deep learning, the construction sector can expect improvements in energy efficiency, worker safety, project management, and overall sustainability (Akinosho et al. 2020).

There is also a growing need for countries to assess sustainability in their construction projects using international metrics. This shall be done by incorporating them in harmonious existence with individual countries' needs and priorities (AlHarmoodi & Dulaimi 2022; Ugwu & Haupt 2007).

Delivering sustainability within the construction sector requires a shift in traditional practices and use of innovative processes, where long-term commitment to an overall vision and related policies and cross-boundary collaboration between and across the supply chain will all play major roles. Innovation in this sector either aims to reduce the consumption of resources, provide better ways of practice, offer improved products, or apply resource management in construction projects. This is usually influenced by the national desire to innovate and the knowledge it diffuses to the sectors and organizations (AlHarmoodi & Dulaimi 2022).

Factors Influencing Innovation in the Construction Industry

Several large-scale empirical studies have found a low level of innovation in the construction industry (Ozorhon et al. 2010). It is commonly accepted that the nature of construction projects is the main obstacle to innovation. A construction project is known for its fragmented supply chains, difficulties with bringing stakeholders to agreements, and poor cross-communication and knowledge management. These factors constrain the generation or adoption, implementation, and diffusion of innovations (Blayse & Manley 2004; Aouad et al. 2010; AlHarmoodi & Dulaimi 2022).

Tabassi et al. (2016) argue that the leadership competencies, transformational leadership, and intellectual competence of project managers play the most significant role in sustainable construction because of their ability to create an innovative culture. Alwan et al. (2017) agreed with previous researchers who have argued that ineffective leadership, ingrained cultures, outdated technologies, and poor logistics are major obstacles to sustainable innovation in construction.

One of the key advances in the modern construction industry is the adoption of innovative technologies, including environmentally sustainable and energy-efficient materials, information technology, cloud computing, and automated equipment (Kuklina et al. 2021; Bello et al. 2021). However, the construction sector remains one of the least digitized industries globally, presenting challenges in addressing its current issues (Abioye et al. 2021).

Building information modeling (BIM) offers promising opportunities for digitalization, enabling the creation of detailed 3D models that represent both the physical and functional characteristics of a project. While it has the potential to improve construction efficiency, such as through enhanced information transfer, barriers to its widespread adoption remain (Qin et al. 2020).

Despite digitization's and AI's potential to serve the construction industry, it still faces challenges like high fragmentation, project-based work, and a skills gap among stakeholders, making standardization and AI integration difficult. This reliance on manual processes adds complexity to project management, hindering innovation (Delgado & Oyedele 2021).

5. INNOVATION AND STAKEHOLDER INTEGRATION

The review of the social network concept has shown that innovation requires the involvement of heterogeneous stakeholders. These can be suppliers, customers, universities, technology centers, trade unions, service providers, financial institutions, and many more, all engaged to achieve a specific outcome. Clients, for instance, can nurture innovation by applying pressure on other stakeholders to improve performance in response to a high standard demand or a novel requirement. While contractors usually implement and exert pressure and encourage subcontractors to actually participate in the innovation, it is often manufacturers that develop the product and/or process innovation. In considering this multidisciplinary, multiparty environment in the construction field, an analysis of innovation at the project level crossing organizations' boundaries could produce more relevant data to assist with identifying the conditions under which innovation can be effective in a construction project setting (Ozorhon et al. 2010; Ozorhon 2013; Murphy et al. 2011). The extent that stakeholders are integrated is important in open innovation in particular. Various scholars have evaluated integration as a strategic capability for the organization that leads to the development of other capabilities, such as environmental protection (Sharma & Vredenburg 1998), organizational learning (Heugens et al. 2002), human resource management (Longo & Mura 2008), and innovation (Brown 2003).

According to Sharma and Vredenburg (1998), stakeholder integration is the ability to create positive collaborative relationships with and between a wide range of stakeholders. These collaborative relationships are the result of a number of factors: (1) knowledge of stakeholders and their needs; (2) interaction between stakeholders and the organization; and (3) taking decisions which take into account stakeholders' demands. To understand how stakeholder integration achieves innovation, it is necessary to know the stakeholders' perspectives, attributes, and interests, their dynamics, roles in innovation, and ways that they can be managed and integrated. Recent research has contributed on the link between innovation and stakeholder engagement along with various mediating factors. Appiah (2023) demonstrated mechanisms and boundary conditions on the relationship between innovation and stakeholder engagement. A positive association between stakeholder engagement and innovation has been observed, while highlighting the pivotal roles of absorptive capacity and risk-taking behavior as mechanisms and boundary conditions within this dynamic (Appiah 2023).

Scholars studying networks argue that network density and centrality are key factors influencing stakeholder behavior in innovation adoption (Rowley 1997; Nambisan & Agarwal 1998). Network density refers to the level of interconnectedness within the network, with higher density fostering greater innovation diffusion (Galaskiewicz & Wasserman 1989). In contrast, centralized networks have fewer active ties, with the innovation champion occupying a strategic position and influencing resource control (Burkhardt & Brass 1990; Liu et al. 2005). Centrality has been empirically linked to innovativeness (Valente 1995).

Reypens et al. (2021) examined multi-stakeholder innovation networks and identified three orchestration practices: connecting stakeholders, facilitating communication, and governing through rules and guidelines. These practices help manage complex, diverse stakeholder groups by switching between dominating and consensus-based modes to tackle challenges. Hybrid orchestration is crucial for successful network management in large, diverse groups (Reypens et al. 2021).

Additionally, understanding stakeholder types, their power, urgency, and legitimacy is essential for strategic analysis, as it informs collaboration and coordination efforts to achieve project and innovation goals.

6. THE EMERGENCE OF OPEN INNOVATION

Construction projects, being multidisciplinary, require effective coordination and communication among various stakeholders to achieve innovation. Historically, innovation in the construction industry followed a "closed innovation" model, where firms generated new ideas internally and invested heavily in R&D while protecting their intellectual property (AlHarmoodi & Dulaimi 2022; Chesbrough 2003). This approach required substantial investments to secure profits. However, with the rise of knowledge mobility, globalization, and venture capital markets, protecting intellectual property became increasingly difficult (Chesbrough 2003; Heap 2010). These factors have encouraged the adoption of open innovation, a model that facilitates cross-boundary collaboration.

While open innovation has long existed in the construction sector, it was often limited to primary stakeholders, ignoring secondary and invisible stakeholders, who can significantly impact project outcomes. Today's open innovation strategies can overcome many of these traditional constraints by promoting collaboration beyond the immediate supply chain.

In the context of sustainability, the UN's 2030 Agenda for Sustainable Development highlighted the importance of sustainability goals, although many industries resist adopting sustainable practices due to maturity and innovation resistance (Ogink et al. 2023). Open innovation plays a crucial role in supporting the transition to sustainability by enabling the exchange of information, resources, and technologies among stakeholders (Carrasco-Carvajal et al. 2023).

The literature on open innovation and sustainability uses various terminologies, such as "open innovation for sustainable innovation," "open eco-innovation mode," and "collaborative circular-oriented innovation," which focus on environmentally friendly knowledge absorption and circular economy practices (Brenner & Drdla 2023). Stakeholders in sustainable open innovation include primary actors (customers, employees) and secondary ones (communities, governments, non-government organizations) (Jesus & Jugend 2023).

Open innovation consists of three processes: the outside–in process (inbound), where external stakeholders increase the organization's knowledge base (Laursen & Salter 2006; Lettl et al. 2006; Piller & Walcher 2006); the inside–out process (outbound), where firms commercialize ideas through licensing, joint ventures, and spinoffs (Enkel et al. 2009); and the coupled process, which combines the two to foster co-creation with complementary partners, such as universities or research organizations (Enkel et al. 2009; Gassmann & Enkel 2004; Lichtenthaler 2011). This model integrates external knowledge with internal capabilities to enhance innovation development and commercialization.

7. INNOVATION AND LEADERSHIP

Leadership and innovation have gained increasing attention in the literature (Yukl 2012; Ozorhon et al. 2010; Aronson et al. 2013; Ozorhon et al. 2014; Eckardt et al. 2021; AlHarmoodi & Dulaimi 2022; Costa et al. 2023). Leadership is one of the most influential factors that affect innovation in organizations and projects. Leadership reflects the goal-directed influence that one person has over other members in an organization or a group in guiding, structuring, and facilitating relationships and activities. Thus, the role of the leader is key for the organization/project to function in an effective and innovative manner (Nam & Tatum 1997). It is the leader's role to contribute actively to the introduction of new ideas, goals, and innovations in an organization/project; therefore, leadership style is considered a crucial attribute in influencing innovation (Aronson et al. 2013; Ozorhon et al. 2014; Bossink 2004; Costa et al. 2023). Creating the conditions that facilitate, motivate, and sustain innovation is one of the greatest challenges facing leaders (Yukl 2010). Transformational leadership (Jung et al. 2003), innovation championing (Howell & Higgins 1990), and leader–member exchange (Graen & Uhl-Bien 1995) have been found to have a positive influence on innovation.

Research links transformational leadership to innovation (Howell & Avolio 1993; Jung et al. 2003). This leadership style enhances innovation by engaging followers, aligning their identities with the organization, and boosting intrinsic motivation. It emphasizes vision and mission, improving performance and willingness to prioritize organizational goals (Howell & Avolio 1993). It also fosters collaboration, trust, empowerment, competence development, and visible support (Jung et al. 2003). Key survey instruments are the Multifactor Leadership Questionnaire (MLQ) (Bass & Avolio 1994) and the Leadership Profile Inventory (LPI) (Kouzes & Posner 1995).

Leadership styles that accommodate change are linked to innovation (Dackert et al. 2004; Scott & Bruce 1994; Tierney & Farmer 2004). Change-oriented leadership, a form of transformational leadership (Gil et al. 2005), is relevant here. Yukl (1999) cautions that transformational leadership may overlook the leader's impact on the team. Leader–member exchange (LMX) theory connects leader–follower relationships directly to innovation, relying on trust and support to encourage risk-taking (Scott & Bruce 1994). The LMX7 survey measures leader–member exchange levels (Graen & Uhl-Bien 1995).

Championing behavior is crucial for innovation. Innovation success often involves champions (Howell & Higgins 1990; Howell & Shea 2001; Nam & Tatum 1997), who influence power distribution, resources, strategic actions, and performance. Champions affect career progression, motivation, retention, communication, and decision-making in innovation projects (Howell & Shea 2001). These leadership traits integrate stakeholders and manage innovation teams effectively.

Hui et al. (2024) discussed the role of AI-based virtual assistants in conjunction with transformational leadership, knowledge-sharing, and absorptive capacity within the construction industry to influence team innovation. Transformational leadership was shown to have a positive association with team innovation and knowledge-sharing, but not with absorptive capacity. Both knowledge-sharing and absorptive capacity were positively correlated with team innovation. Interestingly, AI-based virtual assistants played a significant moderating role in the relationship between knowledge-sharing and team innovation but had an insignificant effect on the relationship between absorptive capacity and team innovation (Hui et al. 2024). Technological-based leadership support systems, such as AI-based virtual assistants, represent novel and promising avenues for fostering innovation within the construction industry.

8. INNOVATION AND TEAM IDENTITY

The concept of identity has a long-standing and active presence in social science literature (Brubaker & Cooper 2000; Gleason 1983). Initially developed by Gleason (1983), the notion of identity has spread across disciplines like psychology and sociology. In social psychology, theories like Tajfel's (1978, 1981) social identity theory and Turner's (1987) self-categorization theory are fundamental. In sociology, identity has been central to the theories of Stryker (1987), Burke (1980), and McCall and Simmons (1978), while it has also become significant in anthropology and cultural studies (Eriksen 2001; Holland 1997). Open innovation at the project level requires informal norms and rules to define roles, unlike closed innovation, where tasks are governed by corporate structures (Rese & Baier 2012). The development of a collective team identity is essential in open innovation contexts, especially to navigate the complexities of stakeholder integration and maintain team cohesion (Postmes et al. 2000; Rese & Baier 2012).

Ashmore et al. (2004) expanded on the idea of collective identity, emphasizing that it is shared among individuals who perceive common characteristics. Collective identity is multidimensional, comprising elements like self-categorization, attachment, and social embeddedness. This strong team identity, supported by effective management, enables innovation through collaboration (Ashmore et al. 2004).

Similarly, Biemans and Huizingh (2023) stress that collaboration within teams enhances the sharing of knowledge and expertise, vital for overcoming challenges. Moreover, Mascareño et al. (2020) show that team identity can be measured through language analysis, where frequent use of "we" signals collective focus and goal alignment, driving innovation. Van der Voet and Steijn (2021) underscore the role of strong team identity and visionary leadership in fostering cohesion and creativity in teams, while Hadjielias et al. (2021) highlight that team identity is integral to digital innovation within multidisciplinary teams. Finally, Ali et al. (2020) demonstrate that shared leadership enhances team creativity by fostering professional identity, contributing to more innovative solutions in industries like construction.

9. A CONCEPTUAL MODEL

The literature review concludes that open innovation can enhance networking and collaboration among different stakeholders in pursuit of innovation goals, though this relationship requires further empirical validation (Ozorhon 2013; Ozorhon et al. 2010; Murphy et al. 2011). Additionally, it was found that a clear team identity and strong membership are critical for successful innovation, and that this is significantly influenced by stakeholder contributions to the project (Ozorhon et al. 2014; Aronson et al. 2013; Yukl 2010).

Based on literature review, it was evident that the type of leadership for innovation and the style of leadership influences both stakeholder integration and team identity (van der Voet & Steijn 2021; AlHarmoodi & Dulaimi 2022; Mudassar et al. 2021). While many scholars have studied leadership for innovation (Mudassar et al. 2021; Yukl 2012; Howell & Shea 2001; Yukl 1999; Nam & Tatum 1997), their focus has primarily been at the organizational level, leaving project-level leadership characteristics underexplored, calling for more empirical research to examine the impact of leadership on innovation at the project level in an open innovation context.

From this analysis, it is evident that leadership for innovation exerts a direct, one-way influence on both stakeholder integration and team identity. Meanwhile, stakeholder integration and team identity interact in a reciprocal, two-way relationship, influencing one another. These constructs – open innovation, leadership for innovation, stakeholder integration, team identity, and innovation effectiveness – are interlinked and collectively contribute to the overall effectiveness of innovation (Figure 16.1).

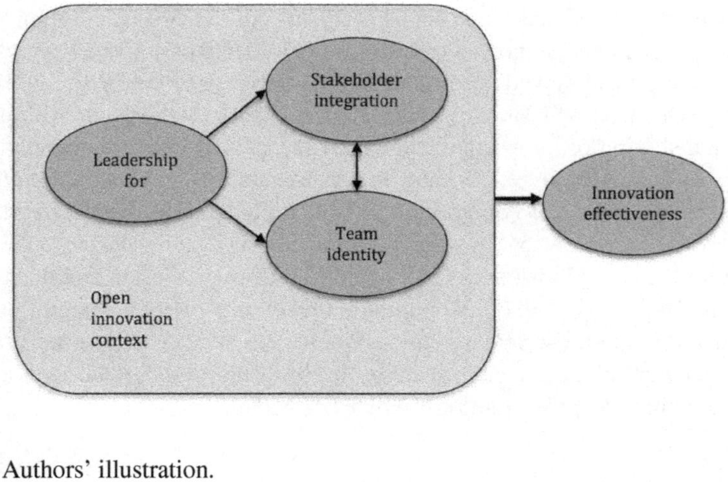

Source: Authors' illustration.

Figure 16.1 Proposed conceptual model

In this model, leadership serves as a central force that shapes both the stakeholders' involvement and the team's sense of identity, fostering better collaboration and alignment toward the innovation goals. Stakeholder integration, which reflects the involvement of diverse external and internal parties, supports team cohesion, while a strong team identity enhances stakeholder collaboration. These interactions, in turn, contribute to the overall effectiveness of the innovation process.

The model underscores that leadership's role in innovation is not only about direct guidance but also about fostering environments where stakeholder involvement and team cohesion feed into each other, ultimately driving innovation success.

The main constructs were defined, and their main dimensions were identified and defined, based on the existing literature (Table 16.1).

10. MODEL EVALUATION

The model evaluation involved two key analysis steps: within-case and cross-case analyses.

Within-case Analysis

The within-case analysis involved collecting and evaluating data from three case studies, comprising 38 interview transcripts and additional relevant documentation like observation notes, firm profiles, and news articles. This data was analyzed using a directed content analysis method, where an initial coding scheme, based on the conceptual thematic framework, was developed and then refined as new codes emerged during the analysis. Frequencies of codes were analyzed across all recording units to understand perceptions about each category and its dimensions. This analysis classified the effect of leadership for innovation on stakeholder integration and team identity, elucidating its influence on innovation effectiveness. The findings were summarized using Miles and Huberman's tabular approach, which separated raw evidence from interpretation, and resulted in narratives highlighting major findings, stakeholder perceptions, and the strengths and weaknesses of each category.

Cross-case Analysis

The cross-case analysis was aimed to validate the case study conclusions by comparing patterns observed across the three cases with theoretical predictions. A pattern-matching technique was employed, where three predicted patterns (high, medium, and low descriptors) for constructs like leadership for innovation, stakeholder integration, team identity, and innovation effectiveness were developed based on interviewee perceptions. The matching criteria ranged from perfect to no match, determining the extent to which case-based patterns aligned with predicted ones. A good or perfect match indicated qualitative validation of the model, while a poor

Table 16.1 The conceptual thematic framework

Conceptual thematic framework	References
Leadership for innovation: Leadership is "the process of influencing others to understand and agree about what needs to be done and how to do it, and the process of facilitating individual and collective efforts to accomplish shared objectives" (based on Yukl 2010, p. 26). Variables: • Encouraging and stimulating innovation • Providing and inspiring vision • Individualized support • Teamwork development • Stakeholder integration	Bass & Avolio (1994), Howell et al. (2005), Yukl et al. (2002), Kouzes & Posner (1995), Skipper & Bell (2006), Podsakoff et al. (1990), Graen & Uhl-Bien (1995), Jung et al. (2003), Mitchell et al. (1997)
Stakeholder integration: Stakeholder integration is "the ability to create positive collaborative relationships with a wide range of stakeholders" (based on Sharma and Vredenburg 1998, p. 735). Variables: • Knowledge of stakeholders • Stakeholders' interactions • Behavior adaptation	Maignan & Ferrell (2004), Mitchell et al. (1997), Friedman & Miles (2002), Elias et al. (2002), Jose et al. (2010), Grafé-Buckens & Hinton (1998), Green & Hunton-Clarke (2003), Sharma & Vredenburg (1998), Lorca & Garcia-Diez (2004), Burchell & Cook (2008)

Conceptual thematic framework	References
Team identity: Team identity is the feeling of "belonging of individuals that create some emotional and value significance to them and a sense of membership" (based on Tajfel & Turner 1979). Variables: Team level: • Boundary clarity • Boundary permeability • Cohesion • Common language and understanding • Cooperative working atmosphere • Cognitive similarity Individual level: • Self-categorization • Evaluation • Importance • Attachment and sense of interdependence • Social embeddedness • Behavioral involvement • Content and meaning	Rese & Baier (2012), Phinney (1996), Ashmore et al. (2004), Henderson-King & Stewart (1994), Abrams et al. (1998), Luhtanen & Crocker (1992), Sellers et al. (1998)

Conceptual thematic framework	References
Innovation effectiveness: Innovation effectiveness is the overall project's outcome that results from implementing the innovation. In the context of the research for this thesis: it is, how the project perceives the overall improvement (including factors such as achieving targets, finance, employee morale), which arises specifically from implementing the innovations (Rese & Baier 2012). Variables: • Effective/efficient innovation capacity • Future/expected innovation potential	Hoegl & Gemuenden (2001), Rese & Baier (2012)

Note: The main constructs of the conceptual model were defined, and their main dimensions were identified and defined based on the existing literature to guide the qualitative analysis.

match suggested the model was not validated by the case studies. This comparison confirmed the robustness and applicability of the conceptual model across different case contexts.

Within-case Analysis Findings

Case studies were conducted with Dubai-based mixed-use construction projects. In total, three firms participated in the study. For the purpose of confidentiality, the three case studies were referred to as SC, AKO, and JFZ to maintain their anonymity. The case study profiles of all projects are summarized in Table 16.2. Overall, the case studies comprised a good mix of projects. As Table 16.2 shows, all of the cases were intended to be sustainable residential projects and the sizes of the projects varied; nevertheless, the researcher interviewed people who worked on similar size clusters for each project to represent similar samples. SC and AKO represented the whole project as a sustainable innovation, whereas JFZ represented a sustainable product innovation within the boundaries of their organization.

In total, there were 38 leaders, team members, and other stakeholders who participated in the interviews. This group of participants consisted of a good combination of stakeholders, such as project leaders, design team members, consultants, contractors, suppliers, and some other administrative positions to complement the information. All of them also possessed at least a bachelor's degree relevant to their field of

Table 16.2 Case study profiles

Case	Description	Project size	Innovation type	Interview participants	Interview location
SC	Sustainable mixed-use residential project	5m square feet	Project-level sustainable innovation	• 2 leaders • 4 design team • 2 consultants • 4 contractors • 1 supplier • 3 other positions	SC head office Onsite location Modular head office
AKO	Sustainable mixed-use residential project	55m square feet	Project-level sustainable innovation	• 2 leaders • 5 design team • 4 consultants • 2 contractors • 1 supplier	AKO head office Onsite location
JFZ	Sustainable residential project	697,210 square feet	Product-level sustainable innovation	• 3 leaders • 3 design team • 1 consultant • 1 contractor	JFZ head office Onsite location

Note: The table presents the profiles of the three cases used in the assessment of the model.

expertise. Therefore, they were considered adequate for this part of the study as they were likely to provide reliable, accurate, and insightful accounts relevant to the topics of investigation.

Table 16.3 summarizes the within-case analysis results, in terms of the qualitative rating assigned for each construct and its associated dimensions, for all of the case

Table 16.3 Within-case analysis results

Construct/dimensions	SC rating	AKO rating	JFZ rating
Leadership for innovation (LI)	**High**	**High**	**Medium**
1.1 Encouraging and stimulating innovation	High	High	Medium
1.2 Providing and inspiring vision	High	Medium	Low
1.3 Individualized support	High	High	Medium
1.4 Teamwork development	High	High	Medium
1.5 Stakeholder integration	High	Medium	Medium
Stakeholder integration (SI)	**High**	**Medium**	**Medium**
2.1 Knowledge of stakeholders	High	Medium	Medium
2.2 Stakeholder interactions	Medium	Medium	Medium
2.3 Behavior adaptation	High	Medium	Medium
Team identity (TI)	**High**	**Medium**	**Low**
3.1 Boundary clarity	High	Medium	Low
3.2 Boundary permeability	High	Medium	Low
3.3 Cohesion	Medium	High	High
3.4 Common language and understanding	Medium	Medium	Low
3.5 Cooperative working atmosphere	High	High	High
3.6 Cognitive similarity	Medium	Low	Low
3.7 Self-categorization	High	Low	Low
3.8 Evaluation	High	Low	Low
3.9 Importance	High	High	High
3.10 Attachment and sense of interdependence	High	High	Low
3.11 Social embeddedness	High	Low	Low
3.12 Behavioral involvement	Low	Medium	Low
3.13 Content and meaning	High	Low	Low
Innovation effectiveness (IE)	**High**	**Medium**	**Medium**
4.1 Effective/efficient innovation capacity	High	Medium	High
4.2 Future/expected innovation potential	High	Medium	Low

Note: The table summarizes the within-case analysis results, in terms of the qualitative rating assigned for each construct and its associated dimensions, for all of the case studies.

studies. The table shows a rating of the dimensions derived from the qualitative findings, which were totaled to represent the overall rating for the main four constructs (LI, SI, TI, IE).

Leadership for Innovation (LI) Construct Rating

LI ratings were high in SC and AKO, and medium in JFZ. SC and AKO had strong leadership, with SC excelling in all five LI dimensions and AKO slightly lower in vision and stakeholder integration. JFZ showed medium ratings in encouraging innovation, individualized support, and stakeholder integration, with high teamwork development but low in inspiring vision. Effective leadership in SC involved active promotion of innovation and stakeholder integration. AKO leaders encouraged innovation but struggled with vision and stakeholder integration. JFZ's leadership was less proactive, leading to a medium overall LI rating.

Stakeholder Integration (SI) Construct Rating

SI ratings were high in SC and medium in AKO and JFZ. SC achieved high SI through extensive stakeholder knowledge and relationship facilitation, despite medium-rated stakeholder interaction. AKO and JFZ had medium ratings due to financial motivations, communication challenges, and limited stakeholder engagement. All firms faced challenges in stakeholder interaction, highlighting the need for improved techniques.

Team Identity (TI) Construct Rating

TI ratings were high in SC and moderate in AKO and JFZ. SC's high TI was due to boundary clarity, cooperative atmosphere, strong self-categorization, and a sense of importance and interdependence. AKO's moderate TI was due to medium boundary clarity, common understanding, and behavioral involvement, with low social embeddedness and self-categorization. JFZ had high feelings of importance and cohesion but low ratings in other dimensions, resulting in moderate TI. Consistent high ratings in cooperative atmosphere and team importance were crucial for team identity.

Innovation Effectiveness (IE) Construct Rating

IE ratings were high in SC and medium in AKO and JFZ. SC had high effective innovation capacity and future potential. AKO had medium ratings in both dimensions, while JFZ had high effective capacity but low future potential, resulting in an overall medium IE rating.

Robustness of Model Dimensions

The model dimensions showed consistent and positive association across most projects. LI dimensions were consistent in all cases, while SI dimensions were well correlated, especially in AKO and JFZ. IE dimensions were consistent in SC and AKO but less so in JFZ. TI dimensions were mostly correlated in SC and JFZ, with some inconsistency in AKO. Overall, the model dimensions proved reliable in capturing the qualitative rating levels, supporting the robustness of the model's structure.

Cross-case Analysis Findings

To perform the cross-case analysis, the relationships between the rated constructs in each case were evaluated against developed predicted patterns. This process contributes to validating the relationships of the model constructs through pattern-matching techniques (McCutcheon & Meredith 1993). Much of the activity of pattern-matching in qualitative analyses involves attempts by the research to link two patterns: one is theoretical and the other is observable or operational. This technique is used to establish links between the collected data and the theoretical propositions by comparing the proposed patterns with the actual values of variables obtained from data collection and analysis (McCutcheon & Meredith 1993; Yin 2003).

The predicted patterns were developed using high, medium and low value descriptors for the constructs and their related dimensions. These descriptors were based on the perceptions and remarks given by the interviewees. Positive remarks reflect a "high" description of the construct, whereas when there is a combination of positive and negative remarks and the positive remarks are dominant, it reflects a "medium" description of the construct, while when the negative remarks are more dominant, it reflects a "low" description of the construct. It is worthwhile to note that the dimensions that were not reflected on or that had a very low number of codes (less than 20% of the sample size) in one case were compared with the similar ones in the other two cases; if one case had a good number of codes reflecting the dimension, it indicates the viability of the dimension and a decision to keep the dimension was made with a description of "low" being given. On the other hand, dimensions that have a very low number of codes or no codes at all in all three cases were neglected since they were considered not viable.

Three main predicted patterns were developed, as presented in Figure 16.2.

Based on the extent to which cases matched the predicted pattern, the developed model can be judged as acceptable or not. If the matching results between the case-based patterns and the predicted patterns have a good or perfect match, it means that the model can be qualitatively validated by the case studies. On the other hand, if most of the results display a poor match, the model cannot be validated by the case studies.

Table 16.4 presents the cross-case analysis of the three case studies, which elaborates on the assessment as to whether the cases match the predicted patterns.

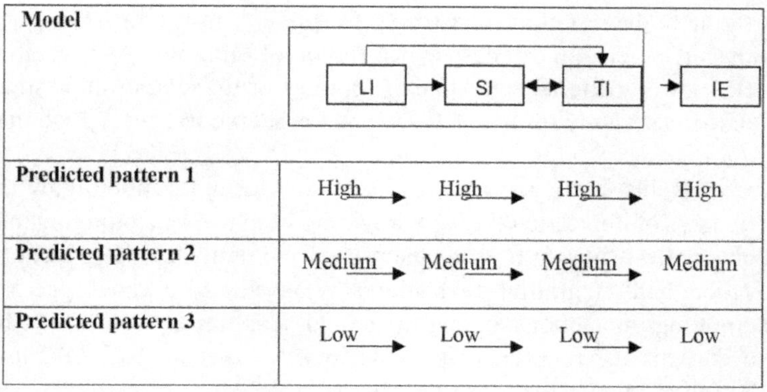

Figure 16.2 *Predicted patterns of relationships between model constructs*

As presented in Table 16.4, the relationship patterns of the rated constructs in the SC case show a perfect match to *Predicted Pattern 1.* The high level of the constructs in turn has a strong association with the level of IE (high).

In the AKO case, LS is rated high; however, it yielded a medium-rank SI and TI, showing that they also do not associate as expected. Nevertheless, the association between SI and TI matches the predictions. In addition to that, the overall medium level of the constructs led to a medium level of the IE, which matches the *Predicted Pattern 2.*

Finally, the results of the JFZ case indicate, as in *Predicted Pattern 2*, that the moderate level of LS led to a moderate level of IE. There is only one discrepancy in the pattern, which is the low rating of the TI construct that does not match the overall

Table 16.4 *Cross-case analysis results*

Case	II rating	LS rating	SI rating	TI rating	IE rating	Matching results
SC	High	High	High	High	High	Perfect match to *Predicted Pattern 1*
AKO	Low	High	Medium	Medium	Medium	Partial match to *Predicted Pattern 2*
JFZ	Medium	Medium	Medium	Low	Medium	Good match to *Predicted Pattern 2*

Note: The table presents the cross-case analysis of the three case studies, which elaborates on the assessment as to whether the cases match the predicted patterns.

medium rating of the rest of the constructs. Despite this minor variation, the remaining relationship patterns in this case match *Predicted Pattern 2*. As a result, since the overall relationship pattern seems to explain most of the relationships stipulated in the model, the case study results of JFZ were considered to have a good match with *Predicted Pattern 2*.

Thus, overall, the findings prove that leadership for innovation highly influences the effectiveness of innovation through providing vision and encouraging innovation, building interested teams with strong team-level and individual-level identity toward the innovation, and integrating stakeholders by developing a knowledge base about them, facilitating interactions among them, and adapting to each other's behaviors (Figure 16.3). This also revealed that clients are key stakeholders driving innovation, especially if they are also the developers of the project. Their capacity to exert influence on individuals and teams in a way that fosters innovation is very significant. They can influence innovation by different means, such as identifying specific novel requirements to be delivered by other stakeholders; exerting pressure on project participants to improve project lifecycle performance, overall characteristics, and

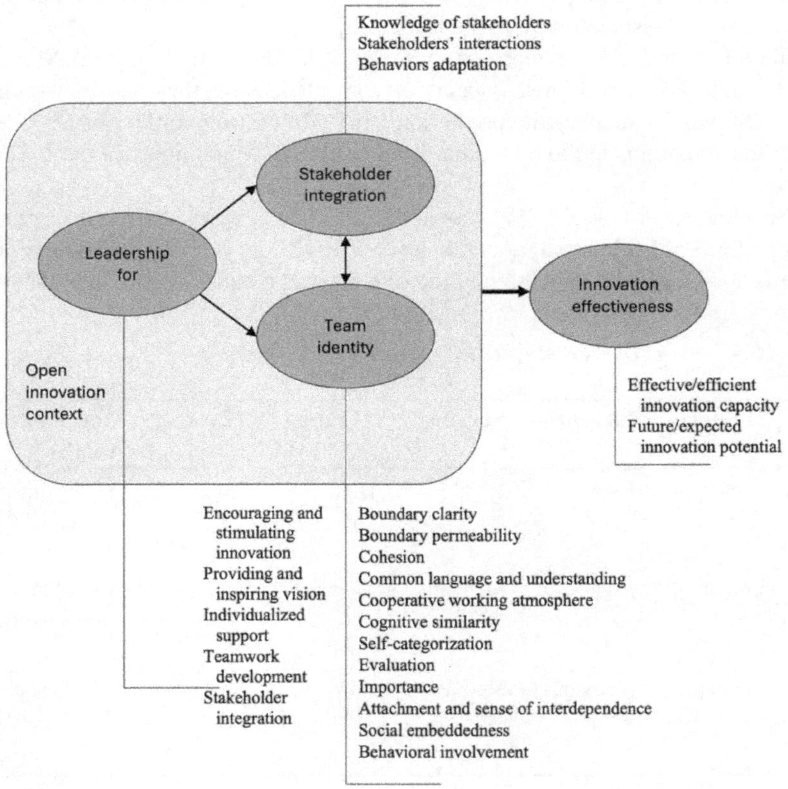

Figure 16.3 *The final model*

project flexibility to cope with unforeseen changes; and in general demand higher standards of work. The more knowledgeable and experienced the client, the more likely it is to stimulate innovation in projects.

This provides empirical evidence that stakeholder integration is critical in facilitating the achievement of effective innovation. The foregoing result underlies the importance of understanding stakeholders, facilitating interactions among them and adapting to their behavior, bearing in mind the different types of internal and external stakeholders and their associated power, legitimacy, and urgency.

Interestingly, it was observed that green certifications, regulations, and policies can have a negative effect on innovation. Although they are often seen to promote innovation, the pressure they exert on the team within a limited timeframe can hinder team members from being creative and innovative, as they give their full attention to meeting the requirements. For the management level, procedures and documentation required for certifications are part of a larger goal. By contrast, other stakeholders see them as extra work with supplementary fees that, together, increase stressful situations and tensions.

Finally, the study provides empirical evidence that integrated teams with a strong sense of membership and identity to the innovation group provide their full effort and potential for the sake of the innovation. These findings make this particular study one of the first which provides empirical evidence that stakeholder integration and innovation team identity are interdependent; therefore, it links social identity theory with the construction and innovation theories. The study further reveals that the integration and identity of the innovation team is highly influenced by leaders. Hence, it adds a new element that leaders of construction projects ought to consider in achieving their innovation goals.

Thus, it should be emphasized that open innovation has to be well thought out and project-managed to integrate the multidisciplinary and multiparty stakeholders who cross boundaries of their organizational settings along with the innovation team to achieve the best outcome of open-source innovation. This can only be achieved with an innovation-oriented leadership that is highly interested in innovation.

11. IMPLICATIONS OF THIS MODEL

Evaluating the main conceptual model that proposes the major aspects that need to be taken into consideration to achieve a successful integration, and – consequently – an effective innovation outcome, a number of implications that could guide construction projects to develop strategies that enhance the effectiveness of the innovation are presented below.

- The client's/owner's leadership should be enhanced to achieve successful innovations, through:
  - building robust relationships with stakeholders and improving coordination, collaboration, and knowledge flows among them;
  - integration of project experiences into continuous business processes to limit the loss of tacit knowledge between projects;
  - promoting innovative procurement systems, including partnering or alliancing, to enhance cooperative problem-solving, the adoption of non-standard solutions, and equitable allocation of risk;
  - strengthening of performance-based regulations and standards, through the enhancement of technical knowledge held by regulators and other key players, and through the formulation of simple enforcement strategies;
  - developing a culture supportive of innovation, enhancing technical competence, supporting innovation champions, and developing an effective innovation strategy.
- Whenever possible, there should be a shift from competitive tendering to partnering and/or alliancing (between the client and the contractor). In addition, those in a position to innovate need to be rewarded for taking such risks. Partnering has to be associated with the use of a range of tools, including charters, workshops, team-building exercises, dispute resolution mechanisms, benchmarking, total quality management, and business process mapping.
- Project owners should selectively appoint an innovation-oriented leader that can successfully integrate the right stakeholders at the right time in the project lifecycle and develop an innovative team with strong identity and membership.
- Apart from urging project owners to assign the right leader, this study provides leaders with techniques that they should focus on for fostering innovative team members and stakeholders, since they play a key role in leading the way to effective innovation.
- Innovation effectiveness should be measured with respect to innovation objectives, without being limited to standard project management performance criteria. Much of the innovation focus revealed by the case studies is environmental sustainability. Therefore, innovation effectiveness should include measures related to environmental drivers, just as other project objectives.

REFERENCES

Abioye, S. O., Oyedele, L. O., Akanbi, L., Ajayi, A., Davila Delgado, J. M., Bilal, M., … Ahmed, A. (2021). Artificial intelligence in the construction industry: A review of present status, opportunities and future challenges. *Journal of Building Engineering*, *44*, 103299. https://doi.org/https://doi.org/10.1016/j.jobe.2021.103299

Abrams, D., Ando, K., & Hinkle, S. (1998). Psychological attachment to the group: Cross-cultural differences in organizational identification and subjective norms as predictors of workers' turnover intentions. *Personality and Social Psychology Bulletin*, *24*, 1027–1039.

Akanbi, L. A., Oyedele, A. O., Oyedele, L. O., & Salami, R. O. (2020). Deep learning model for Demolition Waste Prediction in a circular economy. *Journal of Cleaner Production*, *274*, 122843.

Akinosho, T. D., Oyedele, L. O., Bilal, M., Ajayi, A. O., Delgado, M. D., Akinade, O. O., & Ahmed, A. A. (2020). Deep learning in the construction industry: A review of present status and future innovations. *Journal of Building Engineering*, *32*, 101827. https://doi.org/https://doi.org/10.1016/j.jobe.2020.101827

AlHarmoodi, S. M., & Dulaimi, M. (2022). The alignment of the open innovation process and the project lifecycle. *Built Environment Project and Asset Management*, *12*(4), 556–572. https://doi.org/10.1108/BEPAM-06-2021-0089

Ali, A., Wang, H., Soomro, M. A., & Islam, T. (2020). Shared leadership and team creativity: Construction industry perspective. *Journal of Construction Engineering and Management*, *146*(10), 04020122.

Alwan, Z., Jones, P., & Holgate, P. (2017). Strategic sustainable development in the UK construction industry, through the framework for strategic sustainable development, using building information modelling. *Journal of Cleaner Production*, *140*, 349–358.

Aouad, G., Ozorhon, B., & Abbot, C. (2010). Facilitating innovation in construction: Directions and implications for research and policy. *Construction Innovation*, *10*(4), 374–394.

Appiah, L. O. (2023). Stakeholder engagement for green process innovation: Exploring the link and boundary conditions. *Operations and Supply Chain Management*, *16*(2), 153–163.

Aronson, Z. H., Shenhar, A. J., & Patanakul, P. (2013). Managing the intangible aspects of a project: Quantitative and qualitative evidence on the affect of vision, artifacts, and leader values on project spirit and success in technology driven projects. *Project Management Journal*, *44*(1), 35–58.

Ashmore, R. D., Deaux, K., & McLaughlin-Volpe, T. (2004). An organizing framework for collective identity: Articulation and significance of multidimensionality. *Psychological Bulletin*, *130*, 80–114.

Bass, B. M., & Avolio, B. J. (1994). *Improving organizational effectiveness through transformational leadership*. Thousand Oaks: Sage Publications.

Bello, S. A., Oyedele, L. O., Akinade, O. O., Bilal, M., Delgado, J. M. D., Akanbi, L. A., … Owolabi, H. A. (2021). Cloud computing in construction industry: Use cases, benefits and challenges. *Automation in Construction*, *122*, 103441.

Biemans, W. G., & Huizingh, E. K. (2023). Why so serious? The effects of humour on creativity and innovation. *Creativity and Innovation Management*, *33*(1). doi:10.1111/caim.12587.

Bigliardi, B., Dolci, V., Filippelli, S., Petroni, A., Pini, B., & Tagliente, L. (2023). The adoption of Open Innovation in Manufacturing: A review. *Procedia Computer Science*, *217*, 1765–1774.

Blayse, A. M., & Manley, K. (2004). Key influences on construction innovation. *Construction Innovation*, *4*(3), 143–154.

Bossink, B. A. G. (2004). Effectiveness of innovation leadership styles: A manager's influence on ecological innovation in construction projects. *Construction Innovation*, *4*(4), 211–228.

Brenner, B., & Drdla, D. (2023). Business model innovation toward sustainability and circularity: A systematic review of innovation types. *Sustainability*, *15*(15), 11625.

Brown, N. (2003). Hope against hype: Accountability in biopasts, presents and futures. *Science Studies*, *16*, 3–21.

Brubaker, R., & Cooper, F. (2000). Beyond "Identity". *Theory and Society*, *29*, 1–47.

Burchell, J., & Cook, J. (2008). Stakeholders *Dialogue and Organizational Learning*: Changing relationships between companies and NGOs. *Business Ethics: A European Review*, *17*(1), 35–46.

Burke, P. J. (1980). The self: Measurement implications from a symbolic interactionist perspective. *Social Psychology Quarterly*, *43*, 18–29.

Burkhardt, M. E., & Brass, D. J. (1990). Changing patterns or patterns of change? The effects of a change in technology on social network structure and power. *Administrative Science Quarterly*, *35*(1), 104–127.

Carrasco-Carvajal, O., Castillo-Vergara, M., & García-Pérez-de-Lema, D. (2023). Measuring open innovation in SMEs: An overview of current research. *Review of Managerial Science, 17*(2), 397–442.

Chesbrough, H. W. (2003). *Open innovation: The new imperative for creating and profiting from technology.* Cambridge, MA: Harvard Business School Publishing.

Chistov, V., Aramburu, N., & Carrillo-Hermosilla, J. (2021). Open eco-innovation: A bibliometric review of emerging research. *Journal of Cleaner Production, 311*, 127627.

Costa, J., Padua, M. & Moreira, A. (2023). Leadership styles and innovation management: What Is the role of human capital? *Administrative Sciences, 13*(2), 47. https://doi.org/10.3390/admsci13020047

Dackert, I., Loov, L. A., & Martensson, M. (2004). Leadership and climate for innovation in teams. *Economic and Industrial Democracy, 25*(2), 301–318.

Delgado, J. M. D., & Oyedele, L. (2021). Digital twins for the built environment: Learning from conceptual and process models in manufacturing. *Advanced Engineering Informatics, 49*, 101332.

Deschryvere, M., Mikkola, M., & Conn, S. (2020). On the structural barriers to public innovation support for SMEs and the opportunity COVID-19 can offer to overcome these barriers. *Journal of Innovation Management, 8*(2), 16–25.

Dubouloz, S., Bocquet, R., Equey Balzli, C., Gardet, E., & Gandia, R. (2021). SMEs' open innovation: Applying a barrier approach. *California Management Review, 64*(1), 113–137.

Eaton, D., Akbiyikli, R., & Dickinson, M. (2006). An evaluation of the stimulants and impediments to innovation within PFI/PPP projects. *Construction Innovation, 6*(2), 63–77.

Eckardt, R., Tsai, C. Y., Dionne, S. D., Dunne, D., Spain, S. M., Park, J. W., Cheong, M., Kim, J., Guo, J., Hao, C., & Kim, E. I. (2021). Human capital resource emergence and leadership. *Journal of Organizational Behavior, 42*(2), 269–295. https://doi.org/10.1002/job.2446

Egbu, C. O., Henry, J., Kaye, G. R., Quintas, P., Schumacher, T. R., & Young, B. A. (1998). *Managing organizational innovations in construction: Proceedings of the 14th Annual Conference of the Association of Researchers in Construction Management.* ARCOM ed.

Elias, A. A., Cavana, R. Y., & Jackson, L. S. (2002). Stakeholder analysis for R&D project management. *R&D Management, 32*(4), 301–310.

Enkel, E., Gassmann, O., & Chesbrough, H. (2009). Open R&D and Open Innovation: Exploring the phenomenon. *R&D Management, 39*(4), 311–314.

Eriksen, T. H. (2001). Ethnic identity, national identity, and intergroup conflict: The significance of personal experiences, in R. D. Ashmore, L. Jussim, & D. Wilder (eds), *Social identity, intergroup conflict, and conflict reduction* (Rutgers Series on Self and Social Identity). New York: Oxford University Press, vol. 3, pp. 42–68.

Fatima, S., Tandon, P., & Singh, A. B. (2023). Current state and future directions of sustainability and innovation in finance: A bibliometric review. *International Journal of System Assurance Engineering and Management*, 1–24.

Friedman, A. L., & Miles, S. (2002). Developing stakeholders theory. *Journal of Management Studies, 39*, 1–21.

Galaskiewicz, J., & Wasserman, S. (1989). Mimetic processes within an interorganizational field: An empirical test. *Administrative Science Quarterly, 34*, 354–479.

Gassmann, O. & Enkel, E. (2004). *Towards a theory of open innovation: Three core process archetypes: Proceedings of the R&D Management Conference*. Lisbon, Portugal, 6–9 July.

Gil, F., Rico, R., Alcover, C. M., & Barrasa, A. (2005). Change-oriented leadership satisfaction and performance in work groups: Effects of TAM climate and group potency. *Journal of Managerial Psychology, 20*(3–4), 312–328.

Gleason, P. (1983). Identifying identity: A semantic history. *Journal of American History, 69*, 910–931.

Graen, G. B., & Uhl-Bien, M. (1995). Relationship-based approach to leadership: Development of Leader–Member Exchange (LMX) theory of leadership over 25 years: Applying a multi-level multi-domain perspective. *Leadership Quarterly, 6*(2), 219–247.

Grafé-Buckens, A., & Hinton, A. F. (1998). Engaging the stakeholder: Corporate views and current trends. *Business Strategy and the Environment*, *7*(3), 124–133.

Green, A. O., & Hunton-Clarke, L. (2003). A typology of stakeholder participation for company environmental decision-making. *Business Strategy and the Environment*, *12*(5), 292–299.

Hadjielias, E., Dada, O., Discua Cruz, A., Zekas, S., Christofi, M., & Sakka, G. (2021). How do digital innovation teams function? Understanding the team cognition-process nexus within the context of digital transformation. *Journal of Business Research*, *122*, 373–386. https://doi.org/https://doi.org/10.1016/j.jbusres.2020.08.045

Hagedoorn, J. (1996). Innovation and entrepreneurship: Schumpeter revisited. *Industrial and Corporate Change*, *5*(3), 883–896.

Heap, J. (2010). Open innovation. *Management Services*, *54*(2), 26.

Henderson-King, D., & Stewart, A. (1994). Women or feminists? Assessing women's group consciousness. *Sex Roles*, *31*, 505–516.

Herazo, B., & Lizarralde, G. (2015). The influence of green building certifications in collaboration and innovation processes. *Construction Management and Economics*, *33*(4), 279–298.

Heugens, P., Van den Bosch, F., & Van Riel, C. (2002). Stakeholder integration. *Business & Society*, *41*(1), 36–60.

Hoegl, M., & Gemuenden, H. G. (2001). Teamwork quality and the success of innovative projects: A theoretical concept and empirical evidence. *Organization Science*, *12*(4), 435–449.

Holland, D. (1997). Selves as cultured: As told by an anthropologist who lacks a soul, in R. D. Ashmore & L. J. Jussim (eds), Self and identity: Fundamental issues. Oxford: Oxford University Press, pp. 160–190.

Howell, J. M., & Avolio, B. J. (1993). Transformational leadership, transactional leadership, locus of control, and support for innovation: Key predictors of consolidated-business-unit performance. *Journal of Applied Psychology*, *78*(6), 891–902.

Howell, J. M., & Higgins, C. A. (1990). Champions of technological innovation. *Administrative Science Quarterly*, *35*(2), 317–341.

Howell, J., & Shea, C. (2001). Individual differences, environmental scanning, innovation framing, and champion behavior: Key predictors of project performance. *Journal of Product Innovation Management*, *18*(1), 15–27. doi:10.1016/S0737-6782(00)00067-9.

Howell, J. M., Shea, C. M., & Higgins, C. A. (2005). Champions of product innovations: Defining, developing, and validating a measure of champion behavior. *Journal of Business Venturing*, *20*(5), 641–661.

Hui, Z., Khan, N. A., & Akhtar, M. (2024). AI-based virtual assistant and transformational leadership in social cognitive theory perspective: A study of team innovation in construction industry. *International Journal of Managing Projects in Business*. doi:10.1108/IJMPB-10-2023-0241

Jesus, G. M. K., & Jugend, D. (2023). How can open innovation contribute to circular economy adoption? Insights from a literature review. *European Journal of Innovation Management*, *26*(1), 65–98.

Jose, A., de Burgos-Jimenez, J., & Carmona-Moreno, E. (2010). Measuring stakeholder integration dimensions: Knowledge, interactions, and adaptational behavior dimensions. *Journal of Business Ethics*, *93*(3), 419–442.

Jung, D. I., Chow, C., & Wu, A. (2003). The role of transformational leadership in enhancing organizational innovation: Hypotheses and some preliminary findings. *The Leadership Quarterly*, 14(4–5), 525–544.

Kouzes, J. M., & Posner, B. Z. (1995). *The leadership challenge: How to keep getting extraordinary things done in organizations*. 2nd ed. San Francisco: Jossey-Bass.

Kuklina, M., Rogov, V., Erdinieva, S., & Urazov, I. (2021). Innovation in the construction industry. *IOP Conference Series: Earth and Environmental Science*, *751*, 012101. https://doi.org/10.1088/1755-1315/751/1/012101

Laursen, K., & Salter, A. (2006). Open for innovation: The role of openness in explaining innovation performance among UK manufacturing firms. *Strategic Management Journal*, *27*(2), 131–150.

Lettl, C., Herstatt, C., & Gemuenden, H. G. (2006). Users' contributions to radical innovation: Evidence from four cases in the field of medical equipment technology. *R&D Management*, *36*(3), 251–272.

Lichtenthaler, U. (2011). Open innovation: Past research, current debates, and future directions. *Academy of Management Perspectives*, *25*(1), 75–93.

Liu, B. S. C., Madhavan, R., & Sudharshan, D. (2005). DiffuNET: The impact of network structure on diffusion of innovation. *European Journal of Innovation Management*, *8*(2), 240–262.

Longo, M., & Mura, M. (2008). Stakeholder management and human resources: Development and implementation of a performance measurement system corporate governance. *The International Journal of Business in Society*, *8*(2), 191–213.

Lorca, P., & Garcia-Diez, P. (2004). The relation between firm survival and the achievement of balance among its stakeholders: An analysis. *International Journal of Management*, *21*(1), 93–99.

Luhtanen, R., & Crocker, J. (1992). A collective self-esteem scale: Self-evaluation of one's social identity. *Personality and Social Psychology Bulletin*, *18*, 302–318.

Maignan, I., & Ferrell, O.C. (2004). Corporate social responsibility and marketing: An integrative framework. *Journal of the Academy of Marketing Science*, *32*(1), 3–19. doi:10.1177/0092070303258971

Maldonado-Guzmán, G., Garza-Reyes, J. A., Pinzón-Castro, S. Y., & Kumar, V. (2017). Barriers to innovation in service SMEs: Evidence from Mexico. *Industrial Management & Data Systems*, *117*(8), 1669–1686.

Manuylenko, V. V., Mishchenko, A. A., Bigday, O. B., Putrenok, Y. L., & Savtsova, A. V. (2015). A comprehensive definition of the concept of innovation in Russian and international science. *International Journal of Economics and Financial Issues*, *5*(4), 1029–1037.

Mascareño, J., Rietzschel, E., & Wisse, B. (2020). Envisioning innovation: Does visionary leadership engender team innovative performance through goal alignment? *Creativity and Innovation Management*, *29*(1), 33–48. https://doi.org/https://doi.org/10.1111/caim.12341

McCall, G. J., & Simmons, J. L. (1978). *Identities and interactions: An examination of human associations in everyday life*. New York: Free Press.

McCutcheon, D. M., & Meredith, J. R. (1993). Conducting case study research in operations management. *Journal of Operations Management*, *11*(3), 239–256.

McKinley, W., Latham, S., & Braun, M. (2014). Organizational decline and innovation: Turnarounds and downward spirals. *Academy of Management Review*, *39*(1), 88–110.

Miranda, L. F., Cruz-Cázares, C., & Saunila, M. (2023). Towards a comprehensive framework to analyse the benefits of openness for sustainability-oriented innovation: A systematic literature review. *Sustainable Development*, *31*(6), 4019–4038. doi:10.1002/sd.2531

Mishra, P., Padhy, C., Mishra, N., Roy, D., Prasanna, V., & Madhuri, C. R. (2023). 21st century entrepreneurship challenges and opportunities. *Asian Journal of Agricultural Extension, Economics & Sociology*, *41*(12), 196–202.

Mitchell, R., Agle, B., & Wood, J. (1997). Toward a theory of stakeholder identification and salience: Defining the principle of who and what really counts. *Academy of Management Review*, *22*(4), 853–886.

Modranský, R., Jakabová, S., & Oláh, A. (2020). Innovation management and barriers: Creating space for innovation and organizational change. *Emerging Science Journal*, *4*(5), 345–364.

Mudassar, A., Zhang, L., Dilawar, D., Adnan, S., & Waqas, K. (2021). Goal clarity as a link between humble leadership and project success: The interactive effects of organizational culture. *Baltic Journal of Management*, *16*(3), 407–423.

Murphy, M., Heaney, J., & Perera, S. (2011). A methodology for evaluating construction innovation constraints through project stakeholder competencies and FMEA. *Construction Innovation*, *11*(4), 416–440.

Nam, C. H., & Tatum, C. B. (1997). Leaders and champions for construction innovation. *Construction Management and Economics*, *15*(3), 259–270.

Nambisan, S., & Agarwal, R. (1998). *The adoption and use of national information infrastructure: A social network and stakeholder perspective: Proceedings of the 19th International Conference on Information Systems*. Helsinki, 13–16 December, pp. 313–318.

Ogink, R. H., Goossen, M. C., Romme, A. G. L., & Akkermans, H. (2023). Mechanisms in open innovation: A review and synthesis of the literature. *Technovation*, *119*, 102621.

Ozorhon, B. (2013). Analysis of construction innovation process at project level. *Journal of Management in Engineering*, *29*(4), 455–463.

Ozorhon, B., Abbott, C., & Aouad, G. (2014). Integration and leadership as enablers of innovation in construction: A case study. *Journal of Management in Engineering*, *30*(2), 256–263.

Ozorhon, B., Abbott, C., Aouad, G., & Powell, J. (2010). *Innovation in construction: A project life-cycle approach*. Salford, UK: University of Salford. Available at: https://www.irbnet.de/daten/iconda/CIB16211.pdf

Phinney, J. S. (1996). When we talk about American ethnic groups, what do we mean? *American Psychologist*, *51*, 918–927.

Piller, F. T., & Walcher, D. (2006). Toolkits for idea competitions: A novel method to integrate users in new product development. *R&D Management*, *36*(1), 307–318.

Podsakoff, P. M., MacKenzie, S. B., Moorman, R. H., & Fetter, R. (1990). Transformational leader behaviors and their effects on followers' trust in leader, satisfaction, and organizational citizenship behaviors. *Leadership Quarterly*, *1*(2), 107–142.

Postmes, T., Spears, R., & Lea, M. (2000). The formation of group norms in computer-mediated communication. *Human Communication Research*, *26*(3), 341–371.

Qin, X., Shi, Y., Lyu, K., & Mo, Y. (2020). Using a TAM-TOE model to explore factors of Building Information Modelling (BIM) adoption in the construction industry. *Journal of Civil Engineering and Management*, *26*(3), 259–277.

Rauter, R., Perl-Vorbach, E., & Baumgartner, R. J. (2017). Is open innovation supporting sustainable innovation? Findings based on a systematic, explorative analysis of existing literature. *International Journal of Innovation and Sustainable Development*, *11*(2–3), 249–270.

Rese, A. & Baier, D. (2012). Networks of promotors as success factor for radical innovations: Theoretical considerations and empirical findings, in K. Fichter and S. Beucker (eds), *Innovation communities*. Berlin Heidelberg: Springer-Verlag, pp. 17–56.

Reypens, C., Lievens, A., & Blazevic, V. (2021). Hybrid orchestration in multi-stakeholder innovation networks: Practices of mobilizing multiple, diverse stakeholders across organizational boundaries. *Organization Studies*, *42*(1), 61–83. https://doi.org/10.1177/0170840619868268

Rowley, T. J. (1997). Moving beyond dyadic ties: A network theory of stakeholder influences. *Academy of Management Review*, *22*(4), 887–910.

Sanni, M., & Verdolini, E. (2022). Eco-innovation and openness: Mapping the growth trajectories and the knowledge structure of open eco-innovation. *Sustainable Futures*, *4*, 100067.

Scott, S. G., & Bruce, R. A. (1994). Determinants of innovative behavior: A path model of individual innovation in the workplace. *Academy of Management Journal*, *37*(3), 580–607.

Sellers, R. M., Smith, M. A., Shelton, J. N., Rowley, S. A. J., & Chavous, T. M. (1998). Multidimensional model of racial identity: A reconceptualization of African American racial identity. *Personality and Social Psychology Review*, *2*, 18–39.

Sexton, M., & Barrett, P. (2005). Performance-based building and innovation: Balancing client and industry needs. *Building Research & Information*, *33*(2), 142–148.

Sharma, S., & Vredenburg, H. (1998). Proactive corporate environmental strategy and the development of competitively valuable organizational capabilities. *Strategy Management Journal*, *19*(8), 729–753.

Skipper, C. O., & Bell, L. C. (2006). Assessment with 360° evaluations of leadership behavior in construction project managers. *Journal of Management in Engineering*, *22*(2), 75–80.

Slaughter, E. S. (2000). Implementation of construction innovations. *Building Research and Information*, *28*(1), 2–17.

Stryker, S. (1987). Identity theory: Developments and extensions, in K. Yardley and T. Honess (eds), *Self and identity: Psychosocial perspectives*. New York: Wiley, pp. 89–103.

Tabassi, A. A., Roufechaei, K. M., Ramli, M., Bakar, A. H. A., Ismail, R., & Pakir, A. H. K. (2016). Leadership competences of sustainable construction project managers. *Journal of Cleaner Production*, *124*, 339–349.

Tajfel, H. (1978). Social categorization, social identity, and social comparison, in H. Tajfel (ed.), *Differentiation between social groups: Studies in the social psychology of intergroup relations*. London: Academic Press, pp. 61–76.

Tajfel, H. (1981). *Human groups and social categories*. Cambridge: Cambridge University Press.

Tajfel, H., & Turner, J.C. (1979). An integrative theory of intergroup conflict, in W.G. Austin and S. Worchel (eds), *The Social Psychology of Intergroup Relations*. Monterey, CA: Brooks/Cole, pp. 33–37.

Tierney, P., & Farmer, S. M. (2004). The Pygmalion process and employee creativity. *Journal of Management*, *30*, 413–432.

Toole, T. M., Hallowell, M., & Chinowsky, P. (2013). A tool for enhancing innovation in construction organizations. *Engineering Project Organization Journal*, *3*(1), 32–50.

Turner, J. C. (1987). *Rediscovering the social group: A self-categorization theory*. Oxford: Blackwell.

Ugwu, O. O., & Haupt, T. C. (2007). Key performance indicators and assessment methods for infrastructure sustainability: A South African construction industry perspective. *Building and Environment*, *42*, 665–680.

Valente, T. W. (1995). *Network models of the diffusion of innovations*. Cresskill, NJ: Hampton Press.

van der Voet, J., & Steijn, B. (2021). Team innovation through collaboration: How visionary leadership spurs innovation via team cohesion. *Public Management Review*, *23*(9), 1275–1294. https://doi.org/10.1080/14719037.2020.1743344

Yigitcanlar, T., & Suharto, T. (2015). Rethinking sustainable urban development: Towards an integrated planning and development process. *International Journal of Environmental Science and Technology*, *12*(1), 341–352.

Yin, R. K. (2003). *Case study research: Design and methods*. 3rd ed. Thousand Oaks, CA: Sage Publications.

Yukl, G. (1999). An evaluation of conceptual weaknesses in transformational and charismatic leadership theories. *Leadership Quarterly*, *10*(2), 285–305.

Yukl, G. (2010). *Leadership in organizations*. 7th ed. Englewood Cliffs, NJ: Prentice Hall.

Yukl, G. (2012). Effective leadership behavior: What we know and what questions need more attention. *Academy of Management Perspectives*, *26*(4), 66–85.

Yukl, G., Gordon, A., & Taber, T. (2002). A hierarchical taxonomy of leadership behavior: Integrating a half century of behavior research. *Journal of Leadership and Organizational Studies*, *9*(1), 15–32.

Index